Selected Titles in This Series

18 **Winfried Just and Martin Weese,** Discovering modern set theory. II: Set-theoretic tools for every mathematician, 1997
17 **Henryk Iwaniec,** Topics in classical automorphic forms, 1997
16 **Richard V. Kadison and John R. Ringrose,** Fundamentals of the theory of operator algebras. Volume II: Advanced theory, 1997
15 **Richard V. Kadison and John R. Ringrose,** Fundamentals of the theory of operator algebras. Volume I: Elementary theory, 1997
14 **Elliott H. Lieb and Michael Loss,** Analysis, 1997
13 **Paul C. Shields,** The ergodic theory of discrete sample paths, 1996
12 **N. V. Krylov,** Lectures on elliptic and parabolic equations in Hölder spaces, 1996
11 **Jacques Dixmier,** Enveloping algebras, 1996 Printing
10 **Barry Simon,** Representations of finite and compact groups, 1996
9   **Dino Lorenzini,** An invitation to arithmetic geometry, 1996
8   **Winfried Just and Martin Weese,** Discovering modern set theory. I: The basics, 1996
7   **Gerald J. Janusz,** Algebraic number fields, second edition, 1996
6   **Jens Carsten Jantzen,** Lectures on quantum groups, 1996
5   **Rick Miranda,** Algebraic curves and Riemann surfaces, 1995
4   **Russell A. Gordon,** The integrals of Lebesgue, Denjoy, Perron, and Henstock, 1994
3   **William W. Adams and Philippe Loustaunau,** An introduction to Gröbner bases, 1994
2   **Jack Graver, Brigitte Servatius, and Herman Servatius,** Combinatorial rigidity, 1993
1   **Ethan Akin,** The general topology of dynamical systems, 1993

Discovering Modern Set Theory. II

Set-Theoretic Tools for Every Mathematician

Graduate Studies in Mathematics

Volume 18

Discovering Modern Set Theory. II

Set-Theoretic Tools for Every Mathematician

Winfried Just
Martin Weese

American Mathematical Society

Editorial Board
James E. Humphreys (Chair)
David J. Saltman
David Sattinger
Julius L. Shaneson

1991 *Mathematics Subject Classification.* Primary 04–01, 03E05, 04A20.

ABSTRACT. Short but rigorous introductions to various set-theoretic techniques that have found numerous applications outside of set theory are given. Topics covered include: trees, partition calculus, applications of Martin's Axiom and the ◊-principle, closed unbounded and stationary sets, measurable cardinals, and the use of elementary submodels. This volume is aimed at advanced graduate students and mathematical researchers specializing in areas other than set theory who want to broaden their knowledge of contemporary set theory. It can be studied independently of Volume I of the same text.

Library of Congress Cataloging-in-Publication Data
Just, W. (Winfried)
   Discovering modern set theory / Winfried Just, Martin Weese.
     p. cm. — (Graduate studies in mathematics, ISSN 1065-7339; V. 8)
   Includes bibliographical references and index.
   Contents: 1. The basics
   ISBN 0-8218-0266-6 (v. 1 : hard cover : alk. paper)
   1. Set theory. I. Weese. Martin. II. Title. III. Series.
QA248.J87 1995
511.3'22–dc20   95-44663
                                                                                                                        CIP

Copying and reprinting. Individual readers of this publication, and nonprofit libraries acting for them, are permitted to make fair use of the material, such as to copy a chapter for use in teaching or research. Permission is granted to quote brief passages from this publication in reviews, provided the customary acknowledgment of the source is given.

  Republication, systematic copying, or multiple reproduction of any material in this publication (including abstracts) is permitted only under license from the American Mathematical Society. Requests for such permission should be addressed to the Assistant to the Publisher, American Mathematical Society, P. O. Box 6248, Providence, Rhode Island 02940-6248. Requests can also be made by e-mail to `reprint-permission@ams.org`.

       © 1997 by the American Mathematical Society. All rights reserved.
         The American Mathematical Society retains all rights
         except those granted to the United States Government.
                Printed in the United States of America.

     ∞ The paper used in this book is acid-free and falls within the guidelines
          established to ensure permanence and durability.
         Visit the AMS home page at URL: `http://www.ams.org/`

           10 9 8 7 6 5 4 3 2 1   02 01 00 99 98 97

Contents

Preface	ix
Notation	xi
Chapter 13. Filters and Ideals in Partial Orders	1
13.1. The general concept of a filter	1
13.2. Ultraproducts	8
13.3. A first look at Boolean algebras	12
Mathographical Remarks	24
Chapter 14. Trees	27
Mathographical Remarks	48
Chapter 15. A Little Ramsey Theory	49
Mathographical Remarks	65
Chapter 16. The Δ-System Lemma	67
Chapter 17. Applications of the Continuum Hypothesis	71
17.1. Applications to Lebesgue measure and Baire category	71
17.2. Miscellaneous applications of CH	79
Mathographical Remarks	85
Chapter 18. From the Rasiowa-Sikorski Lemma to Martin's Axiom	87
Mathographical Remarks	94
Chapter 19. Martin's Axiom	95
19.1. MA essentials	95
19.2. MA and cardinal invariants of the continuum	102
19.3. Ultrafilters on ω	110
Mathographical Remarks	116
Chapter 20. Hausdorff Gaps	117
Mathographical Remarks	122
Chapter 21. Closed Unbounded Sets and Stationary Sets	123
21.1. Closed unbounded and stationary sets of ordinals	123
21.2. Closed unbounded and stationary subsets of $[X]^{<\kappa}$	131
Chapter 22. The $\Diamond$-principle	139
Mathographical Remarks	146

Chapter 23. Measurable Cardinals	147
Mathographical Remarks	157
Chapter 24. Elementary Submodels	159
24.1. Elementary facts about elementary submodels	159
24.2. Applications of elementary submodels in set theory	167
Mathographical Remarks	185
Chapter 25. Boolean Algebras	187
Mathographical Remark	205
Chapter 26. Appendix: Some General Topology	207
Index	217
Index of Symbols	223

Preface

This is the second volume of a graduate course in set theory. Volume I covered the basics of modern set theory and was primarily aimed at beginning graduate students. Volume II is aimed at more advanced graduate students and research mathematicians specializing in fields other than set theory. It contains short but rigorous introductions to various set-theoretic techniques that have found applications outside of set theory. Although we think of Volume II as a natural continuation of Volume I,[1] each volume is sufficiently self-contained to be studied separately.

The main prerequisite for Volume II is a knowledge of basic naive and axiomatic set theory.[2] Moreover, some knowledge of mathematical logic and general topology is indispensible for reading this volume. A minicourse in mathematical logic was given in Chapters 5 and 6 of Volume I, and we include an appendix on general topology at the end of this volume. Our terminology is fairly standard. For the benefit of those readers who learned their basic set theory from a different source than our Volume I we include a short section on somewhat idiosyncratic notations introduced in Volume I. In particular, some of the material on mathematical logic covered in Chapter 5 is briefly reviewed. The book can be used as a text in the classroom as well as for self-study.

We tried to keep the length of the text moderate. This may explain the absence of many a worthy theorem from this book. Our most important criterion for inclusion of an item was frequency of use outside of pure set theory. We want to emphasize that "item" may mean either an important concept (like "equiconsistency with the existence of a measurable cardinal"), a theorem (like Ramsey's Theorem), or a proof technique (like the craft of using Martin's Axiom). Therefore, we occasionally illustrate a technique by proving a somewhat marginal theorem. Of course, the "frequency of use outside set theory" is based on our subjective perceptions.

At the end of most chapters there are "Mathographical Remarks." Their purpose is to show where the material fits in the history and literature of the subject. We hope they will provide some guidance for further reading in set theory. They should not be mistaken for "scholarly remarks" though. We did not make any effort whatsoever to trace the theorems of this book to their origins. However, each of the theorems presented here can also be found in at least one of the more specialized texts reviewed in the "Mathographical Remarks." Therefore, we do not feel guilty of severing chains of historical evidence.

---

[1] This is the reason why the present volume starts with Chapter 13.
[2] Possible alternatives to our Volume I are such texts as: K. Devlin, *The Joy of Sets. Fundamentals of Contemporary Set Theory*, Springer Verlag, 1993; A. Lévy; *Basic Set Theory*, Springer Verlag, 1979; or J. Roitman, *Introduction to Modern Set Theory*, John Wiley, 1990.

Much of this book is written like a dialogue between the authors and the reader. This is intended to model the practice of creative mathematical thinking, which more often than not takes on the form of an inner dialogue in a mathematician's mind. You will quickly notice that this text contains many question marks. This reflects our conviction that in the mathematical thought process it is at least as important to have a knack for asking the right questions at the right time as it is to know some of the answers.

You will benefit from this format only if you do your part and actively participate in the dialogue. This means in particular: Whenever we pose a rhetorical question, pause for a moment and ponder the question before you read our answer. Sometimes we put a little more pressure on you and call our rhetorical questions EXERCISES. Not all exercises are rhetorical questions that will be answered a few lines later. Often the completion of a proof is left as an exercise. We also may ask you to supply the entire proof of an interesting theorem, or an important example. Nevertheless, we recommend that you attempt the exercises right away, especially all the easier ones. Most of the time it will be easier to digest the ensuing text if you have worked on the exercise, even if you were unable to solve it.

We often make references to solutions of exercises from earlier chapters. Sometimes the new material will make an old and originally quite hard exercise seem trivial, and sometimes a new question can be answered by modifying the solution to a previous problem. Therefore it is a good idea to collect your solutions and even your failed attempts at solutions in a folder where you can look them up later.

The level of difficulty of our exercises varies greatly. To help the reader save time, we rated each exercise according to what we perceive as its level of difficulty. The rating system is the same as used by American movie theatres. Everybody should attempt the exercises rated G (general audience). Beginners are encouraged to also attempt exercises rated PG (parental guidance), but may sometimes want to consult their instructor for a hint. Exercises rated R (restricted) are intended for mature audiences. The X–rated problems must not be attempted by anyone easily offended or discouraged.

In Chapters 17, 18, 19, and 22 we will discuss consequences of statements that are relatively consistent with, but no provable in ZFC: the Continuum Hypothesis (abbreviated CH), Martin's Axiom (abbreviated MA), and the Diamond Principle (abbreviated $\Diamond$). We will write "THEOREM n.m (CH)" in order to indicate that Theorem n.m is provable in the theory ZFC + CH rather than ZFC alone.

We are greatly indebted to Mary Anne Swardson of Ohio University for reading the very first draft of this book and generously applying her red pencil to it. Special thanks are due to Ewelina Skoracka-Just for her beautiful typesetting of this volume.

Notation

Here is a list of somewhat ideosyncratic symbols that will be used in this volume:

$f\restriction A$ — restriction of a function f to a subset A of its domain;
$f[A]$ — image of a set A under a function f;
Δ — symmetric difference of two sets;
$\vec{a}$ — abbreviation for $\langle a_0, \ldots, a_n \rangle$;
$^{<\kappa}X$ — the set $\bigcup_{\alpha<\kappa} {}^{\alpha}X$ of all functions from some ordinal $\alpha < \kappa$ into X;
$\lambda^{<\kappa}$ — the cardinality of the set $^{<\kappa}\lambda$;
$s{\frown}i$ — (where $s \in {}^{<\omega}I$ and $i \in I$) denotes the function $s \cup \langle dom(s), i\rangle$;
$TC(x)$ — the transitive closure of a set x, i.e., the smallest set $y \supseteq x$ such that $z \subseteq y$ for all $z \in y$;
H_λ — the family of all sets hereditarily of cardinality less than λ, i.e., the family $\{x : |TC(x)| < \lambda\}$;
$[X]^\kappa$ — the family of all subsets of X of size κ;
$[X]^{<\kappa}$ — the family of all subsets of X of size less than κ;
Fin — denotes $[\omega]^{<\aleph_0}$, i.e., the family of finite subsets of ω;
V — the set-theoretical universe, i.e., the class of all sets;
ON — the class of all ordinals;
LIM — the class of all limit ordinals;
Card — the class of all cardinals;
L — the constructible universe, i.e., the class of all constructible sets;
V_α — the α-th level of the cumulative hierarchy, where $\mathbf{V} = \bigcup_{\alpha \in \mathbf{ON}} V_\alpha$;
L_α — the α-th level of the constructible hierarchy, where $\mathbf{L} = \bigcup_{\alpha \in \mathbf{ON}} L_\alpha$;
$(\alpha, \beta]$ — the set of ordinals $\{\gamma \in \mathbf{ON} : \alpha < \gamma \leq \beta\}$;
$\alpha^{\cdot\beta}$ — ordinal exponentiation is written with a dot in front of the exponent in order to distinguish it from cardinal exponentiation;
p.o. — abbreviates "partial order;"
l.o. — abbreviates "linear order;"
w.o. — abbreviates "wellorder;"
CH — abbreviates the Continuum Hypothesis;
GCH — abbreviates the generalized Continuum Hypothesis.

Now let us review the rudiments of mathematical logic that were introduced in Chapter 5.

The *logical symbols* of a first-order language L are $\land$, $\neg$, $\exists$, $=$, brackets, and variable symbols v_i for every $i \in \omega$. The symbols $\lor$, $\to$, $\leftrightarrow$, $\forall$ are considered abbreviations. Each language L also has *nonlogical symbols:* A set $\{r_i : i \in I\}$ of relational symbols, a set $\{f_j : j \in J\}$ of functional symbols, and a set $\{c_k : k \in K\}$

of constant symbols. The sets I, J, K may be empty. The arity of r_i is denoted by $\tau_0(i)$, the arity of f_j by $\tau_1(j)$.

Here are some prominent examples of first-order languages: The language L_G of group theory has no relational symbols, one functional symbol $*$ of arity 2, and one constant symbol e. The language L_S of set theory has no functional or constant symbols, and only one relational symbol $\in$ of arity 2. Similarly, the only nonlogical symbol of the language $L_\leq$ is the relational symbol $\leq$ of arity 2. In principle, set theory could be developed in $L_\leq$ and the theory of partial orders could be expressed in L_S; the difference is purely a matter of convention.

Given a language L, one defines the set $Term_L$ of all *terms* of L as the smallest set T of finite strings of symbols of L that satisfies the following conditions:

(i) $v_i \in T$ for all $i \in \omega$;
(ii) $c_k \in T$ for all $k \in K$;
(iii) If $t_0, ..., t_{n-1} \in T$, $j \in J$, and $\tau_1(j) = n$, then $f_j(t_0, ..., t_{n-1}) \in T$.

The set $Form_L$ of all *formulas* of L is the smallest set of finite strings of symbols of L that satisfies the following conditions:

(i) If $s, t \in Term$ then $s = t \in F$;
(ii) If $t_0, ..., t_{n-1} \in Term$, $i \in I$, and $\tau_0(i) = n$, then $r_i(t_0, ..., t_{n-1}) \in F$;
(iii) If φ and ψ are in F, then so are $(\varphi \wedge \psi)$, $(\neg \varphi)$, and $(\exists v_i \, \varphi)$.

The above definition of $Term_L$ and $Form_L$ allows us to use induction and recursion over the wellfounded relations of being a subterm or a subformula. These techniques are commonly referred to as induction or recursion over the *length* of a term or formula.

The occurrence of a variable v_i in a formula φ is *bound* if it happens within the range of a quantifier $\exists v_i$. Otherwise it is called *free*. A formula without free variables is called a *sentence*.

Let L be a first-order language with nonlogical symbols $\{r_i : i \in I\} \cup \{f_j : j \in J\} \cup \{c_k : k \in K\}$. A *model of L* or an *L-structure* is a structure of the form $\mathfrak{A} = \langle A, (R_i)_{i \in I}, (F_j)_{j \in J}, (C_k)_{k \in K} \rangle$. The set A is called the *universe* or *underlying set* of the model $\mathfrak{A}$. For each $i \in I$, R_i is a relation on A of arity $\tau_0(i)$ (i.e., $R_i \subseteq A^{\tau_0(i)}$); for each $j \in J$, F_j is a function from $A^{\tau_1(j)}$ into A; and for each $k \in K$, C_k is an element of A. For example, if M is a set and $\bar\in$ denotes the membership relation restricted to elements of M, then $\langle M, \bar\in \rangle$ is a model of L_S.

Now let us recall what it means for a model of L to satisfy a sentence of L.

A *valuation* (of the variables) is a function s that assigns an element of A to each natural number. Given s, we assign to each $t \in Term$ some $t^s \in A$ as follows:

$$v_i^s = s(i);$$
$$c_k^s = C_k;$$
$$f_j^s(t_0, ..., t_{\tau_1(j)-1}) = F_j(t_0^s, ..., t_{\tau_1(j)-1}^s).$$

By recursion over the length of formulas we define a relation

$$\mathfrak{A} \models_s \varphi$$

(read as: "$\mathfrak{A}$ *satisfies* φ under the valuation s." The relation $\models_s$ is called the *satisfaction relation*.)

$\mathfrak{A} \models_s t_0 = t_1$ iff $t_0^s = t_1^s$;
$\mathfrak{A} \models_s r_i(t_0, ..., t_{\tau_0(i)-1})$ iff $(t_0^s, ..., t_{\tau_0(i)-1}^s) \in R_i$;
$\mathfrak{A} \models_s \neg \varphi$ iff it is not the case that $\mathfrak{A} \models_s \varphi$;

$\mathfrak{A} \models_s \varphi \wedge \psi$ iff $\mathfrak{A} \models_s \varphi$ and $\mathfrak{A} \models_s \psi$;
$\mathfrak{A} \models_s \exists v_i \varphi$ iff there exists a valuation s^* such that $\mathfrak{A} \models_{s^*} \varphi$ and
$s(k) = s^*(k)$ for all $k \neq i$.

Let $\varphi \in Form$, let $a_{i_0}, ..., a_{i_{n-1}} \in A$, and let s be a valuation. We write
$$\mathfrak{A} \models_s \varphi[a_{i_0}, ..., a_{i_{n-1}}]$$
if $\mathfrak{A} \models_{s^*} \varphi$ for the valuation s^* that is obtained from s by replacing $s(j)$ with $s^*(j) = a_j$ for $j = i_0, \ldots, i_{n-1}$ and leaving $s^*(j) = s(j)$ for $j \neq i_0, \ldots, i_{n-1}$. One can show that if all free variables of φ are among $v_{i_0}, \ldots, v_{i_{n-1}}$ and s_0, s_1 are valuations, then
$$\mathfrak{A} \models_{s_0} \varphi[a_{i_0}, \ldots, a_{i_{n-1}}] \text{ iff } \mathfrak{A} \models_{s_0} \varphi[a_{i_0}, \ldots, a_{i_{n-1}}].$$
In particular, if φ is a sentence, then the satisfaction relation $\mathfrak{A} \models_s \varphi$ does not depend on s. If $\mathfrak{A} \models_s \varphi$ for every valuation s, then we write $\mathfrak{A} \models \varphi$.

When working with specific formulas, we shall often use more suggestive, self-explanatory terminology. For example, if $\mathfrak{M} = \langle M, E \rangle$ is a model of L_S and $a \in M$, then we shall write "$\mathfrak{M} \models \exists x (x \in a)$" instead of "$\mathfrak{M} \models \exists v_1 (v_1 \in v_0)[a]$."

A *theory* in a language L is any set of sentences of this language. For example, ZFC is a theory in L_S.

Let T be a theory in a language L. We say that a model $\mathfrak{A}$ of L is a *model of* T and write $\mathfrak{A} \models T$ if $\mathfrak{A} \models \varphi$ for each $\varphi \in T$. A *theorem* of T is a sentence φ of L that has a formal proof from the sentences of L. We write $T \vdash \varphi$ to indicate that φ is a theorem of T. By Gödel's Completeness Theorem, φ is a theorem of T if and only if $\mathfrak{A} \models \varphi$ for all models $\mathfrak{A}$ of T.

CHAPTER 13

Filters and Ideals in Partial Orders

This chapter contains assorted results on filters and partial orders. The first section is devoted to the general notion of a filter in a p.o. This material will be used in several later chapters and probably should be read up front. Section 13.2 is devoted to applications of ultrafilters to an important construction in model theory. This material will not be needed until Chapter 24. In Section 13.3 we develop the theory of Boolean algebras (which can be looked at as the theory of a very special kind of partial order) to the extent required for the proofs of Lemmas 19.7 and 19.8.

13.1. The general concept of a filter

Recall that if X is a nonempty set, then a *filter on* X is a subfamily $\mathcal{F}$ of $\mathcal{P}(X)$ such that

(i) $\mathcal{F}$ is closed under supersets, i.e., $\forall Y \in \mathcal{F} \, \forall Z \subseteq X \, (Y \subseteq Z \to Z \in \mathcal{F})$;
(ii) $\mathcal{F}$ is closed under finite intersections, i.e., $\bigcap H \in \mathcal{F}$ for all $H \in [\mathcal{F}]^{<\omega}$.

Recall also that a filter $\mathcal{F}$ is *proper* if $\emptyset \notin \mathcal{F}$. A maximal proper filter on X is called an *ultrafilter on* X.

EXAMPLE 13.1. If $a \in X$ then $\mathcal{F}_a = \{Y \subseteq X : a \in Y\}$ is a proper filter on X. It is called the *principal ultrafilter determined by* a.

EXAMPLE 13.2. The subsets of the unit interval $[0, 1]$ of Lebesgue measure one form a proper filter $\mathcal{F}_1$ on $[0, 1]$.

EXAMPLE 13.3. If $\langle X, \tau \rangle$ is a topological space and $x \in X$, then the family

$$\mathcal{N}_x = \{Y \subseteq X : \exists U \in \tau \, (x \in U \subseteq Y)\}$$

forms a proper filter on X, called the *neighborhood filter of* x or *neighborhood system at* x.

A filter $\mathcal{F}$ on X is *uniform* if $|Y| = |X|$ for all $Y \in \mathcal{F}$. The filter in Example 13.1 is uniform only if $|X| = 1$. Thus, every uniform ultrafilter on an infinite set X is nonprincipal. Moreover, if X is denumerable, then the notions of a nonprincipal and a uniform ultrafilter on X coincide. An ultrafilter on a set X of cardinality $\aleph_1$ is uniform if and only if it contains all *cocountable* subsets of X, i.e., all subsets of X with countable complement. The filter of Example 13.2 is uniform. Neighborhood filters as in Example 13.3 are uniform in some but not all topological spaces.

Given a proper filter $\mathcal{F}$ on a set X, the *dual ideal to* $\mathcal{F}$ is the family

$$\mathcal{F}^* = \{Y \subseteq X : X \backslash Y \in \mathcal{F}\}.$$

The sets in $\mathcal{P}(X) \backslash \mathcal{F}^*$ are called the *stationary sets with respect to* $\mathcal{F}$, or simply the $\mathcal{F}$-*stationary* sets.

EXERCISE 13.1 (G). Let $\mathcal{F}$ be a proper filter on a nonempty set X.
(a) Show that if $\mathcal{I} = \mathcal{F}^*$, then $\mathcal{I}$ satisfies the following conditions:
   (i)* $\mathcal{I}$ is closed under subsets, i.e., $\forall Y \in \mathcal{I} \forall Z \subseteq Y \, (Z \in \mathcal{I})$;
   (ii)* $\mathcal{I}$ is closed under finite unions, i.e., $\bigcup H \in \mathcal{I}$ for all $H \in [\mathcal{I}]^{<\omega}$.
(b) Show that if a family $\mathcal{I} \subseteq \mathcal{P}(X)$ satisfies canditions (i)* and (ii)*, then the family $\mathcal{I}^* = \{Y \subseteq X : X \backslash Y \in \mathcal{I}\}$ is a filter on X.
(c) Show that $\mathcal{F}^{**} = \mathcal{F}$.
(d) Show that a set $Y \subseteq X$ is $\mathcal{F}$-stationary if and only if $Y \cap Z \neq \emptyset$ for all $Z \in \mathcal{F}$.

A family $\mathcal{I} \subseteq \mathcal{P}(X)$ that satisfies conditions (i)* and (ii)* of Exercise 13.1(a) is called an *ideal on* X. An ideal $\mathcal{I}$ on X is *proper* if $\mathcal{I} \neq \mathcal{P}(X)$, i.e., if $X \notin \mathcal{I}$. If $\mathcal{I}$ is a proper ideal on X, then $\mathcal{I}^*$ is a proper filter on X. The family of $\mathcal{I}^*$-stationary sets is often denoted by $\mathcal{I}^+$.

One can think of a proper filter on X as a family of "large" subsets of X; the dual ideal would then be the family of "small" subsets of X, and the stationary sets would be "not too small" in a sense. Example 13.2 nicely illustrates this point of view: $\mathcal{F}_1^*$ consists of all subsets of the unit interval of measure zero, whereas the $\mathcal{F}_1$-stationary sets are precisely the sets of positive outer measure.

In Example 13.3, a subset Y of X is $\mathcal{N}_x$-stationary if and only if x is in the closure of Y.

In Example 13.1, the family of $\mathcal{F}_a$-stationary sets is $\mathcal{F}_a$ itself. The reason is that $\mathcal{F}_a$ is an ultrafilter. In our new terminology, Theorem 9.10 can be expressed as follows:

CLAIM 13.1. *Let $\mathcal{F}$ be a proper filter on a nonempty set X. Then $\mathcal{F}$ is an ultrafilter if and only if $\mathcal{F}$ coincides with the family of $\mathcal{F}$-stationary sets.*

In the proof of Theorem 9.10, we constructed a filter that contained a given subfamily of $\mathcal{P}(X)$. Let us now review this construction. For $\mathcal{A} \subseteq \mathcal{P}(X)$, let

$$flt(\mathcal{A}) = \{Y \subseteq X : \exists k \in \omega \exists A_0, \ldots, A_{k-1} \in \mathcal{A} \, (Y \supseteq A_0 \cap \cdots \cap A_{k-1})\}.$$

In particular, if $\mathcal{F}$ is a filter on X and $Y \subseteq X$, then the set $flt(\mathcal{F}, Y)$ defined in Chapter 9 is the same as $flt(\mathcal{F} \cup \{Y\})$ in our new terminology. We say that a subfamily $\mathcal{A}$ of $\mathcal{P}(X)$ has the *finite intersection property* (abbreviated *fip*) if $A_0 \cap \cdots \cap A_k \neq \emptyset$ for every finite subset $\{A_0, \ldots, A_k\}$ of $\mathcal{A}$. We call $\mathcal{A}$ a *filter base* if for every finite subset $\{A_0, \ldots, A_k\}$ of $\mathcal{A}$ there exists $A \in \mathcal{A}$ such that $A \subseteq A_0 \cap \cdots \cap A_k$.

EXERCISE 13.2 (G). Let $X \neq \emptyset$ and $\mathcal{A} \subseteq \mathcal{P}(X)$.
(a) Show that the family $flt(\mathcal{A})$ is the smallest filter on X that contains $\mathcal{A}$.
(b) Show that $flt(\mathcal{A})$ is proper iff $\mathcal{A}$ has the fip.
(c) Show that $flt(\mathcal{A}) = \{Y \subseteq X : \exists A \in \mathcal{A} \, (A \subseteq Y)\}$ iff $\mathcal{A}$ is a filter base.
(d) Show that if X is infinite, then every uniform filter on X is contained in a uniform ultrafilter on X.

We call $flt(\mathcal{A})$ the *filter generated by* $\mathcal{A}$. Note that if $\mathcal{N}_x$ is as in Example 13.3 and $\mathcal{B} \subseteq \mathcal{P}(X)$, then $\mathcal{B}$ is a filter base with $flt(\mathcal{B}) = \mathcal{N}_x$ if and only if $\mathcal{B}$ is a neighborhood base at x.

Lemma 9.11(b) says that if $\mathcal{F}$ is a filter on X and $Y \subseteq X$, then $flt(\mathcal{F} \cup \{Y\})$ is proper if and only if $X \backslash Y \notin \mathcal{F}$. This translates into the following characterization of $\mathcal{F}$-stationary sets:

CLAIM 13.2. *Let $\mathcal{F}$ be a proper filter on a set X, and let $Y \subseteq X$. Then the family $\mathcal{F} \cup \{Y\}$ generates a proper filter if and only if Y is $\mathcal{F}$-stationary.*

Let κ be a regular uncountable cardinal, and let $\mathcal{F}$ be a filter on a nonempty set X. We say that $\mathcal{F}$ is κ-*complete* or κ-*closed* if $\bigcap \mathcal{A} \in \mathcal{F}$ for every $\mathcal{A} \in [\mathcal{F}]^{<\kappa}$. An $\aleph_1$-complete filter is also called *countably complete*.

Every principal ultrafilter $\mathcal{F}_a$ is κ-complete for all κ. The filter $\mathcal{N}_x$ may or may not be countably complete. If it is, then x is said to be a *P-point in X*. The filter $\mathcal{F}_1$ of Example 13.2 is countably complete but not $(2^{\aleph_0})^+$-complete, since the family $\mathcal{A} = \{[0,1]\backslash\{x\} : 0 \leq x \leq 1\} \subseteq \mathcal{F}_1$ is of cardinality $2^{\aleph_0}$, but $\bigcap \mathcal{A} = \emptyset \notin \mathcal{F}_1$.

EXERCISE 13.3 (G). (a) Show that if $\mathcal{F}$ is a nonprincipal ultrafilter on a set X with $|X| = \kappa$, then $\mathcal{F}$ is not κ^+-complete.

(b) Show that if an ultrafilter $\mathcal{F}$ on a set X is not countably complete, then there exists a denumerable family $\{Y_n : n \in \omega\} \subseteq \mathcal{F}$ such that $\bigcap_{n \in \omega} Y_n = \emptyset$.

If there exists a nonprincipal κ-complete ultrafilter on a set of size κ, then κ is called a *measurable cardinal*. Measurable cardinals will be studied in Chapter 23.

So far we have been talking only about filters on a set X, i.e., filters consisting of arbitrary subsets of X. But if you know some general topology, you will certainly have noticed that topologists speak about "filters of closed sets," "ultrafilters of zero sets," etc. These objects are usually not filters of the type we have discussed so far. In particular, our filters are closed under arbitrary supersets, and a superset of a closed set is not in general closed. We need a more general concept of filters.

DEFINITION 13.3. Let $\langle \mathbb{P}, \leq \rangle$ be a p.o., and let $A \subseteq \mathbb{P}$. An element p of $\mathbb{P}$ is a *lower bound* for A if $p \leq q$ for every $q \in A$. We say that p and q are *compatible in A* if A contains a lower bound for the set $\{p, q\}$. If p and q are compatible in $\mathbb{P}$, then we just say that they are *compatible* and write $p \not\perp q$; otherwise we say that p and q are *incompatible* and write $p \perp q$. A subset $\mathbb{F}$ of $\mathbb{P}$ is called a *filter in $\langle \mathbb{P}, \leq \rangle$* (or just a *filter in $\mathbb{P}$* if the p.o. relation is implied by the context), if

(I) $\mathbb{F}$ is closed upwards, i.e., $\forall p \in \mathbb{F} \, \forall q \in \mathbb{P} \, (p \leq q \rightarrow q \in \mathbb{F})$;
(II)   Every finite subset of $\mathbb{F}$ has a lower bound in $\mathbb{F}$.

Note that we speak about filters *in* $\mathbb{P}$ to avoid confusion with filters *on* X.

EXERCISE 13.4 (G). Convince yourself that if we replace (II) in the above definition by the apparently weaker demand that every two elements of $\mathbb{F}$ are compatible in $\mathbb{F}$, then we get an equivalent definition.

EXAMPLE 13.4. Let $\mathbb{P}$ be the family of all open subsets of the unit interval of positive Lebesgue measure, and consider the p.o. $\langle \mathbb{P}, \subseteq \rangle$. Let $\mathbb{A}(0.5)$ be the family of all subsets of $\mathbb{P}$ of measure > 0.5. Then every two elements of $\mathbb{A}(0.5)$ are compatible in $\mathbb{P}$, but not necessarily in $\mathbb{A}(0.5)$. Moreover, not every finite subset of $\mathbb{A}(0.5)$ has a lower bound in $\mathbb{P}$.

EXAMPLE 13.5. A filter on X is the same thing as a filter in $\langle \mathcal{P}(X), \subseteq \rangle$, and a proper filter on X is the same as a filter in $\langle \mathcal{P}(X)\backslash\{\emptyset\}, \subseteq \rangle$. Thus there is no need to introduce a separate notion of proper filter in a p.o.

EXAMPLE 13.6. Let X be a topological space, and let $\mathcal{K}(X) = \{K \subseteq X : K \text{ is closed and nonempty}\}$. A filter $\mathbb{F}$ in $\mathcal{K}(X)$ will be referred to as a *filter of closed sets* (the adjective "nonempty" is usually dropped in this context).

EXAMPLE 13.7. Suppose $\langle \mathbb{P}, \leq \rangle$ is a l.o. Then $\mathbb{F}$ is a filter in $\langle \mathbb{P}, \leq \rangle$ if and only if it is a final segment of $\langle \mathbb{P}, \leq \rangle$, i.e., iff $\forall p \in \mathbb{F} \, \forall q \in \mathbb{P} \, (p \leq q \rightarrow q \in \mathbb{F})$.

Now that we are equipped with the general notion of a filter, let us see how the other concepts from the beginning of this section translate into the broader context. An *ultrafilter in* $\langle \mathbb{P}, \leq \rangle$ is a maximal filter in $\langle \mathbb{P}, \leq \rangle$. Thus an ultrafilter in $\langle \mathcal{P}(X) \setminus \emptyset, \subseteq \rangle$ is the same thing as an ultrafilter on X, but the only ultrafilter in $\langle \mathcal{P}(X), \subseteq \rangle$ is $\mathcal{P}(X)$ itself. Similarly, if $\langle \mathbb{P}, \leq \rangle$ is a l.o., then the only ultrafilter in $\langle \mathbb{P}, \leq \rangle$ is $\mathbb{P}$ itself.

EXERCISE 13.5 (G). Show that if $\mathbb{F}$ is an ultrafilter in a p.o. $\langle \mathbb{P}, \leq \rangle$, then $\mathbb{F} \neq \mathbb{P}$ if and only if $\mathbb{P}$ contains two incompatible elements.

Are there ultrafilters in every p.o.? Yes.

THEOREM 13.4. *If* $\langle \mathbb{P}, \leq \rangle$ *is a p.o., then there exists an ultrafilter* $\mathbb{F} \subseteq \mathbb{P}$. *Moreover, every filter* $\mathbb{F}$ *in* $\langle \mathbb{P}, \leq \rangle$ *can be extended to an ultrafilter in* $\langle \mathbb{P}, \leq \rangle$.

EXERCISE 13.6 (PG). Prove Theorem 13.4.

EXAMPLE 13.8. Let I, J be nonempty sets. Define:

$$Fn(I, J) = \{p : p \text{ is a function from a finite subset of } I \text{ into } J\}.$$

EXERCISE 13.7 (G). (a) Convince yourself that p and q are compatible in $\langle Fn(I, J), \supseteq \rangle$ iff $p \cup q$ is a function.

(b) Prove that a filter $\mathbb{F}$ in $\langle Fn(I, J), \supseteq \rangle$ is an ultrafilter in this p.o. iff $\bigcup \mathbb{F}$ is a function from I into J.

EXAMPLE 13.9. If a is a minimal element in a p.o. $\langle \mathbb{P}, \leq \rangle$, then the set $\mathbb{F}_a = \{p \in \mathbb{P} : a \leq p\}$ is an ultrafilter in $\langle \mathbb{P}, \leq \rangle$. Ultrafilters of the form $\mathbb{F}_a$ will be called *principal ultrafilters* or *fixed ultrafilters*. Ultrafilters which do not contain a smallest element will be called *free ultrafilters*.

Let us now try to generalize the other notions introduced for filters on X. As you might have guessed, an *ideal in a p.o.* $\langle \mathbb{P}, \leq \rangle$ is a subset $\mathbb{I}$ of $\mathbb{P}$ such that

(I)* $\mathbb{I}$ is closed downwards, i.e., $\forall p \in \mathbb{I} \, \forall q \in \mathbb{P} \, (q \leq p \rightarrow q \in \mathbb{I})$;

(II)*   Every finite subset of $\mathbb{I}$ has an upper bound in $\mathbb{I}$.

Subsets of p.o.'s that are closed downwards are also called *open*. The following exercise explains why.

EXERCISE 13.8 (G). Let $\langle \mathbb{P}, \leq \rangle$ be a p.o. Then $\{\mathbb{I} \subseteq \mathbb{P} : \mathbb{I} \text{ satisfies } (I)^*\}$ is a topology on $\mathbb{P}$.

The notion of the dual ideal of a filter has no generalization to the context of all p.o.'s, because not every partial order admits a meaningful notion of the *complement* of an element p of $\mathbb{P}$. One important class of p.o.'s for which such a notion is defined, the class of Boolean algebras, will be introduced in Section 13.3 and studied in detail in Chapter 25.

The concept of a filter base has a straightforward generalization: A subset B of $\mathbb{P}$ is a *filter base* in $\langle \mathbb{P}, \leq \rangle$ if it satisfies condition (II) of the definition of a filter, i.e., if every finite subset of B has a lower bound in B. It is easy to see that the family $\mathbb{F}(B) = \{q \in \mathbb{P} : \exists p \in B \, (p \leq q)\}$ is a filter in $\langle \mathbb{P}, \leq \rangle$ if and only if B is a filter base in $\langle \mathbb{P}, \leq \rangle$.

The generalization of the finite intersection property is a bit trickier. As a candidate, one might consider the following notion: If $\langle \mathbb{P}, \leq \rangle$ is a p.o. and A is a subset of $\mathbb{P}$, then A is *centered* if every finite subset of A has a lower bound in $\mathbb{P}$. Unfortunately, it is not always true that every centered subset of a p.o. is contained in a filter.

EXERCISE 13.9 (G). Let $\mathbb{P} = \{\{n\} : n \in \omega\} \cup \{\omega \backslash \{n\} : n \in \omega\}$. Show that $A = \{\omega \backslash \{n\} : n \in \omega\}$ is a centered subset of $\langle \mathbb{P}, \subseteq \rangle$, but A is not contained in any filter in $\langle \mathbb{P}, \subseteq \rangle$.

The trouble with generalizing the fip is caused by the lack of "intersections" in some p.o.'s. If $\langle \mathbb{P}, \leq \rangle$ is a p.o. and $p, q \in \mathbb{P}$, then an element r of $\mathbb{P}$ is called the *greatest lower bound* or *meet* of p and q if r is a lower bound of $\{p, q\}$ and $s \leq r$ for every lower bound s of $\{p, q\}$. A p.o. $\langle \mathbb{P}, \leq \rangle$ is called *well-met* if every two compatible elements of $\mathbb{P}$ have a greatest lower bound in $\mathbb{P}$.[1]

EXERCISE 13.10 (G). Convince yourself that the p.o. of Exercise 13.9 is not well-met, but the p.o.'s of Examples 13.4—13.8 are.

For well-met p.o.'s, the analogue of Exercise 13.2(a) holds.

THEOREM 13.5. *Let $\langle \mathbb{P}, \leq \rangle$ be a well-met p.o., and let A be a centered subset of $\mathbb{P}$. Then there exists a smallest filter $\mathbb{F}$ in $\langle \mathbb{P}, \leq \rangle$ such that $A \subseteq \mathbb{F}$.*

EXERCISE 13.11 (G). Prove Theorem 13.5.

Of course, the smallest filter containing A will be called the *filter generated by A*. By Theorems 13.4 and 13.5, every centered subset of a well-met p.o. $\langle \mathbb{P}, \leq \rangle$ is contained in an ultrafilter in this p.o.

Let us now present two important applications of ultrafilters in topology.

THEOREM 13.6. *Let X be a topological space. The following are equivalent:*
(i)   *X is compact;*
(ii) *Every filter of closed subsets of X has nonempty intersection;*
(iii) *Every ultrafilter of closed subsets of X is fixed.*

PROOF. Let X be a fixed topological space. For any filter $\mathbb{F}$ of closed subsets of X, let $\mathbb{F}^\bullet = \{X \backslash Y : Y \in \mathbb{F}\}$. Of course, $\mathbb{F}^\bullet$ is a family of open sets.

EXERCISE 13.12 (G). Show that for every filter in $\mathcal{K}(X)$ the following are equivalent:
(1) $\bigcap \mathbb{F} = \emptyset$;
(2) $\mathbb{F}^\bullet$ is an open cover of X without finite subcover.

Moreover, show that if $\mathbb{F}$ is an ultrafilter in $\mathcal{K}(X)$, then $\mathbb{F}$ is free if and only if it satisfies conditions (1) and (2).

The implications (i) $\Rightarrow$ (ii) and (ii) $\Rightarrow$ (iii) follow immediately from Exercise 13.12. For the implication (iii) $\Rightarrow$ (i), assume that X is not compact, and let $\mathcal{U}$ be an open cover of X without finite subcover. Then the set $\mathcal{U}^\bullet = \{X \backslash U : U \in \mathcal{U}\}$ is a centered subset of $\mathcal{K}(X)$, and since $\langle \mathcal{K}(X), \subseteq \rangle$ is a well-met p.o., by Theorem 13.5 there exists an ultrafilter $\mathbb{F}$ of closed subsets of X with $\mathcal{U}^\bullet \subseteq \mathbb{F}$. Since $\bigcap \mathcal{U}^\bullet = \emptyset$, also $\bigcap \mathbb{F} = \emptyset$, and Exercise 13.12 implies that $\mathbb{F}$ is not fixed. □

---
[1] A well-met p.o. with no incompatible elements is called a *lower semilattice*.

THEOREM 13.7 (Tychonoff's Theorem). *If $\{X_i : i \in I\}$ is any family of compact topological spaces, then $\prod_{i \in I} X_i$ is compact.*

PROOF. Let $\{X_i : i \in I\}$ be as in the assumption, and let $\mathbb{F}$ be an ultrafilter of closed subsets of $\prod_{i \in I} X_i$. By Theorem 13.6 and Exercise 13.11, it suffices to show that $\bigcap \mathbb{F} \neq \emptyset$. For each $i \in I$, let $\mathbb{F}_i = \{Y \subseteq X_i : \pi_i^{-1} Y \in \mathbb{F}\}$. Since the projection π_i on the i-th coordinate is a continuous function and since inverse images preserve set-theoretic operations, $\mathbb{F}_i$ is a filter of closed subsets of X_i. By compactness of X_i, $\bigcap \mathbb{F}_i \neq \emptyset$ for every $i \in I$. Now the following exercise concludes the proof of Theorem 13.7.

EXERCISE 13.13 (R). Show that if $f \in \prod_{i \in I} X_i$ is such that $f(i) \in \bigcap \mathbb{F}_i$ for every $i \in I$, then $f \in \bigcap \mathbb{F}$. □

REMARK 13.8. In Chapter 9 we showed that Tychonoff's Theorem implies the Axiom of Choice. Note that (AC) is necessary in the proof of Theorem 13.7 in order to guarantee the existence of a function f as in Exercise 13.13. But if each of the spaces X_i is Hausdorff, then the function f is uniquely determined.

EXERCISE 13.14 (G). Show that if X_i is Hausdorff and $\mathbb{F}_i$ is defined as in the proof of Theorem 13.7, then $|\bigcap \mathbb{F}_i| = 1$.

However, this does not mean that the restriction of Tychonoff's Theorem to the class of Hausdorff spaces is provable in ZF alone. We also used the fact that there exists an ultrafilter of closed sets in $\prod_{i \in I} X_i$, and this fact is not provable in ZF alone, although it follows from the so-called *Prime Ideal Theorem*[2] which is weaker than the full Axiom of Choice (see T. Jech, *The Axiom of Choice*, North Holland, Amsterdam, 1973).

The lack of "intersections" in arbitrary p.o.'s also dictates some caution in generalizing the notion of κ-closedness. Let κ be a regular uncountable cardinal. A p.o. $\langle \mathbb{P}, \leq \rangle$ is κ-*closed* if every decreasing sequence $\langle p_\xi : \xi < \lambda \rangle$ of elements of $\mathbb{P}$ of length $\lambda < \kappa$ has a lower bound in $\mathbb{P}$. A p.o. $\langle \mathbb{P}, \leq \rangle$ is κ-*directed closed* if every filter base $B \subseteq \mathbb{P}$ of size less than κ has a lower bound in $\mathbb{P}$. Since every chain in a p.o. is a filter base in this p.o., κ-directed closed p.o.'s are κ-closed. However, the converse is not true.

EXERCISE 13.15 (PG). (a) Consider the following strict partial order relation on $[\omega_1]^{\leq \aleph_0}$: $X < Y$ iff Y is a proper subset of X and Y is finite. Show that the corresponding p.o. $\langle [\omega_1]^{\leq \aleph_0}, \leq \rangle$ is $\aleph_2$-closed but not $\aleph_2$-directed closed.

(b) Show that a p.o. is countably closed (i.e., $\aleph_1$-closed) if and only if it is $\aleph_1$-directed closed.

A filter $\mathbb{F}$ in a p.o. $\langle \mathbb{P}, \leq \rangle$ will be called κ-*closed* if the p.o. $\langle \mathbb{F}, \leq \rangle$ is κ-closed.

EXERCISE 13.16 (PG). Show that if $\mathbb{F}$ is a filter in a p.o. $\langle \mathbb{P}, \leq \rangle$, then $\langle \mathbb{F}, \leq \rangle$ is κ-closed iff $\langle \mathbb{F}, \leq \rangle$ is κ-directed closed.

Let us conclude this section by introducing a few concepts that will play an important role in Chapters 18 and 19. A subset $A \subseteq \mathbb{P}$ is an *antichain* in a p.o. $\langle \mathbb{P}, \leq \rangle$ if every two elements of A are incompatible. The set A is a *maximal*

---

[2]The Prime Ideal Theorem is a statement very similar to our Theorem 13.4. It states that every Boolean algebra has a prime ideal.

13.1. THE GENERAL CONCEPT OF A FILTER

antichain in $\langle \mathbb{P}, \leq \rangle$ if it is not properly contained in any other antichain of the same p.o.

EXAMPLE 13.10. Let $A = \{\{n\} : n \in \omega\}$. Then A is an antichain both of $\langle \mathcal{P}(\omega)\backslash\{\emptyset\}, \subseteq \rangle$ and $\langle \mathcal{P}(\omega_1)\backslash\{\emptyset\}, \subseteq \rangle$, but not of $\langle \mathcal{P}(\omega), \subseteq \rangle$. The antichain A is maximal in $\langle \mathcal{P}(\omega)\backslash\{\emptyset\}, \subseteq \rangle$, but not in $\langle \mathcal{P}(\omega_1)\backslash\{\emptyset\}, \subseteq \rangle$.

A subset $D \subseteq \mathbb{P}$ is *dense* in the p.o. $\langle \mathbb{P}, \leq \rangle$ if $\forall p \in \mathbb{P} \, \exists q \in D \, (q \leq p)$. Note that the set A of Example 13.10 is dense in $\langle \mathcal{P}(\omega)\backslash\{\emptyset\}, \subseteq \rangle$, but is not dense in $\langle \mathcal{P}(\omega), \subseteq \rangle$.

EXERCISE 13.17 (G). Show that if $D \subseteq \mathbb{P}$ is a dense subset in a p.o. $\langle \mathbb{P}, \leq \rangle$, then there exists a maximal antichain A in this p.o. such that $A \subseteq D$.

A subset $C \subseteq \mathbb{P}$ is *predense* in a p.o. $\langle \mathbb{P}, \leq \rangle$ if $\forall p \in \mathbb{P} \, \exists q \in C \, (p \not\perp q)$. Note that both dense subsets and maximal antichains of $\langle \mathbb{P}, \leq \rangle$ are examples of predense subsets of $\langle \mathbb{P}, \leq \rangle$.

EXERCISE 13.18 (G). (a) Show that if C is a finite predense subset of a p.o. $\langle \mathbb{P}, \leq \rangle$ and $\mathbb{F}$ is an ultrafilter in $\langle \mathbb{P}, \leq \rangle$, then $\mathbb{F} \cap C \neq \emptyset$.
(b) Show that if $\langle \mathbb{P}, \leq \rangle$ has the property that for all $p \in \mathbb{P}$ there exist $q, r \leq p$ such that $q \perp r$, then for every ultrafilter $\mathbb{F}$ in $\langle \mathbb{P}, \leq \rangle$, the set $\mathbb{P}\backslash\mathbb{F}$ is dense in $\langle \mathbb{P}, \leq \rangle$.

By point (b) of the last exercise, as long as the partial order $\langle \mathbb{P}, \leq \rangle$ has enough incompatible elements, no ultrafilter in $\langle \mathbb{P}, \leq \rangle$ intersects every predense subset of $\mathbb{P}$. However, it is often useful to know that there exist ultrafilters that intersect every set in a given large family of predense subsets of $\mathbb{P}$. Sufficient conditions for the existence of such ultrafilters will be investigated in Chapters 18 and 19.

Exercise 13.18(a) can be generalized a little if the filter in question has certain completeness properties.

EXERCISE 13.19 (G). Show that if κ is a regular infinite cardinal, C is a predense subset of a p.o. $\langle \mathbb{P}, \leq \rangle$ such that $|C| < \kappa$, and $\mathbb{F}$ is a κ-complete ultrafilter in $\langle \mathbb{P}, \leq \rangle$, then $\mathbb{F} \cap C \neq \emptyset$.

A p.o. $\langle \mathbb{P}, \leq \rangle$ satisfies the *κ-chain condition* (abbreviated κ-c.c.), if every antichain in $\langle \mathbb{P}, \leq \rangle$ has cardinality less than κ. The $\aleph_1$-c.c. is also called the *countable chain condition* and abbreviated *c.c.c.*[3] By Exercise 13.17, if $\langle \mathbb{P}, \leq \rangle$ has the κ-c.c., then every predense subset of $\mathbb{P}$ contains a predense subset of size less than κ.

EXAMPLE 13.11. If $\kappa > 0$ is a cardinal, then $\langle \mathcal{P}(\kappa)\backslash\emptyset, \subseteq \rangle$ has the κ^+-c.c., but does not have the κ-c.c.

EXAMPLE 13.12. A topological space $\langle X, \tau \rangle$ has the c.c.c. if and only if the p.o. $\langle \tau\backslash\{\emptyset\}, \subseteq \rangle$ has the c.c.c.

EXERCISE 13.20 (PG). Let κ be an infinite cardinal. Suppose $\langle \mathbb{P}, \leq \rangle$ is a p.o. with the κ-c.c. and $\mathbb{F}$ is a κ-complete ultrafilter in $\langle \mathbb{P}, \leq \rangle$. Show that there exists $p \in \mathbb{F}$ with no incompatible elements below p, i.e., such that $\forall q, r \leq p \, (q \not\perp r)$.

---

[3] Of course, this should be called the *countable antichain condition*, but our inappropriate terminology is used throughout the literature.

13.2. Ultraproducts

We are going to discuss a fundamental construction in model theory that utilizes ultrafilters. Throughout this section, let L denote a first-order language with nonlogical symbols $\{r_i : i \in I\} \cup \{f_j : j \in J\} \cup \{c_k : k \in K\}$. Let $X \neq \emptyset$, and suppose we are given an indexed family $\{\mathfrak{A}_\xi : \xi \in X\}$ of models of L. For $\xi \in X$, let $\mathfrak{A}_\xi = \langle A_\xi, (R_{i,\xi})_{i \in I}, (F_{j,\xi})_{j \in J}, (C_{k,\xi})_{k \in K}\rangle$. Let $\mathcal{U}$ be a proper ultrafilter on X. We are going to define a model $\prod_{\xi \in X} \mathfrak{A}_\xi / \mathcal{U}$ of L, called the *ultraproduct of* $\{\mathfrak{A}_\xi : \xi \in X\}$ *by* $\mathcal{U}$. First, we define a relation $\sim_\mathcal{U}$ on $\prod_{\xi \in X} A_\xi$ by:

$$g \sim_\mathcal{U} h \text{ iff } \{\xi \in X : g(\xi) = h(\xi)\} \in \mathcal{U}.$$

If $g \sim_\mathcal{U} h$, then we say that g *is equal to* h *modulo* $\mathcal{U}$.

EXERCISE 13.21 (G). Show that $\sim_\mathcal{U}$ is an equivalence relation on $\prod_{\xi \in X} A_\xi$.

For $g \in \prod_{\xi \in X} A_\xi$ let $g/\mathcal{U}$ denote the equivalence class $\{h \in \prod_{\xi \in X} A_\xi : g \sim_\mathcal{U} h\}$, and let $\prod_{\xi \in X} A_\xi / \mathcal{U} = \{g/\mathcal{U} : g \in \prod_{\xi \in X} A_\xi\}$. The set $\prod_{\xi \in X} A_\xi / \mathcal{U}$ is called the *reduced product* of the A_ξ's.

EXERCISE 13.22 (PG). (a) Show that if for some fixed $n \in \omega$, $|A_\xi| \leq n$ for all $\xi \in X$, then $|\prod_{\xi \in X} A_\xi / \mathcal{U}| \leq n$.

(b) Assume $X = \omega$, and $n \leq |A_n| < \aleph_0$ for each $n \in X$. Show that $\prod_{\xi \in X} A_\xi / \mathcal{U}$ is finite iff $\mathcal{U}$ is principal.

Now we define relations R_i, functions F_j, and constants C_k on $\prod_{\xi \in X} A_\xi / \mathcal{U}$. Suppose r_i and f_j have arity n, and let $g_0, \ldots, g_{n-1}, g_n \in \prod_{\xi \in X} A_\xi$. Then we define

(1) $\langle g_{0/\mathcal{U}}, \ldots, g_{n-1/\mathcal{U}}\rangle \in R_i$ iff $\{\xi \in X : \langle g_0(\xi), \ldots, g_{n-1}(\xi)\rangle \in R_{i,\xi}\} \in \mathcal{U}$;

(2) $g_{n/\mathcal{U}} = F_j(g_{0/\mathcal{U}}, \ldots, g_{n-1/\mathcal{U}})$ iff $\{\xi : g_n(\xi) = F_{j,\xi}(g_0(\xi), \ldots, g_{n-1}(\xi))\} \in \mathcal{U}$.

Note that R_i and F_j were defined in terms of representatives. We need to verify that (1) indeed defines a subset of $(\prod_{\xi \in X} A_\xi / \mathcal{U})^n$ and that (2) defines a function from $(\prod_{\xi \in X} A_\xi / \mathcal{U})^n$ into $\prod_{\xi \in X} A_\xi / \mathcal{U}$.

EXERCISE 13.23 (G). (a) Show that if $g_\ell \sim_\mathcal{U} h_\ell$ for $\ell \leq n$, then it follows that $\{\xi : \langle g_0(\xi), \ldots, g_{n-1}(\xi)\rangle \in R_{i,\xi}\} \in \mathcal{U}$ iff $\{\xi : \langle h_0(\xi), \ldots, h_{n-1}(\xi)\rangle \in R_{i,\xi}\} \in \mathcal{U}$, and $\{\xi : g_n(\xi) = F_{j,\xi}(g_0(\xi), \ldots, g_{n-1}(\xi))\} \in \mathcal{U}$ iff $\{\xi : h_n(\xi) = F_{j,\xi}(h_0(\xi), \ldots, h_{n-1}(\xi))\} \in \mathcal{U}$.

(b) Show that if $g_{n/\mathcal{U}} = F_j(g_{0/\mathcal{U}}, \ldots, g_{n-1/\mathcal{U}}) = g'_{n/\mathcal{U}}$, then $g_n \sim_\mathcal{U} g'_n$.

EXERCISE 13.24 (PG). Define interpretations $C_k \in \prod_{\xi \in X} A_\xi / \mathcal{U}$ of the constant symbols c_k in the spirit of (1) and (2), and verify that your C_k's are well defined.

Now let $\prod_{\xi \in X} \mathfrak{A}_\xi / \mathcal{U} = \langle \prod_{\xi \in X} A_\xi / \mathcal{U}, (R_i)_{i \in I}, (F_j)_{j \in J}, (C_k)_{k \in K}\rangle$. This is the ultraproduct of $\{\mathfrak{A}_\xi : \xi \in X\}$ by $\mathcal{U}$.

One would expect that ultraproducts are big and complicated structures. This is usually the case, but there are exceptions. For example, by Exercise 13.22(a), $\prod_{\xi \in X} \mathfrak{A}_\xi / \mathcal{U}$ may be finite, even if X is infinite and $\mathcal{U}$ is nonprincipal. Moreover, as the next exercise shows, it can happen that $\prod_{\xi \in X} \mathfrak{A}_\xi / \mathcal{U}$ is isomorphic to one of the $\mathfrak{A}_\xi$'s.

EXERCISE 13.25 (PG). Let $\{\mathfrak{A}_\xi : \xi \in X\}$ be as above. Show that if $\mathcal{U}$ is a principal ultrafilter on X, then there exists a $\xi_0 \in X$ such that $\prod_{\xi \in X} \mathfrak{A}_\xi / \mathcal{U} \cong \mathfrak{A}_{\xi_0}$.

The following theorem is called *Łoś' Theorem*[4] or the *Fundamental Theorem on Ultraproducts*.

THEOREM 13.9. *Let $\{\mathfrak{A}_\xi : \xi \in X\}$ be an indexed family of models of a first order language L, and let $\mathcal{U}$ be an ultrafilter on X. If Θ is a formula of L with variables among $v_0, \ldots, v_{n-1}$, and if $g_0, \ldots, g_{n-1} \in \prod_{\xi \in X} A_\xi$, then*

$$\prod_{\xi \in X} \mathfrak{A}_\xi / \mathcal{U} \models \Theta[g_{0/\mathcal{U}}, \ldots, g_{n-1/\mathcal{U}}] \quad \text{iff} \quad \{\xi : \mathfrak{A}_\xi \models \Theta[g_0(\xi), \ldots, g_{n-1}(\xi)]\} \in \mathcal{U}.$$

PROOF. Fix $\Theta, g_0, \ldots, g_{n-1}$ as above. Let us make the following simplifying assumption:

(=) If $g_i \sim_\mathcal{U} g_j$ for some $i < j < n$, then $g_i = g_j$.

Fix a function $r : \prod_{\xi \in X} A_\xi / \mathcal{U} \to \prod_{\xi \in X} A_\xi$ such that $r(g_{/\mathcal{U}}) \in g_{/\mathcal{U}}$ for all $g_{/\mathcal{U}} \in \prod_{\xi \in X} A_\xi / \mathcal{U}$. By (=) we may also require that $r(g_{i/\mathcal{U}}) = g_i$ for all $i < n$. For each valuation[5] $s : \omega \to \prod_{\xi \in X} A_\xi / \mathcal{U}$ and each $\xi \in X$, let $s_\xi = \pi_\xi \circ r \circ s$, where $\pi_\xi : \prod_{\xi \in X} A_\xi \to A_\xi$ is the projection on the ξ-th coordinate (i.e., $s_\xi(m) = r(s(m))(\xi)$). Then $s_\xi : \omega \to A_\xi$ is a valuation for the model $\mathfrak{A}_\xi$. Note that in this terminology, if t is a term of L and $\xi \in X$, then the interpretations t^s and t^{s_ξ} will be members of $\prod_{\xi \in X} A_\xi / \mathcal{U}$ and A_ξ respectively.

LEMMA 13.10. *Let s, s_ξ be as above, and let t be a term of L. Then*

$(3)_t \quad \forall g \in \prod_{\xi \in X} A_\xi \, (g_{/\mathcal{U}} = t^s \leftrightarrow \{\xi : g(\xi) = t^{s_\xi}\} \in \mathcal{U}).$

PROOF. By induction over the length of t, i.e., by induction over the well-founded relation of being a subterm.

If $t = v_m$, then $t^s = s(m)$, $t^{s_\xi} = s_\xi(m)$, and $(3)_t$ just expresses a necessary and sufficient condition for g to be equal to $r(s(m))$ modulo $\mathcal{U}$.

If $t = c_k$ for some k, then $t^s = C_k$; and $(3)_t$ should follow from your solution to Exercise 13.24. If it doesn't, you'd better return to Exercise 13.24 and redo it.

The inductive step is covered by the next exercise.

EXERCISE 13.26 (PG). Suppose $t = f_j(t_0, \ldots, t_{n-1})$, and $(3)_{t_\ell}$ holds for every $\ell < n$. Show that $(3)_t$ holds. *Hint:* Use (2) and the definition $t^s = F_j(t_0^s, \ldots, t_{n-1}^s)$. □

Now we are ready to prove Theorem 13.9. For each formula φ of L, consider the statement

$(4)_\varphi \quad \forall s : \omega \to \prod_{\xi \in X} A_\xi / \mathcal{U} \, (\prod_{\xi \in X} \mathfrak{A}_\xi / \mathcal{U} \models_s \varphi \leftrightarrow \{\xi : \mathfrak{A}_\xi \models_{s_\xi} \varphi\} \in \mathcal{U}).$

We are going to prove by induction over the length of φ that $(4)_\varphi$ holds for all formulas φ. In particular, this will imply that $(4)_\Theta$ holds. By the choice of r, if $s(i) = g_{i/\mathcal{U}}$ for some $i < n$, then $s_\xi(i) = g_i(\xi)$. Thus $(4)_\Theta$ yields the last line of Theorem 13.9 for arbitrary Θ and $g_0, \ldots, g_{n-1}$ that satisfy (=).

---

[4] Named after the Polish mathematician Jerzy Łoś who discovered it. The Polish name Łoś is pronounced approximately like the English word "wash."

[5] Valuations and other concepts used in this proof were defined in Chapter 5. These definitions are reviewed at the beginning of this volume.

EXERCISE 13.27 (PG). Deduce Theorem 13.9 in its full generality, i.e., without assuming (=).

To get started with the inductive proof of $(4)_\varphi$, suppose φ is "$t_0 = t_1$" for some terms t_0, t_1. Then $\prod_{\xi \in X} \mathfrak{A}_\xi/\mathcal{U} \models_s \varphi$ if and only if $t_0^s = t_1^s$.

By $(3)_{t_0}$ applied to $g \in t_1^s$, the latter is true if and only if $\{\xi : t_0^{s_\xi} = t_1^{s_\xi}\} \in \mathcal{U}$, and $(4)_\varphi$ follows.

Similarly, if φ is "$r_i(t_0, \ldots, t_{n-1})$," then an argument as in the solution to Exercise 13.26 shows that $\langle t_0^s, \ldots, t_{n-1}^s \rangle \in R_i$ iff $\{\xi : \langle t_0^{s_\xi}, \ldots, t_{n-1}^{s_\xi} \rangle \in R_{i,\xi}\}$, and $(4)_\varphi$ follows.

The induction step involves negation, conjunction, and the existential quantifier.

EXERCISE 13.28 (PG). Suppose that φ and ψ are formulas of L such that $(4)_\varphi$ and $(4)_\psi$ hold. Show that $(4)_{\neg\varphi}$ and $(4)_{\varphi \wedge \psi}$ also hold.

Now suppose that $(4)_\varphi$ holds, where φ is a formula with free variable v_ℓ. We show that $(4)_{\exists v_\ell \varphi}$ holds. Let $s : \omega \to \prod_{\xi \in X} A_\xi/\mathcal{U}$ be a valuation. For the implication from the left to the right, assume that $\prod_{\xi \in X} \mathfrak{A}_\xi/\mathcal{U} \models_s \exists v_\ell \varphi$. By the definition of $\models_s$, this means that there exists a valuation s^* such that $s^*(m) = s(m)$ for all $m \neq \ell$, and $\prod_{\xi \in X} \mathfrak{A}_\xi/\mathcal{U} \models_{s^*} \varphi$. Let $Y = \{\xi : \mathfrak{A}_\xi \models_{(s^*)_\xi} \varphi\}$. By $(4)_\varphi$, $Y \in \mathcal{U}$. Note that for each $\xi \in X$, $(s^*)_\xi : \omega \to A_\xi$ is a valuation such that $(s^*)_\xi(m) = s_\xi(m)$ for all $m \neq \ell$. Therefore, if $\xi \in Y$, then $\mathfrak{A}_\xi \models_{s_\xi} \exists v_\ell \varphi$, and it follows that $\{\xi : \mathfrak{A}_\xi \models_{s_\xi} \exists v_\ell \varphi\} \in \mathcal{U}$.

For the converse, let $Z = \{\xi : \mathfrak{A}_\xi \models_{s_\xi} \exists v_\ell \varphi\}$, and assume $Z \in \mathcal{U}$. For each $\xi \in Z$, choose a valuation $s_\xi^* : \omega \to A_\xi$ such that $s_\xi^*(m) = s_\xi(m)$ whenever $m \neq \ell$, and $\mathfrak{A}_\xi \models_{s_\xi^*} \varphi$. For $\xi \in X \setminus Z$, let $s_\xi^* = s_\xi$. Define a function $g \in \prod_{\xi \in X} A_\xi$ by $g(\xi) = s_\xi^*(\ell)$ for all $\xi \in X$, and let s^* be such that $s^*(m) = s(m)$ if $m \neq \ell$, and $s^*(\ell) = g_{/\mathcal{U}}$. Then $(s^*)_\xi = s_\xi^*$ for all $\xi \in X$, hence $\{\xi : \mathfrak{A}_\xi \models_{(s^*)_\xi} \varphi\} \in \mathcal{U}$. By $(4)_\varphi$, the latter implies that $\mathfrak{A}_\xi \models_{s^*} \varphi$. Since $s^*(m) = s(m)$ for all $m \neq \ell$, we have $\mathfrak{A}_\xi \models_s \exists v_\ell \varphi$. This completes the proof of $(4)_{\exists v_\ell \varphi}$, and simultaneously the proof of Theorem 13.9. $\square$

COROLLARY 13.11. *If T is a theory in L and $\mathfrak{A}_\xi \models T$ for every $\xi \in X$, then $\prod_{\xi \in X} \mathfrak{A}_\xi/\mathcal{U} \models T$.*

Now we can give a simple proof of Theorem 6.7.

COROLLARY 13.12. *Suppose T is a theory in L with the following property: For every natural number n, there exists a natural number $m \geq n$ such that T has an m-element model. Then T has an infinite model.*

PROOF. Let $X = \omega$, and for each $n \in \omega$, let $\mathfrak{A}_n$ be a finite model of T such that $|A_n| \geq n$. Let $\mathcal{U}$ be a nonprincipal ultrafilter on ω. By Corollary 13.11, $\prod_{\xi \in X} \mathfrak{A}_\xi/\mathcal{U} \models T$, and by Exercise 13.22(b), $\prod_{\xi \in X} A_\xi/\mathcal{U}$ is infinite. $\square$

In Exercise 6.9, we suggested that you might use the Compactness Theorem to prove Corollary 13.12. It is perhaps not too surprising that the Compactness Theorem itself also follows from Theorem 13.9.

COROLLARY 13.13. *Let T be a theory in L such that every finite subset $T^- \subseteq T$ has a model. Then T itself has a model.*

13.2. ULTRAPRODUCTS

PROOF. First we prove the corollary for countable T. If T is finite, there is nothing to prove. So suppose T is denumerable, and let $T = \{\varphi_n : n \in \omega\}$. For each $n \in \omega$, choose a model $\mathfrak{A}_n$ of $\{\varphi_0, \ldots, \varphi_{n-1}\}$, let $\mathcal{U}$ be a nonprincipal ultrafilter on ω, and consider $\prod_{n \in \omega} \mathfrak{A}_n/\mathcal{U}$. If $m \in \omega$, then $\{n \in \omega : \mathfrak{A}_n \models \varphi_m\} \supseteq \{n \in \omega : n > m\} \in \mathcal{U}$. By Theorem 13.9, $\prod_{n \in \omega} \mathfrak{A}_n/\mathcal{U} \models \varphi_m$. Since this is true for every $\varphi_m \in T$, we have found a model of T.

Now assume $|T| = \aleph_1$, let $T = \{\varphi_\eta : \eta < \omega_1\}$, and for each $\xi < \omega_1$, let $T_\xi = \{\varphi_\eta : \eta < \xi\}$. We have already shown that each T_ξ has a model $\mathfrak{A}_\xi$. Fix an ultrafilter $\mathcal{U}$ on ω_1 such that $\omega_1 \setminus \eta \in \mathcal{U}$ for every $\eta < \omega_1$.

EXERCISE 13.29 (G). (a) Show that if T, $\{\mathfrak{A}_\xi : \xi < \omega_1\}$, and $\mathcal{U}$ are as above, then $\prod_{\xi \in \omega_1} \mathfrak{A}_\xi/\mathcal{U} \models T$.
(b) Prove Corollary 13.13 for theories T of arbitrary cardinality. *Hint:* Use transfinite induction over $|T|$. □

What consequences does Theorem 13.9 have for models of set theory? Of course, as soon as we want to talk about models of set theory, we run into the metamathematical difficulties connected with Gödels's Second Incompleteness Theorem. For the present discussion it will be best to sidestep these difficulties by assuming that there exists a κ such that $\langle V_\kappa, \bar{\in}\rangle \models \text{ZFC}$.[6] As in Chapters 7 and 12, we will occasionally use the symbol "$\bar{\in}$" for the standard interpretation of the relational symbol "$\in$."

So let κ be as above, let X be a nonempty set, let $\mathfrak{A}_\xi = \langle V_\kappa, \bar{\in}\rangle$ for each $\xi \in X$, and let $\mathcal{U}$ be an ultrafilter on X. As in Chapter 12, we will not make a notational distinction between $\langle V_\kappa, \bar{\in}\rangle$ and V_κ. The ultraproduct $\prod_{\xi \in X} \mathfrak{A}_\xi/\mathcal{U}$ is denoted by $^X V_\kappa/\mathcal{U}$ and called the *ultrapower* of V_κ by $\mathcal{U}$. By Corollary 13.11, $^X V_\kappa/\mathcal{U} \models \text{ZFC}$, and moreover, $^X V_\kappa/\mathcal{U}$ and V_κ are elementarily equivalent (see Definition 6.3). For example, if $V_\kappa \models \text{GCH}$, then also $^X V_\kappa/\mathcal{U} \models \text{GCH}$. In the terminology of Chapter 7, V_κ is a standard model of L_S, and $^X V_\kappa/\mathcal{U}$ is a nonstandard model. The interpretation of the membership relation in V_κ is wellfounded. How about the membership relation in $^X V_\kappa/\mathcal{U}$?

THEOREM 13.14. *Let α be an infinite ordinal, let $X \neq \emptyset$, and let $\langle ^X V_\alpha/\mathcal{U}, E\rangle$ be the ultrapower of $\langle V_\alpha, \bar{\in}\rangle$ by $\mathcal{U}$. Then the relation E is wellfounded if and only if $\mathcal{U}$ is countably complete.*

PROOF. First assume towards a contradiction that $\mathcal{U}$ is countably complete and that $(g_n)_{n \in \omega}$ is a sequence of elements of $^X V_\alpha$ with $\langle g_{n+1}/\mathcal{U}, g_n/\mathcal{U}\rangle \in E$ for all $n \in \omega$. For each n, let $X_n = \{\xi : g_{n+1}(\xi) \in g_n(\xi)\}$. Since $\mathcal{U}$ is countably complete, the intersection $\bigcap_{n \in \omega} X_n$ is nonempty. Let $\xi \in \bigcap_{n \in \omega} X_n$. Then $g_{n+1}(\xi) \bar{\in} g_n(\xi)$ for every $n \in \omega$. But this is impossible, since $\bar{\in}$ is wellfounded.

Now suppose that $\mathcal{U}$ is not countably complete. By Exercise 13.3(b), there exists a decreasing sequence $(X_m)_{m \in \omega}$ of elements of $\mathcal{U}$ such that $\bigcap_{m \in \omega} X_m = \emptyset$. For each $n \in \omega$, define $g_n \in {}^X V_\alpha$ by:

$$g_n(\xi) = \min\{m \in \omega : \xi \notin X_{m+n}\}.$$

EXERCISE 13.30 (G). Convince yourself that $g_{n+1}(\xi) = g_n(\xi) - 1$ for each $n \in \omega$ and $\xi \in X_{n+1}$.

---

[6] In Chapter 12 it was shown that every strongly inaccessible κ has this property. A more sophisticated way of sidestepping the metamathematical difficulties is outlined in the Mathographical Remarks. In Chapter 24, we will use yet another approach to get around the problem.

Since each X_{n+1} belongs to $\mathcal{U}$ and $m - 1 \in m$ for $m \in \omega\setminus\{0\}$, Exercise 13.30 together with condition (1) implies that $\langle g_{n+1/\mathcal{U}}, g_{n/\mathcal{U}}\rangle \in E$ for all $n \in \omega$, and hence E is not a wellfounded relation. □

Let V_α, X, $\mathcal{U}$ be as in the assumptions of Theorem 13.14. If $\mathcal{U}$ is principal, then $\mathcal{U}$ is countably complete. By Exercise 13.25, ${}^X V_\alpha/\mathcal{U} \cong V_\alpha$. Can one get an ultrapower of V_α that is wellfounded but *not* isomorphic to V_α? By Theorem 13.14, this boils down to the following question:

QUESTION 13.15. Are there nonprincipal, countably complete ultrafilters on any set X?

Question 13.15 was first posed and studied by the Polish mathematician Stanisław Ulam 25 years before Łoś' Theorem was discovered. Ulam was motivated by considerations in measure theory. Question 13.15 has such far-reaching consequences for set theory that we devote the whole Chapter 23 to it. For now, let us just divulge that it has a positive answer if and only if there exists a measurable cardinal.

Now let us suppose that $V_\kappa \models \text{ZFC}$, and consider the question: "How does ${}^X V_\kappa/\mathcal{U}$ look if $\mathcal{U}$ is *not* countably complete?" The answer is: "Weird." For example, let $(g_n)_{n\in\omega}$ be the sequence of functions constructed in the second part of the proof of Theorem 13.14. Fix $n \in \omega$. For each $\xi \in X$,

(13.1) $\qquad\qquad\qquad V_\kappa \models g_n(\xi)$ is a finite ordinal.

Thus, by Theorem 13.9,

(13.2) $\qquad\qquad\qquad {}^X V_\kappa/\mathcal{U} \models g_{n/\mathcal{U}}$ is a finite ordinal.

On the other hand, as shown in the proof of Theorem 13.14,

(13.3) $\qquad\qquad\qquad {}^X V_\kappa/\mathcal{U} \models g_{n+1/\mathcal{U}} = g_{n/\mathcal{U}} - 1$.

Now a simple inductive argument shows that $g_{n/\mathcal{U}}$ cannot be ${}^X V_\kappa/\mathcal{U}$'s version of the smallest finite ordinal 0, the fifth finite ordinal 5, or any other "standard" natural number. The "finite ordinal" $g_{n/\mathcal{U}}$ must be larger than any of these. An object like $g_{n/\mathcal{U}}$ is called a *nonstandard natural number*. ${}^X V_\kappa/\mathcal{U}$ believes that $g_{n/\mathcal{U}}$ is finite, but there are infinitely many $g_{/\mathcal{U}}$'s in ${}^X V_\kappa/\mathcal{U}$ such that ${}^X V_\kappa/\mathcal{U} \models g_{/\mathcal{U}} \in g_{n/\mathcal{U}}$. This gives an alternative method for deriving a negative answer to Question 6.20.

EXERCISE 13.31 (R). Assume that $\mathcal{U}$ is not countably complete. If $V_\kappa \models \text{ZFC}$, then for each $n \in \omega$ there is $\bar{n} \in {}^X V_\kappa/\mathcal{U}$ such that

$$ {}^X V_\kappa/\mathcal{U} \models \bar{n} \text{ is the } n\text{-th natural number.} $$

Show that $\{\bar{n} : n \in \omega\} \notin {}^X V_\kappa/\mathcal{U}$.[7]

13.3. A first look at Boolean algebras

Let A be an arbitrary set. The *power set algebra* of A is defined as the structure $\langle \mathcal{P}(A), \cup, \cap, -, \emptyset, A\rangle$. Here $\cup$, $\cap$, and $\emptyset$ have their regular meaning, and $-$ stands for the complement $A \setminus X$ of a subset X of A. A *set algebra* or *field of sets* is a substructure of a power set algebra, i.e., F is a field of sets if there is a suitable set A such that $F \subseteq \mathcal{P}(A)$ and

---

[7] The set $\{\bar{n} : n \in \omega\}$ is called the *standard part* of $\omega^{{}^X V_\kappa/\mathcal{U}}$.

(i) $\emptyset, A \in F$;
(ii) If $X, Y \in F$, then $X \cup Y \in F$;
(iii) If $X \in F$, then $-X \in F$.

We also say that F is a *subfield* of $\mathcal{P}(A)$.

EXERCISE 13.32 (G). (a) Show that if F is a subfield of $\mathcal{P}(A)$ and $\mathcal{P}(B)$, then $A = B$.
(b) Show that if F is a field of sets, then $X \cap Y \in F$ for all $X, Y \in F$.

EXAMPLE 13.13. Let $A \neq \emptyset$, $F = \{\emptyset, A\}$. Then F is a field of sets that contains exactly two elements.

EXAMPLE 13.14. For an infinite set A the family $Finco(A) = \{X \subseteq A : |X| < \aleph_0 \vee |-X| < \aleph_0\}$ of all finite and all cofinite subsets of A is a subfield of $\mathcal{P}(A)$.

EXAMPLE 13.15. Let A be an arbitrary set, κ an infinite cardinal. Define
$$F_{<\kappa}(A) = \{X \subseteq A : |X| < \kappa \vee |-X| < \kappa\}.$$
Then $F_{<\kappa}$ is a subfield of $\mathcal{P}(A)$. Note that if $\kappa > |A|$, then $F_{<\kappa} = \mathcal{P}(A)$, and if $\kappa = \omega$, then $F_{<\omega}(A) = Finco(A)$.

A set algebra F is κ-*complete* if $\forall \mathcal{X} \subseteq F(|\mathcal{X}| < \kappa \to \bigcup \mathcal{X} \in F)$. Note that a power set algebra is κ-complete for every κ.

EXERCISE 13.33 (PG). (a) Let κ be a regular infinite cardinal. Show that $F_{<\kappa}(A)$ is κ-complete for every set A.
(b) Let A be a set with $|A| \geq \aleph_\omega$. Show that $F_{\aleph_\omega}(A)$ is not $\aleph_\omega$-complete.
(c) Let F be a κ-complete field of sets, and let $\mathcal{X} \subseteq F$ be such that $|\mathcal{X}| < \kappa$. Show that $\bigcap \mathcal{X} \in F$.
(d) Let A be an arbitrary set, let $\mathcal{X} \subseteq \mathcal{P}(A)$, and let κ be an infinite cardinal. Show that there exists a smallest κ-complete set algebra F such that $\mathcal{X} \subseteq F$.

An $\aleph_1$-complete field of sets F is also called σ-*complete* or a σ-*field*.

Let $\langle X, \tau \rangle$ be a topological space, and let $\mathcal{B}$ be the smallest σ-field of subsets of X that contains τ. The elements of $\mathcal{B}$ are called *Borel sets*. In particular, all open and all closed subsets of X are Borel sets. For $\alpha > 0$, we define recursively families $\mathbf{\Sigma}^0_\alpha$ and $\mathbf{\Pi}^0_\alpha$ of subsets of X as follows:

$\mathbf{\Sigma}^0_1 = \tau$, $\qquad\qquad\qquad\qquad \mathbf{\Pi}^0_1 = \{X \setminus A : A \in \mathbf{\Sigma}^0_1\}$,
$\mathbf{\Sigma}^0_\alpha = \{\bigcup M : M \in [\bigcup_{\beta < \alpha} \mathbf{\Pi}^0_\beta]^{\leq \aleph_0}\}$, $\qquad \mathbf{\Pi}^0_\alpha = \{X \setminus A : A \in \mathbf{\Sigma}^0_\alpha\}$.

Note that, in particular, $\mathbf{\Pi}^0_1$ is the family of closed sets, $\mathbf{\Sigma}^0_2$ is the family of F_σ-sets, and $\mathbf{\Pi}^0_2$ is the family of G_δ-subsets of X.[8]

EXERCISE 13.34 (PG). (a) Show that $\mathbf{\Sigma}^0_\alpha = \mathbf{\Sigma}^0_{\omega_1}$ for all $\alpha \geq \omega_1$.
(b) Show that if X is a Polish space and $\alpha < \beta$, then $\mathbf{\Sigma}^0_\alpha \subseteq \mathbf{\Sigma}^0_\beta, \mathbf{\Pi}^0_\alpha \subseteq \mathbf{\Pi}^0_\beta$.

A set $A \subseteq X$ is *nowhere dense* if every nonempty open set U contains a nonempty open subset $V \subseteq U$ such that $V \cap A = \emptyset$. Let $\mathcal{L}$ be the family of nowhere dense subsets of X, and let $\mathcal{M} = \{\bigcup C : C \subseteq \mathcal{L} \wedge |C| \leq \aleph_0\}$. The sets

---

[8]Descriptive set theorists distinguish between "lightface" and "boldface" versions of the families $\mathbf{\Sigma}^0_\alpha$ and $\mathbf{\Pi}^0_\alpha$. The difference is that arbitrary subsets of ω are allowed as parameters in the definitions of the sets appearing in the boldface families, but not in the lightface families. The families defined here are the "boldface" classes.

in $\mathcal{M}$ are called *meager*. Note that every meager subset of a topological space is included in a meager F_σ-set. A set $A \subseteq X$ has the *Baire property*, if there exists an open set $B \subseteq X$ such that the symmetric difference $A \triangle B$ is meager.

EXERCISE 13.35 (PG). (a) Show that if A has the Baire property then $X \setminus A$ also has the Baire property.

(b) Show that the family of subsets of X that has the Baire property forms a σ-algebra.

(c) Conclude that every Borel set has the Baire property.

Closely connected with set algebras are *Boolean algebras*. A Boolean algebra is a structure $\mathfrak{B} = \langle B, +, \cdot, -, 0, 1 \rangle$, where $+, \cdot$ are binary functions, $-$ is a unary function, $0, 1$ are constants such that the following equalities hold for all $a, b, c \in B$:

(Ba1) $\quad (a + b) + c = a + (b + c), \qquad (a \cdot b) \cdot c = a \cdot (b \cdot c),$

(Ba2) $\quad a + b = b + a, \qquad a \cdot b = b \cdot a,$

(Ba3) $\quad a + (a \cdot b) = a, \qquad a \cdot (a + b) = a,$

(Ba4) $\quad a \cdot (b + c) = a \cdot b + a \cdot c, \qquad a + (b \cdot c) = (a + b) \cdot (a + c),$

(Ba5) $\quad a + (-a) = 1, \qquad a \cdot (-a) = 0.$

A structure $\mathfrak{A} = \langle A, +, \cdot \rangle$ that satisfies axioms (Ba1)–(Ba3) is called a *lattice*.

A casual look at the axioms (Ba1)–(Ba5) reveals that for every axiom also the "dual" axiom appears, i.e., the axiom obtained by substituting "$\cdot$" for "$+$," "0" for "1," and vice versa. It follows that if φ is any theorem of the theory of Boolean algebras, then one can prove the dual theorem by switching "$+$" and "$\cdot$" as well as "0" and "1" throughout the proof of φ.

REMARK 13.16. Since $-$ is a unary function, we can simply write $a \cdot -b$ instead of $a \cdot (-b)$. We shall also frequently write B instead of $\langle B, +, \cdot, 0, 1 \rangle$ if the operations are implied by the context.

Obviously, every set algebra is a Boolean algebra. We do not require that $0 \neq 1$. In particular, the set algebra $\langle \{\emptyset\}, \cup, \cap, -, \emptyset, \emptyset \rangle$ is the (unique up to isomorphism) one-element Boolean algebra. We refer to $\langle \{\emptyset\}, \cup, \cap, -, \emptyset, \emptyset \rangle$ as the *trivial Boolean algebra*.

EXERCISE 13.36 (PG). Show that every two-element Boolean algebra is isomorphic to the set algebra of Example 13.13.

EXAMPLE 13.16. Let $\langle X, \tau \rangle$ be a topological space, and let $A \subseteq X$. We say that $A \subseteq X$ is *clopen* if both $A, X \setminus A \in \tau$. Let $Clop(X)$ be the family of clopen subsets of X. Then $\langle Clop(X), \cap, \cup, -, \emptyset, X \rangle$ is a Boolean algebra.

LEMMA 13.17. *Let $\langle B, +, \cdot \rangle$ be a lattice, and let $a \in B$. Then*
$$a + a = a, \qquad a \cdot a = a.$$

PROOF. By (Ba3) we have
$$a = a + a \cdot (a + a) = a + a.$$

The second equality follows from duality. $\square$

LEMMA 13.18. *Let B be a lattice, $a, b \in B$. Then*
$$b = a + b \;\; \textit{iff} \;\; a = a \cdot b.$$

PROOF. Let $b = a + b$. Then (Ba3) implies that $a \cdot b = a \cdot (a + b) = a$. The other direction follows from duality. □

Let B be a lattice. We define a binary relation R on B by letting aRb if $a \cdot b = a$. By Lemma 13.18, this is equivalent to the equality $a + b = b$. Note that if B is a set algebra, then aRb if and only if $a \subseteq b$.

LEMMA 13.19. *Let B be a lattice and let R be the relation defined above. Then $\langle B, R \rangle$ is a p.o.*

PROOF. Clearly, R is reflexive.

If aRb and bRa, then $a = a \cdot b$ and $b = a \cdot b$; hence (Ba2) implies that $a = b$. This shows that R is antisymmetric.

For transitivity, let $a, b, c \in B$, and suppose aRb and bRc. Then $a \cdot c = (a \cdot b) \cdot c = a \cdot (b \cdot c) = a \cdot b = a$. □

We write $a \leq_B b$ (or simply $a \leq b$ if B is implied by the context) instead of aRb. If B is a Boolean algebra, then we call $\leq_B$ the *Boolean order*. The duality principle can be extended to formulas that contain $\leq$: In order to obtain the dual statement to φ, exchange $\leq$ with $\geq$, as well as $+$ with $\cdot$ and 0 with 1.

It follows immediately from (Ba3) that $a \leq a + b$ for all $a, b \in B$. By duality, $a \geq a \cdot b$ for all $a, b \in B$.

LEMMA 13.20. *Let B be a lattice and let $a, b, c \in B$. If $a \leq b$, then $a + c \leq b + c$.*

PROOF. Let $a \leq b$. Then $a + b = b$, and by regrouping we get $(a+c)+(b+c) = (a+b)+(c+c) = b+c$. □

EXERCISE 13.37 (G). Show that for each Boolean algebra B and all $a, b, c \in B$, the inequality $a \leq b$ implies $a \cdot c \leq b \cdot c$.

Now that we have introduced a natural p.o. relation on every lattice, it makes sense to talk about such order concepts as greatest lower bound (glb), least upper bound (lub), minimum element, etc.

LEMMA 13.21. *Let B be a lattice and let $a, b \in B$. Then $a + b = lub\{a, b\}$ and $a \cdot b = glb\{a, b\}$.*

PROOF. We only show that $a + b = lub\{a, b\}$; the other part follows from duality.

We know already that $a \leq a + b$ and $b \leq a + b$; hence $a + b$ is an upper bound of $\{a, b\}$.

Now suppose x is an upper bound of $\{a, b\}$. Then $a + x = x$ and $b + x = x$. Hence $(a + b) + x = a + (b + x) = a + x = x$, which implies that $a + b \leq x$. □

EXERCISE 13.38 (G). Suppose $\langle L, \leq \rangle$ is a p.o. such that every two-element subset of L has a least upper bound and a greatest lower bound. For $a, b \in L$ define: $a + b = lub\{a, b\}$ and $a \cdot b = glb\{a, b\}$. Show that the structure $\langle L, +, \cdot \rangle$ is a lattice.

LEMMA 13.22. *Let B be a lattice. Then $\forall a, b, c \in B(a \cdot (b + c) = a \cdot b + a \cdot c)$ if and only if $\forall a, b, c \in B(a + (b \cdot c) = (a + b) \cdot (a + c))$.*

PROOF. We only prove the implication from the left to the right; the other implication follows from duality. Let $a, b, c \in B$. We have

$$\begin{aligned}(a+b)\cdot(a+c) &= (a+b)\cdot a + (a+b)\cdot c \\ &= a + (a+b)\cdot c \\ &= a + (a\cdot c + b\cdot c) \\ &= a + b\cdot c\end{aligned}$$

The first and third equalities follow from the assumption $\forall a,b,c \in B(a\cdot(b+c) = a\cdot b + a\cdot c)$; the second and fourth ones follow from (Ba3). □

LEMMA 13.23. *Let $\langle B,+,\cdot,-,0,1\rangle$ be a Boolean algebra. Then 0 is the smallest and 1 is the largest element of B with respect to the Boolean order.*

PROOF. (Ba3) and (Ba5) imply that $a = a + a\cdot -a = a + 0$ for all $a \in B$, and hence $0 \leq a$. The dual argument shows that 1 is the largest element of B. □

Let B be a Boolean algebra; $a,b \in B$. On the one hand, we have $a = 1 \cdot a = (-a + -(-a))\cdot a = -(-a)\cdot a$, and hence $a \leq -(-a)$. On the other hand, $-(-a)\cdot 1 = -(-a)(a + -a) = -(-a)\cdot a$, and hence $-(-a) \leq a$. Thus, $a = -(-a)$. Moreover, $(a+b) + (-a\cdot -b) = (a + -a\cdot -b) + b = (a\cdot a + a\cdot -b + 0 + -a\cdot -b) + b = (a + -a)\cdot(a + -b) + b = 1\cdot(a + -b) + b = a + -b + b = 1$.

EXERCISE 13.39 (G). Show that if B is a Boolean algebra and $a,b \in B$, then $(a+b)\cdot(-a\cdot -b) = 0$.

The last exercise and duality yield the following:

LEMMA 13.24 (de Morgan's Laws). *Let B be a Boolean algebra; $a,b \in B$. Then $-(a+b) = -a\cdot -b$ and $-(a\cdot b) = -a + -b$.*

EXERCISE 13.40 (G). Suppose A and B are Boolean algebras and $\varphi : A \to B$ is such that $\varphi(0_A) = 0_B$, $\varphi(a\cdot b) = \varphi(a)\cdot\varphi(b)$, and $\varphi(-a) = -\varphi(a)$ for all $a,b \in A$. Use de Morgan's Laws to show that φ is a homomorphism.

EXERCISE 13.41 (PG). Prove that if B is a Boolean algebra and $a,b \in B$, then
(i) $a\cdot b = 0$ if and only if $a\cdot b = a$;
(ii) $a\cdot b = a$ if and only if $-a\cdot -b = -b$;
(iii) $a \leq b$ if and only if $a\cdot -b = 0$; and
(iv) $a \leq b$ if and only if $-a + b = 1$.

Let B be a Boolean algebra, and let $a,b \in B$ be such that $a\cdot b = 0$. Then $b \leq -a$, i.e., $-a$ is the largest element $b \in B$ with $a\cdot b = 0$.

We say that $-a$ is the *complement of a*. By Exercise 13.41(iii)(iv), $-a$ is the unique element b of B with $a\cdot b = 0$ and $a + b = 1$. In terms of the Boolean order, $-a$ is the largest element b of B such that a and b are incompatible in the p.o. $\langle B^+,\leq\rangle$, where B^+ is an abbreviation for $B\setminus\{0\}$.

For every Boolean algebra $\mathfrak{B} = \langle B,+,\cdot,-,0,1\rangle$, let $\mathfrak{B}^* = \langle B,\cdot,+,-,1,0\rangle$. Since the dual of every axiom of the theory of Boolean algebras is also an axiom of this theory, it follows that $\mathfrak{B}^*$ is a Boolean algebra. $\mathfrak{B}^*$ is called the *dual Boolean algebra* of $\mathfrak{B}$.

EXERCISE 13.42 (G). Show that for each Boolean algebra $\mathfrak{B} = \langle B,+,\cdot,-,0,1\rangle$ the function $\pi : B \to B$ defined by $\pi(a) = -a$ is an isomorphism from $\mathfrak{B}$ onto $\mathfrak{B}^*$.

Exercise 13.42 allows us to strengthen the duality principle to sentences that are not necessarily theorems of the theory of Boolean algebras. If φ is any sentence of the language of Boolean algebra, φ^* is the dual of φ, and $\mathfrak{B}$ is a Boolean algebra, then $\mathfrak{B} \models \varphi$ if and only if $\mathfrak{B}^* \models \varphi^*$. Since $\mathfrak{B} \cong \mathfrak{B}^*$, we can infer that $\mathfrak{B} \models \varphi$ if and only if $\mathfrak{B} \models \varphi^*$.

Now we are going to investigate the relation between set algebras and Boolean algebras in general. We are going to show that every Boolean algebra is isomorphic to a set algebra.

Let B be a Boolean algebra. A *filter in B* is a filter in the p.o. $\langle B^+, \leq_B \rangle$. If F is a filter and $a, b \in F$ then $a \cdot b$ is also an element of F, since a filter contains a lower bound for every two of its elements and $a \cdot b$ is the greatest lower bound of a and b.

LEMMA 13.25. *Let B be a Boolean algebra. A filter F in B is an ultrafilter if and only if for each $a \in B$ exactly one of the elements $a, -a$ is in F.*

PROOF. Let F be a filter in B and let $a \in B$ be such that $a, -a \notin F$. Consider the sets
$$F_0 = \{a \cdot b : b \in F\}, \qquad F_1 = \{-a \cdot b : b \in F\}.$$
At least one of these sets does not contain 0 as an element, because otherwise there would exist $b_0, b_1 \in F$ such that $a \cdot b_0 = -a \cdot b_1 = 0$. But then $c = b_0 \cdot b_1 \in F$, and since $a \cdot c = -a \cdot c = 0$, we would have $c = 0$. Let us assume that $0 \notin F_0$, and let $G = \{b \in B : \exists c \in F_0\, (c \leq b)\}$. Then G is a filter in B with $F \subset G$ and $a \in G \setminus F$; hence F is not an ultrafilter.

On the other hand, note that a and $-a$ cannot belong simultaneously to F, since in this case $0 = a \cdot -a$ would also have to be an element of F, but 0 is not an element of B^+. □

Let B be a Boolean algebra, and let $\mathit{Ult}\,B$ denote the set of all ultrafilters in B. For $a \in B$ define
$$U_a = \{F \in \mathit{Ult}\,B : a \in F\},$$
and let
$$B' = \{U_a : a \in B\}.$$
Consider $a, b \in B$ and $F \in \mathit{Ult}\,B$. Then $a \in F$ und $b \in F$ if and only if $a \cdot b \in F$; and $a \in F$ or $b \in F$ if and only if $a + b \in F$. Hence,
$$U_a \cap U_b = U_{a \cdot b} \quad \text{and} \quad U_a \cup U_b = U_{a+b}.$$
Moreover, since each ultrafilter contains exactly one of the elements $a, -a$, we have
$$U_{-a} = X \setminus U_a.$$
Thus $\langle B', \cup, \cap, -, \emptyset, B' \rangle$ is a set algebra. Define a function $\pi : B \to B'$ by $\pi(a) = U_a$. Since $\pi(a \cdot b) = \pi(a) \cdot \pi(b)$ and $\pi(-a) = -\pi(a)$ for all $a, b \in B$, the function π is a homomorphism from B onto B'. It remains to show that π is one-to-one. Let $a, b \in B$ be such that $a \neq b$. Exercise 13.41(iii) implies that $a \cdot -b \neq 0$ or $-a \cdot b \neq 0$ (otherwise we would have $a \leq b$ and $b \leq a$, and thus $a = b$). Without loss of generality assume $a \cdot -b \neq 0$. By Theorem 13.4 there exists an ultrafilter F that contains $a \cdot -b$. But then $F \in U_a \setminus U_b$, and hence $U_a \neq U_b$. We have shown that π is one-to-one and have thereby proved the following theorem:

THEOREM 13.26. *For every Boolean algebra B there exists a set algebra B' such that $B \cong B'$.*

Next we want to show that the Boolean algebra B' of Theorem 13.26 can always be represented as the algebra of clopen subsets of a suitable topological space.

We need a few preliminaries. A topological space $\langle X, \tau \rangle$ is *totally disconnected* if every connected subspace of X has at most one point. X is a *Boolean space* if X is a compact Hausdorff space that has a base consisting of clopen sets.

EXERCISE 13.43 (PG). (a) Prove that $\langle X, \tau \rangle$ is a Boolean space if and only if it is a compact, totally disconnected Hausdorff space.

(b) Show that the product of an arbitrary family of Boolean spaces is again a Boolean space.

EXAMPLE 13.17. Let κ be a cardinal. The *discrete space* with κ points, i.e., the space $\langle \kappa, \mathcal{P}(\kappa) \rangle$, is denoted by $D(\kappa)$. For $\kappa < \omega$ the space $D(\kappa)$ is Boolean. By Exercise 13.43(b), $(D(2))^\lambda$ is a Boolean space for every cardinal λ. Such spaces are called *Cantor spaces*.

EXERCISE 13.44 (PG). Let $D_0 = I$ be the closed unit interval, let $D_1 = [0, \frac{1}{3}] \cup [\frac{2}{3}, 1]$, and $D_2 = [1, \frac{1}{9}] \cup [\frac{2}{9}, \frac{1}{3}] \cup [\frac{2}{3}, \frac{7}{9}] \cup [\frac{8}{9}, 1]$. In general, construct D_{k+1} by removing the middle third from each interval of D_k. Let $D = \bigcap_{n<\omega} D_n$.

Show that D with the topology of a subspace of I is homeomorphic to $(D(2))^\omega$. This space is called the *Cantor discontinuum* or simply the *Cantor set*.

Let B be a Boolean algebra, and let τ be the topology on $Ult\, B$ generated by $\mathcal{U} = \{U_a : a \in B\}$. Note that $\mathcal{U}$ is a base for this topology, $\mathcal{U}$ consists of clopen sets, and the resulting space is Hausdorff. We show that $\langle Ult\, B, \tau \rangle$ is compact. Consider an open cover $\mathcal{V} = \{V_i : i \in I\}$ of $Ult\, B$. Without loss of generality we may assume that the V_i's are elements of $\mathcal{U}$. Then for each $i \in I$ there exists an $a_i \in B$ such that $V_i = U_{a_i}$. Suppose $\mathcal{V}$ does not contain a finite subcover. Then $Ult\, B \neq \bigcup_{i \in H} U_{a_i}$ for each $H \in [I]^{<\aleph_0}$. Thus $F = \{-a_i : i \in I\}$ generates a filter on B. By Theorem 13.5 there exists an ultrafilter G in B such that $F \subseteq G$. But then $G \notin \bigcup \mathcal{V}$, contradicting our assumption that $\mathcal{V}$ was a cover of $\langle Ult\, B, \tau \rangle$.

The space $\langle Ult\, B, \tau \rangle$ is called the *Stone space* of B. It will be denoted by $st\, B$.

Let us show that $\mathcal{U}$ is the family of clopen sets in $st\, B$. Suppose $V \subseteq st\, B$ is clopen. Consider the family $\mathcal{A} = \{U \in \mathcal{U} : U \subseteq V\}$. Then $\mathcal{A}$ is a cover of the compact set V, and thus has a finite subcover $\{U_{a_0}, \ldots, U_{a_{n-1}}\}$. Let $a = a_0 + a_1 + \cdots + a_{n-1}$. Then $U_a = \bigcup_{i<n} U_{a_i} = V$, and hence $V \in \mathcal{U}$. Thus $B \cong Clop(st\, B)$.

THEOREM 13.27 (Stone Duality Theorem). (a) *Let B be a Boolean algebra. Then $Clop\,(st\, B) \cong B$.*

(b) *Let X be a Boolean space. Then X is homeomorphic to $st\,(Clop\, X)$.*

PROOF. Point (a) has already been shown. For the proof of point (b) let X be a Boolean space. The function that maps each $x \in X$ to $\{U \in Clop\, X : x \in U\}$ is a homeomorphism between X and $st\,(Clop\, X)$. □

Let us give some examples of Stone spaces.

If $B = \mathcal{P}(n)$, then there are exactly n ultrafilters in B and $st\, B$ is homeomorphic to $D(n)$.

Now consider $B = Finco\,(\omega)$. For each $n \in \omega$ the set $F_n = \{A \in Finco\,(\omega) : n \in A\}$ is an ultrafilter, and so is $F_\omega = \{A \subseteq \omega : |\omega \setminus A| < \omega\}$. It is not hard to see that $Ult\, B = \{F_\alpha : \alpha \leq \omega\}$, and $\mathcal{U} = \{\{F_n\} : n < \omega\} \cup \{\{F_\alpha : n \leq \alpha \leq \omega\} : n < \omega\}$

is a base of the topology of $st\, B$. Thus $st\, B$ is homeomorphic to the one-point compactification of $D(\omega)$.

Let A be a Boolean algebra. A subset $X \subseteq A$ is *dense in* A if for every $a \in A^+$ there exists $x \in X$ with $0 < x \le a$. Thus X is dense in A if and only if $X \setminus \{0\}$ is dense in the partial order $\langle A^+, \le_A \rangle$.

EXERCISE 13.45 (G). Let B be a Boolean algebra, and let A be a dense subset of B^+. Show that $\bigcup \{U_a : a \in A\}$ is a dense open subset of $st\, B$.

Let κ be an infinite cardinal, B a Boolean algebra. We say that B satisfies the κ-*chain condition* (κ-c.c.) if the p.o. $\langle B^+, \le_B \rangle$ satisfies the κ-c.c. The Stone Duality Theorem implies that B has the κ-c.c. if and only if $st(B)$ has the κ-c.c.

For set algebras, we introduced the notion of κ-completeness. This notion can be generalized to the class of all Boolean algebras: A Boolean algebra B is κ-*complete* if for every $A \in [B]^{<\kappa}$, $\sup A$ exists in the Boolean order $\langle B, \le \rangle$. We say that B is *complete* if B is $|B|^+$-complete.

If B is a Boolean algebra, $X \subseteq B$, and $\sup X$ or $\inf X$ exists, then we write $\sum X$ for $\sup X$ and $\prod X$ for $\inf X$. We can treat $\sum X$ and $\prod X$ as infinitary operations. Let us prove a distributivity law for $\sum X$.

LEMMA 13.28. *Let B be a Boolean algebra, $a \in B$, $X \subseteq B$, and assume that $\sum X$ exists. Then $a \cdot \sum X = \sum \{a \cdot x : x \in X\}$.*

PROOF. Let $s = \sum X$. We show that $a \cdot s$ is the least upper bound of $\{a \cdot x : x \in X\}$. First note that $a \cdot x \le a \cdot s$ for every $x \in X$; hence $a \cdot s$ is an upper bound of $\{a \cdot x : x \in X\}$.

Now let d be an arbitrary upper bound of $\{a \cdot x : x \in X\}$. Then $a \cdot x \le d$ for each $x \in X$. The inequality $a \cdot x \le d$ implies $x \le d + -a$. Since this holds for all $x \in X$, we conclude that $s \le d + -a$. Multiplying by a we get $s \cdot a \le d \cdot a \le d$, and hence $s \cdot a$ is the least upper bound of $\{a \cdot x : x \in X\}$. $\square$

EXERCISE 13.46 (PG). Let B be a Boolean algebra, $X, Y \subseteq B$, and assume that $\sum X, \sum Y$ exist. Show that
(a) $-\sum X = \prod \{-x : x \in X\}$;
(b) $\sum X \cdot \sum Y = \sum \{x \cdot y : x \in X,\ y \in Y\}$.

We want to show that every Boolean algebra can be embedded as a dense subalgebra into a complete Boolean algebra. Let us first consider the analogue for linear orders.

Recall that a *Dedekind cut* in a l.o. $\langle L, \le \rangle$ is a pair $\langle A, B \rangle$ of nonempty subsets of L such that $A \cap B = \emptyset$, $A \cup B = L$, and $a < b$ whenever $a \in A$ and $b \in B$. A Dedekind cut $\langle A, B \rangle$ is a *gap* if A has no largest element and B has no smallest element. Let us now use Dedekind cuts to show that every l.o. $\langle L, \le \rangle$ can be extended to a complete l.o. $\langle L_1, \le_1 \rangle$ so that L is a dense subset of L_1 and, moreover, $\forall a \in L_1\, \exists b, c \in L\, (b \le a \wedge a \le c)$. Such a l.o. $\langle L_1, \le_1 \rangle$ will be called a *completion* of $\langle L, \le \rangle$.

EXERCISE 13.47 (PG). Show that a l.o. $\langle L, \le \rangle$ has, up to isomorphism, at most one completion.

Exercise 13.47 entitles us to speak from now on about *the* completion of a linear order. The next theorem is a generalization of Lemma 4.17.

THEOREM 13.29. *Every l.o.* $\langle L, \leq \rangle$ *has a completion* $\langle L_1, \leq_1 \rangle$.

PROOF. Let $\langle L, \leq \rangle$ be a linear order. For $a \in L$, let $\hat{a} = \{b \in L : b < a\}$. The Dedekind cut $\langle \hat{a}, L \setminus \hat{a} \rangle$ will be denoted by S_a. Let $\mathcal{L}$ be the set of all gaps in L. Finally, let $\mathcal{C} = \{S_a : a \in L\} \cup \mathcal{L}$. Thus $\mathcal{C}$ contains all Dedekind cuts $\langle A, B \rangle$ such that A does not have a largest element or B does have a smallest element.

If $\langle A_0, B_0 \rangle$ and $\langle A_1, B_1 \rangle$ are Dedekind cuts such that $A_1 \not\subseteq A_0$, then A_0 is an initial segment of A_1. Thus we can define a linear order relation $\leq_{\mathcal{C}}$ on $\mathcal{C}$ as follows: $\langle A_0, B_0 \rangle \leq_{\mathcal{C}} \langle A_1, B_1 \rangle$ if and only if $A_0 \subseteq A_1$.

The function $i : L \to \mathcal{C}$ defined by $i(a) = S_a$ is an embedding of L into $\mathcal{C}$, and it is easy to see that $i[L]$ is dense in $\mathcal{C}$.

Now we show that the l.o. $\langle \mathcal{C}, \leq_{\mathcal{C}} \rangle$ is complete. Consider a Dedekind cut $\langle X, Y \rangle$ in $\mathcal{C}$, where $X = \{\langle A_i, B_i \rangle : i \in I\}$ and $Y = \{\langle D_j, E_j \rangle : j \in J\}$. Let

$$A' = \bigcup_{i \in I} A_i, \qquad B' = \bigcap_{i \in I} B_i.$$

Then $\langle A', B' \rangle \in \mathcal{C}$ and $X \leq_{\mathcal{C}} \langle A', B' \rangle \leq_{\mathcal{C}} Y$. Thus $\langle \mathcal{C}, \leq_{\mathcal{C}} \rangle$ is complete. It is not hard to see that $\mathcal{C}$ has a minimum if and only if L does, and that a minimum element of C will necessarily be in the range of i. The same is true for maxima. This proves that $\forall a \in \mathcal{C} \exists b, c \in L\, (i(b) \leq_{\mathcal{C}} a \wedge a \leq_{\mathcal{C}} i(c))$. □

Now let us generalize Theorem 13.29 to arbitrary p.o.'s. First we consider p.o.'s with a special property.

We say that a p.o. $\langle \mathbb{P}, \leq \rangle$ is *separative* if for all $p, q \in \mathbb{P}$ with $p \not\leq q$ there exists $r \in \mathbb{P}$ such that $r \leq p$ and $r \perp q$.

Note that if B is a Boolean algebra, then B^+ is separative. On the other hand, a linear order is separative if and only if it has at most one element.

EXERCISE 13.48 (G). Let $\langle \mathbb{P}, \leq_{\mathbb{P}} \rangle$ and $\langle \mathbb{Q}, \leq_{\mathbb{Q}} \rangle$ be p.o.'s; assume $\langle \mathbb{P}, \leq_{\mathbb{P}} \rangle$ is separative, and $\pi : \mathbb{P} \to \mathbb{Q}$ is a homomorphism from $\mathbb{P}$ onto $\mathbb{Q}$ such that $\pi(p) \perp \pi(q)$ for all $p, q \in \mathbb{P}$ with $p \perp q$. Show that π is an isomorphism.

LEMMA 13.30. *For every p.o.* $\langle \mathbb{P}, \leq_{\mathbb{P}} \rangle$ *there exist a separative p.o.* $\langle \mathbb{Q}, \leq_{\mathbb{Q}} \rangle$ *and a surjection* $i : \mathbb{P} \to \mathbb{Q}$ *such that*
  (i) *If* $p \leq_{\mathbb{P}} q$ *then* $i(p) \leq_{\mathbb{Q}} i(q)$;
  (ii) *If* $p \perp q$ *then* $i(p) \perp i(q)$.
The p.o. $\langle \mathbb{Q}, \leq_{\mathbb{Q}} \rangle$ *is unique up to isomorphism.*

PROOF. We define a binary relation $\preceq$ on $\mathbb{P}$ by:

$$p \preceq q \text{ if and only if } \forall r \in \mathbb{P}\, (r \not\perp p \to r \not\perp q).$$

The relation $\preceq$ is transitive and reflexive. We let

$$p \approx q \text{ if and only if } p \preceq q \text{ and } q \preceq p.$$

Then $\approx$ is an equivalence relation on $\mathbb{P}$. For $p \in \mathbb{P}$ let $[p] = \{q \in \mathbb{P} : p \approx q\}$, i.e., let $[p]$ be the equivalence class of p. Let $\mathbb{Q} = \{[p] : p \in \mathbb{P}\}$, and define a relation $\leq_{\mathbb{Q}}$ by: $[p] \leq_{\mathbb{Q}} [q]$ if and only if $p \preceq q$. Define a function $i : \mathbb{P} \to \mathbb{Q}$ by $i(p) = [p]$.

EXERCISE 13.49 (G). Convince yourself that this function i satisfies conditions (i) and (ii).

It remains to show that $\langle \mathbb{Q}, \leq_{\mathbb{Q}} \rangle$ is unique up to isomorphism. Assume that $\langle \mathbb{R}, \leq_{\mathbb{R}} \rangle$ is a separative p.o. and $j : \mathbb{P} \to \mathbb{R}$ is a surjection such that

(i) If $p \leq_{\mathbb{P}} q$ then $j(p) \leq_{\mathbb{R}} j(q)$;
(ii) If $p \perp q$ then $j(p) \perp j(q)$.

CLAIM 13.31. *If $\forall r \, (r \not\perp p \to r \not\perp q)$ then $j(p) \leq_{\mathbb{R}} j(q)$.*

PROOF OF CLAIM 13.31. If $j(p) \not\leq_{\mathbb{R}} j(q)$ then there exists $s^* \in \mathbb{R}$ with $s^* \leq_{\mathbb{R}} j(p)$ and $s^* \perp j(q)$. Let $s \in j^{-1}(s^*)$. Note that $s \not\perp p$ and $s \perp q$. □

CLAIM 13.32. *If $\exists r \, (r \not\perp p \wedge r \perp q)$ then $j(p) \not\leq_{\mathbb{R}} j(q)$.*

PROOF OF CLAIM 13.32. If $r \not\perp p \wedge r \perp q$ then there exists $s \in \mathbb{P}$ with $s \leq r, p$. Since $r \perp q$, we have $s \perp q$. If $j(p) \leq_{\mathbb{R}} j(q)$ then $j(s) \leq_{\mathbb{R}} j(p) \leq_{\mathbb{R}} j(q)$, and hence $j(s) \not\perp j(q)$, which contradicts the assumption. □

EXERCISE 13.50 (G). Deduce from Claims 13.31 and 13.32 that $\langle \mathbb{Q}, \leq_{\mathbb{Q}} \rangle \cong \langle \mathbb{R}, \leq_{\mathbb{R}} \rangle$. □

EXERCISE 13.51 (G). Show that in the formulation of Lemma 13.30 point (ii) cannot be replaced by the condition: $i(p) = i(q)$ iff $\forall r \in \mathbb{P} \, (r \not\perp p \leftrightarrow r \not\perp q)$. *Hint:* Consider the p.o.'s $\langle \mathbb{P}_0, \leq_0 \rangle$ and $\langle \mathbb{P}_1, \leq_1 \rangle$, where $\mathbb{P}_0 = \mathbb{P}_1 = \{0, 1, 2\}$, and the corresponding strict p.o. relations are $<_0 = \emptyset$, $<_1 = \{(0, 1), (2, 1)\}$.

Let $\preceq$ be the relation introduced in the proof of Lemma 13.30. Our next example demonstrates that $[p] \preceq [q]$ does not necessarily imply the existence of p_1, q_1 such that $[p] = [p_1]$, $[q] = [q_1]$, and $p_1 \leq q_1$.

EXAMPLE 13.18. Let $\mathbb{P} = \{a_0, a_1, a_2, b, c\}$, and consider the strict p.o. relation $<_{\mathbb{P}} = \{\langle a_0, b \rangle, \langle a_1, b \rangle, \langle a_0, c \rangle, \langle a_1, c \rangle, \langle a_2, c \rangle\}$. Then $[b] = \{b\}$, $[c] = \{c\}$, and $[b] \preceq [c]$, but b and c are not comparable in $\langle \mathbb{P}, \leq_{\mathbb{P}} \rangle$.

If $\langle A, B \rangle$ is a Dedekind cut in a l.o. $\langle \mathbb{P}, \leq \rangle$, then A is an open subset of $\mathbb{P}$ in the sense of the definition preceding Exercise 13.8. It also works the other way round: If A is an open subset of $\mathbb{P}$, then there exists a unique $B \subseteq \mathbb{P}$ such that $\langle A, B \rangle$ is a Dedekind cut. So let us for the purpose of the next proof refer to open subsets of arbitrary p.o.'s as *cuts*. That is, if $\langle \mathbb{P}, \leq \rangle$ is a p.o., then $U \subseteq \mathbb{P}$ will be called a cut in $\mathbb{P}$ if $\forall p, q \in \mathbb{P} \, (p \leq q \wedge q \in U \to p \in U)$. Note that if $\mathcal{X}$ is a nonempty set of cuts in $\mathbb{P}$, then $\bigcup \mathcal{X}$ and $\bigcap \mathcal{X}$ are also cuts in $\mathbb{P}$.

For $p \in \mathbb{P}$ let $U_p = \hat{p} \cup \{p\}$. A cut U is *regular* if for every $p \notin U$ there exists $q \leq p$ such that $U_q \cap U = \emptyset$. Note that in a separative p.o. every cut of the form U_p is regular. Moreover, the intersection of an arbitrary family of regular cuts is a regular cut.

LEMMA 13.33. *Let $\langle \mathbb{P}, \leq \rangle$ be a separative p.o. Then there exists a complete Boolean algebra B such that $\mathbb{P} \subseteq B^+$, $\leq \, = \, \leq_B \cap \, \mathbb{P}^2$, and $\mathbb{P}$ is dense in B^+. This Boolean algebra B is uniquely determined up to isomorphism.*

PROOF. The construction is similar to the one used in the proof of Theorem 13.29.

Let B denote the set of all regular cuts in $\mathbb{P}$. For $U, V \in B$ let $U \cdot V = U \cap V$, $U + V = \bigcap \{W \in B : U, V \subseteq W\}$. Finally, let $-U = \{p \in P : U_p \cap U = \emptyset\}$. Note that if $p \notin -U$, then $U_p \cap U \neq \emptyset$, and hence there exists a $q \in U_p$ with $U_q \subseteq U$. Then $U_q \cap W = \emptyset$, and hence $-U$ is a regular cut.

EXERCISE 13.52 (G). (a) Verify that $\langle B, +, \cdot, -, \emptyset, \mathbb{P} \rangle$ is a Boolean algebra.

(b) Verify that the set-theoretic intersection of an arbitrary subset of B is the greatest lower bound for this subset in the sense of $\leq_B$. Conclude that B is a complete Boolean algebra.

(c) Show that the function $i : \mathbb{P} \to B^+$ defined by $i(p) = U_p$ is a dense embedding of $\mathbb{P}$ into B^+.

(d) Show that if $\mathfrak{C} = \langle C, \ldots \rangle$ is a complete Boolean algebra with $\mathbb{P} \subseteq C$, $\leq \, = \, \leq_C \cap \mathbb{P}^2$, and if $\mathbb{P}$ is a dense subset of C^+, then $\mathfrak{C} \cong \langle B, +, \cdot, -, \emptyset, \mathbb{P} \rangle$. $\square$

Lemmas 13.30 and 13.33 yield the following:

THEOREM 13.34. *Let $\langle \mathbb{P}, \leq \rangle$ be an arbitrary p.o. Then there exist a complete Boolean algebra B and a function $i : \mathbb{P} \to B^+$ such that*

(i) $\quad rng(i)$ *is dense in* B^+;
(ii) $\quad \forall p, q \in \mathbb{P} \, (p \leq q \to p \leq_B q)$;
(iii) $\quad \forall p, q \in \mathbb{P} \, (p \perp q \leftrightarrow i(p) \perp i(q))$.

Since for every Boolean algebra B the p.o. $\langle B^+, \leq_B \rangle$ is separative, Lemma 13.33 also implies:

COROLLARY 13.35. *For each Boolean algebra B there exists a unique (up to isomorphism) complete Boolean algebra cB such that B^+ is dense in cB.*

The Boolean algebra cB of Corollary 13.35 is called the *completion* of B. Note that since B^+ is dense in cB, every dense subset X of B is also a dense subset of cB.

In Example 13.16 we saw that the clopen subsets of any topological space form a set algebra, and hence a Boolean algebra. The Stone Duality Theorem shows that every Boolean algebra can be represented in this way. We conclude this section with a construction that associates a complete Boolean algebra with every topological space.

Let $\langle X, \tau \rangle$ be a topological space, $A \subseteq X$. We define the *regularization* of A by:
$$rA = \operatorname{int} \operatorname{cl} A,$$
and call A *regular open* if $rA = A$. The family of all regular open subsets of X will be denoted by $RO(X)$.

EXAMPLE 13.19. Open intervals are regular open subsets of $\mathbb{R}$ with the usual topology. The set $(0,1) \cup (2,3)$ is also regular open, but the set $(0,1) \cup (1,2)$ is not.

EXERCISE 13.53 (G). Let $\langle \mathbb{P}, \leq \rangle$ be a p.o. and let τ be the topology on $\mathbb{P}$ described in Exercise 13.8. Show that an open set in this topology is regular if and only if it is a regular cut in the sense of the proof of Lemma 13.33.

EXERCISE 13.54 (G). Prove that the operation r satisfies for each $A, B \subseteq X$ and each open $U \subseteq X$:

(13.4) $\qquad\qquad\qquad \forall A, B \subseteq X (A \subseteq B \to rA \subseteq rB);$

(13.5) $\qquad\qquad\qquad U \subseteq rU.$

Let U and V be regular open. By 13.5, we have $U \cap V \subseteq r(U \cap V)$. Moreover, since $r(U \cap V) \subseteq rU = U$ and $r(U \cap V) \subseteq rV = V$, we also have $r(U \cap V) \subseteq U \cap V$. Thus

(13.6) $\qquad\qquad\qquad \forall U, V \in RO(X) \, (U \cap V \in RO(X)).$

Let A be an arbitrary subset of X. Then $rA = int\,cl\,A \subseteq cl\,A$; hence also $cl\,r\,A \subseteq cl\,cl\,A = cl\,A$, and thus $r\,r\,A = int\,cl\,r\,A \subseteq int\,cl\,A = rA$. Together with 13.5 this implies

(13.7) $$\forall A \subseteq X(r\,r\,A = r\,A).$$

In other words, $r\,A \in RO(X)$ for each $A \subseteq X$.

If U is open, then 13.5 implies that $U \subseteq rU$. If $V \in RO(X)$ and $U \subseteq V$, then 13.4 implies that $rU \subseteq rV$. Thus rU is the smallest regular open set that contains U, i.e.,

(13.8) $$\forall V \in RO(X)\,(U \subseteq V \to rU \subseteq V).$$

Now suppose $U \in \tau$ and $A \subseteq X$. Then $U \cap cl\,A \subseteq cl\,(U \cap A)$. Thus $U \cap rA = int\,U \cap int\,cl\,A = int\,(U \cap cl\,A) \subseteq int\,cl\,(U \cap A) = r\,(U \cap A)$. We have shown that

(13.9) $$\forall U, A\,(U \in \tau \to U \cap r\,A \subseteq r\,(U \cap A)).$$

If U and V are disjoint open sets, then also $cl\,U \cap V = \emptyset$, and hence $r\,U \cap V = \emptyset$. Repeating this argument we get

(13.10) $$\forall U, V \in \tau\,(U \cap V = \emptyset \to rU \cap rV = \emptyset).$$

Now let $U \in \tau$, $V = int\,(X \setminus U)$, and $x \in X \setminus (U \cup V)$. Then every neighborhood of x intersects U, thus $x \in cl\,U$, and hence $cl\,(U \cup V) = X$.

For $U \in RO(X)$ we define
$$neg\,U = int\,(X \setminus U).$$

Since $X \setminus U$ is closed, 13.7 implies that $neg\,U \in RO(X)$. Moreover, we have

(13.11) $$\forall U \in \tau\,(r\,(U \cup neg\,U) = X \wedge r\,(U \cap neg\,U) = \emptyset).$$

Let us summarize what we have shown so far.

LEMMA 13.36. *Let $\langle X, \tau \rangle$ be a topological space. Then*
(i) $\emptyset, X \in RO(X)$;
(ii) $RO(X)$ *is closed under* $\cap$ *and* neg;
(iii) $\forall U \in RO(X)(U \cap neg\,U = \emptyset \wedge r\,(U \cup neg\,U) = X)$;
(iv) $\forall U, V \in RO(X)\exists W \in RO(X)\,(U \cup V \subseteq W$
$\wedge \forall W_1 \in RO(X)\,(U \cup V \subseteq W_1 \to W \subseteq W_1))$.

We define operations $+, \cdot, -$ on $RO(X)$ as follows: $A + B = r\,(A \cup B)$, $A \cdot B = A \cap B$, $-A = neg\,A$. The structure $\langle RO(X), +, \cdot, -, \emptyset, X \rangle$ is called the *algebra of regular open subsets of X*.

THEOREM 13.37. *For every topological space X the structure $\langle RO(X), +, \cdot, -, \emptyset, X \rangle$ is a complete Boolean algebra.*

PROOF. To show the first part of (Ba1), associativity of addition, let $U, V, W \in RO(X)$. Then $U + (V + W) = r\,(U \cup r\,(V \cup W)) \subseteq r\,(U \cup r\,(U \cup V \cup W)) = r\,r\,(U \cup V \cup W) = r\,(U \cup V \cup W)$, and thus $U \cup V \cup W \subseteq r\,(U \cup r\,(V \cup W)) \subseteq r\,(U \cup V \cup W)$. Applying the operation r to all of these sets we get $r\,(U \cup V \cup W) \subseteq r\,(U \cup r\,(V \cup W)) \subseteq r\,(U \cup V \cup W)$, and hence $U + (V + W) = r\,(U \cup V \cup W)$. A similar argument shows that $(U + V) + W = r\,(U \cup V \cup W)$, and the equality $U + (V + W) = (U + V) + W$ follows.

The second part of (Ba1) and the axioms (Ba2), (Ba3) follow immediately from the definitions. Axiom (Ba5) follows from Lemma 13.36(iii). For the proof of (Ba4) it suffices by Lemma 13.22 to show the first equality.

Let $U, V, W \in RO(X)$. Then $U \subseteq U+V$, and thus $W \cdot U \subseteq W \cdot (U+V)$. A similar argument shows that $W \cdot V \subseteq W \cdot (U+V)$, and hence $W \cdot U + W \cdot V \subseteq W \cdot (U+V)$. On the other hand, we have

$$\begin{aligned} W \cdot (U+V) &= W \cap r(U \cup V) \\ &\subseteq r(W \cap (U \cup V)) \quad \text{(by 13.9)} \\ &= r((W \cap U) \cup (W \cap V)) \\ &= W \cdot U + W \cdot V. \end{aligned}$$

Thus $RO(X)$ is a Boolean algebra. Now let $M \subseteq RO(X)$. Then $r(\bigcup M)$ is the smallest regular open set that contains all sets from M. We have shown that $RO(X)$ is a complete Boolean algebra. □

EXERCISE 13.55 (PG). Show that the complete Boolean algebra constructed in the proof of Lemma 13.33 is the algebra of regular open subsets of the space $X = \mathbb{P}$ with the topology of Exercise 13.8.

Note that Exercise 13.55 implies together with Lemma 13.33 that for every complete Boolean algebra B there exists a topological space X such that B is isomorphic to the algebra of regular open subsets of X.

Mathographical Remarks

One is naturally tempted to try forming ultrapowers $^X\mathbf{V}/\mathcal{U}$ of the whole universe $\mathbf{V}$. Supposedly, such ultrapowers would be proper class models of all axioms of ZFC. The drawback of this approach is that one has to check very carefully that all objects under consideration are lawfully formed classes and not illegitimate concoctions like $\{\mathbf{V}, \mathbf{ON}\}$. Now $^X\mathbf{V}$ and $\sim_\mathcal{U}$ can easily be construed as classes. There is a problem with $^X\mathbf{V}/\mathcal{U}$ though: If $g \in {}^X\mathbf{V}$, then $g_{/\mathcal{U}}$ is usually a proper class, and thus cannot be a member of the class $^X\mathbf{V}/\mathcal{U}$. However, this problem can be overcome by using the same trick as in Definition 12.12. So it is not surprising that set theorists do study ultrapowers of $\mathbf{V}$ a lot. If $\mathcal{U}$ is nonprincipal and countably complete, then $^X\mathbf{V}/\mathcal{U}$ is wellfounded (i.e., the associated relational class that interprets $\in$ is), and moreover, it satisfies the assumptions of a generalization of Theorem 4.39 to proper relational classes. Hence, there exists a proper class $\mathbf{M}$ (the Mostowski collapse of $^X\mathbf{V}/\mathcal{U}$) such that $\langle \mathbf{M}, \bar{\in} \rangle \cong {}^X\mathbf{V}/\mathcal{U}$ (the notation we are using here is of course politically incorrect, but it is more expressive than its translation into L_S). Moreover, there exists an *elementary embedding* $j_\mathcal{U} : \mathbf{V} \to \mathbf{M}$ (elementary embeddings will be defined and studied in Chapter 24). It turns out that there is a close correspondence between combinatorial properties of $\mathcal{U}$ and properties of the associated elementary embedding $j_\mathcal{U}$. This makes it possible to characterize large cardinal properties in terms of elementary embeddings; an approach that has been tremendously fruitful.

In this text we will not formally develop the machinery of elementary embeddings of proper classes. An introduction to these techniques can be found in T. Jech's *Set Theory*, Academic Press, 1978. The most up-to-date survey of large cardinals and their associated elementary embeddings is contained in the monograph *The Higher Infinite*, by A. Kanamori, Springer Verlag, Vol. I: 1995, Vol. II: in preparation.

CHAPTER 14

Trees

A *tree* is a p.o. $\langle T, \leq_T \rangle$ such that for every $t \in T$ the initial segment $\hat{t} = \{s \in T : s <_T t\}$ is wellordered by the relation $\leq_T$. If it is more convenient to consider the corresponding strict p.o. relation, we will speak of the tree $\langle T, <_T \rangle$. If the relation is implied by the context, we will just speak of the tree T. Whenever T is a tree and $S \subseteq T$, then the suborder $\langle S, \leq_T \rangle$ is also a tree. We call S a *subtree* of T. The elements of a tree T are often called the *nodes* of the tree. Most trees considered in this chapter will have a minimum element. The minimum element of T, if it exists, is called the *root* of T. To get used to this terminology, let us contemplate a few examples:

EXAMPLE 14.1. See Figure 1.

EXAMPLE 14.2. See Figure 2.

EXAMPLE 14.3. See Figure 3.

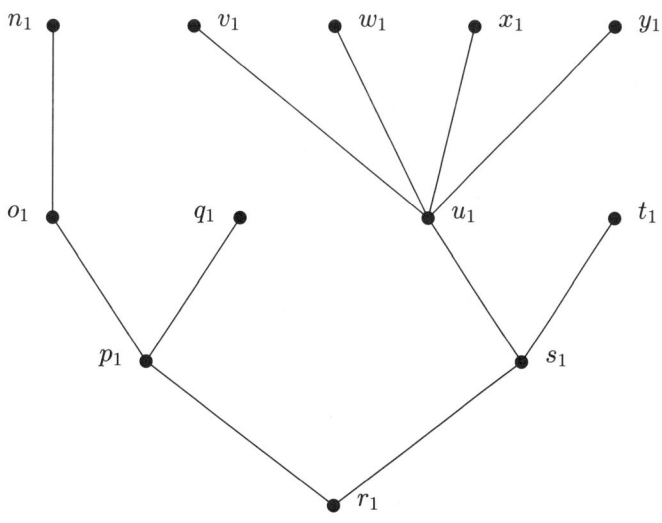

FIGURE 1

FIGURE 2

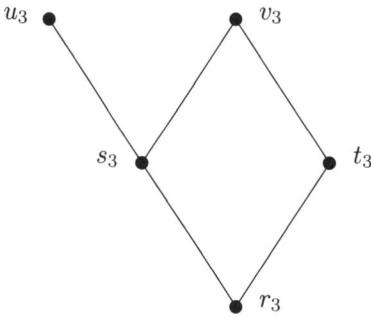

FIGURE 3

The p.o.'s in 14.1 and 14.2 are trees with roots r_1 and r_2. Moreover, $\langle T_1 \cup T_2, \leq_1 \cup \leq_2 \rangle$ is also a tree, but one without root.[1] The p.o. in Example 14.3 is not a tree, since the initial segment $\hat{v}_3 = \{r_3, s_3, t_3\}$ is not wellordered.

EXAMPLE 14.4. If α is an ordinal, then $\langle \alpha, \in \rangle$ is a tree. Note that the tree of Example 14.2 is order-isomorphic to $\langle 4, \in \rangle$.

EXAMPLE 14.5. Let κ, λ be cardinals > 0, and let $^{<\kappa}\lambda = \bigcup \{^{\alpha}\lambda : \alpha < \kappa\}$. Then $\langle ^{<\kappa}\lambda, \subseteq \rangle$ is a tree. If $\lambda = 2$, this tree is called the *full binary tree of height* κ.

A node r of a tree T is *splitting* if there are nodes $s, t \in T$ such that $r <_T s$, $r <_T t$, and s, t are incomparable in $\langle T, \leq_T \rangle$. In Example 14.1, the nodes r_1 and p_1 are splitting, whereas the nodes o_1 and n_1 are not. If every node of T is splitting, then we say that *the tree T is splitting*. The trees in Examples 14.1, 14.2, 14.4 are not splitting. The tree in Example 14.5 is splitting if and only if κ is infinite.

The *height* of a node $t \in T$, denoted by $ht(t)$, is the order type of $\hat{t}$. The *height of the tree T* is the ordinal $ht(T) = \sup\{ht(t) + 1 : t \in T\}$. For an ordinal α we

---

[1]Some authors would call such a structure a *forest* and reserve the word *tree* for forests with a minimal element.

define:
$$T(\alpha) = \{t \in T : ht(t) = \alpha\};$$
$$T_{(\alpha)} = \{t \in T : ht(t) < \alpha\}.$$

$T(\alpha)$ is called the α-th level of T, whereas $T_{(\alpha)}$ is the *initial part of height α of T*. To illustrate these concepts, consider Example 14.1. $T_1(2) = \{o_1, q_1, u_1, t_1\}$, whereas $(T_1)_{(2)} = \{p_1, s_1, r_1\}$. The height of both T_1 and T_2 is 4. If α is an ordinal, then the height of $\langle \alpha, \in \rangle$ is α, every $\beta < \alpha$ has height β in this tree, and $\alpha_{(\beta)}$ is simply β.

In Example 14.5, if $\alpha < \kappa$ and $f \in {}^\alpha\lambda$, then f has height α in ${}^{<\kappa}\lambda$. Moreover, $ht({}^{<\kappa}\lambda) = \kappa$.

EXERCISE 14.1 (PG). (a) Show that if T is any tree, then $ht(T) = \min\{\alpha \in \mathbf{ON} : T(\alpha) = \emptyset\}$ and $T = \bigcup_{\alpha < ht(T)} T(\alpha)$.

(b) Show that if $\beta < \alpha < ht(T)$ and $t \in T(\alpha)$, then there exists exactly one $s \in T(\beta)$ such that $s <_T t$.

Let T be a tree. We say that a subset P of T is a *path* in T if P is a chain (i.e., linearly ordered by $\leq_T$) such that $\hat{t} \subseteq P$ for all $t \in P$. A *branch* is a maximal path. A branch B is *cofinal* in T if B intersects every nonempty level of T, i.e., if $ot(\langle B, \leq_T \rangle) = ht(T)$. For example, in $\langle T_1, \leq_1 \rangle$, the chain $\{p_1, q_1\}$ is not a path, $\{r_1, p_1\}$ is a path but not a branch, $\{r_1, p_1, q_1\}$ is a branch, but not cofinal, and $\{r_1, s_1, u_1, x_1\}$ is a cofinal branch. The only branch in $\langle \alpha, \in \rangle$ is α itself, and this branch is cofinal. The branches of $\langle {}^{<\kappa}\lambda, \subseteq \rangle$ are sets of the form $B_f = \{f\lceil\alpha : \alpha < \kappa\}$, where $f \in {}^\kappa\lambda$, and every such branch is automatically cofinal.

EXERCISE 14.2 (G). Show that if T is a tree, then every chain $C \subseteq T$ is contained in a branch B of T.

While the above exercise implies that every tree has a branch, it is not the case that every tree has a cofinal branch.

EXAMPLE 14.6. Let $\gamma > 0$ be a limit ordinal, and let $G = \{\langle \alpha, \beta \rangle : \alpha < \beta < \gamma\}$. Define a p.o. relation $\leq_G$ on G by:
$$\langle \alpha, \beta \rangle \leq_G \langle \alpha', \beta' \rangle \text{ if and only if } \alpha \leq \alpha' \text{ and } \beta = \beta'.$$

EXERCISE 14.3 (G). Convince yourself that the p.o. $\langle G, \leq_G \rangle$ of Example 14.6 is a tree of height γ without cofinal branch.

For each tree $\langle T, \leq_T \rangle$ the relation $\leq_T$ is wellfounded. This allows us to construct mathematical objects by recursion over a given tree $\langle T, \leq_T \rangle$, i.e., by recursion over the wellfounded relation $\leq_T$. The proof of Claim 11.5 is a nice illustration of this technique. We still owe you this proof, so let us give it now. First let us reproduce the claim.

CLAIM 14.1. *Every closed subset of a Polish space is either countable or of cardinality $2^{\aleph_0}$.*

PROOF. Since singletons are closed subsets of Polish spaces, it follows from Exercise 11.9 that every Polish space has at most $2^{\aleph_0}$ points. Hence it suffices to show that each uncountable closed subset of a Polish space contains a subset of cardinality $2^{\aleph_0}$. So let $\langle X, \tau \rangle$ be a Polish space and assume that ϱ is a complete metric on X that induces the topology τ. Consider an uncountable closed subset F of X, and let $H = \{x \in F : \exists \varepsilon > 0\, (|B(x, \varepsilon) \cap F| \leq \aleph_0)\}$, where $B(x, \varepsilon)$ denotes the open ball with center x and radius ε.

EXERCISE 14.4 (G). Show that
(a) $F \setminus H$ is uncountable, and
(b) $|B(x,\varepsilon) \cap (F \setminus H)| > \aleph_0$ for each $x \in F \setminus H$ and $\varepsilon > 0$.

Let $T = {}^{<\omega}2$. Then $\langle T, \subseteq \rangle$ is the full binary tree of height ω. By recursion over this tree, we construct functions $f : T \to F \setminus H$ and $g : T \to \tau \setminus \{\emptyset\}$ as follows:

Let $f(\emptyset)$ be an arbitrary element of $F \setminus H$, and let $G(\emptyset)$ be an open neighborhood of $f(\emptyset)$ of diameter at most 1.

Since T doesn't have elements at level ω or higher, we only need to specify how $f(s{}^\frown i)$ and $g(s{}^\frown i)$ are being constructed when $f(s)$ and $g(s)$ are given and $i \in \{0,1\}$.

So suppose $s \in T$, and $f(s), g(s)$ have been defined in such a way that $g(s) \in \tau$ and $f(s) \in F \setminus H \cap g(s)$. We let $f(s{}^\frown 0)$, $f(s{}^\frown 1)$ be two different points in $g(s)$. Such points can be found by Exercise 14.4(b). Now we choose open sets $g(s{}^\frown 0)$, $g(s{}^\frown 1)$ such that for $i \in \{0, 1\}$:

(i) $f(s{}^\frown i) \in g(s{}^\frown i) \subset g(s)$;
(ii) The diameter of $g(s{}^\frown i)$ is at most 2^{-s-1}; and
(iii) $cl\,(g(s{}^\frown 0)) \cap cl\,(g(s{}^\frown 1)) = \emptyset$.

EXERCISE 14.5 (G). Show by induction over T (i.e., by induction over the wellfounded relation $\subseteq \cap T^2$) that for all $s, t \in T$, if $s \subseteq t$, then $\varrho(f(s), f(t)) \leq 2^{-|s|}$.

Now consider a function $b : \omega \to 2$. The set $\{b\lceil n : n \in \omega\}$ forms an infinite branch of T. For $n \in \omega$, we define $x_n^b = f(b\lceil n)$. By Exercise 14.5, the sequence $(x_n)_{n \in \omega}$ is a Cauchy sequence. By completeness of ϱ, this sequence must converge to a limit point $x^b \in X$. Since each x_n^b is in F and F is closed, x^b must be an element of F.

Let $b, c : \omega \to 2$ be two different functions, and let k be the smallest integer such that $b(k) \neq c(k)$. To be specific, let us assume $b(k) = 0$ and $c(k) = 1$. By induction over T one can deduce from (i) that $x_n^b \in g(b\lceil k{}^\frown 0)$ and $x_n^c \in g(b\lceil k{}^\frown 1)$ for all $n > k$. Hence $x^b \in cl(g(b\lceil k{}^\frown 0))$ and $x^c \in cl(g(b\lceil k{}^\frown 1))$. By (iii), $x^b \neq x^c$.

Now consider the function $\varphi : {}^{\omega}2 \to F$ that assigns x^b to each b in its domain. We have just shown that this function is injective. Since ${}^{\omega}2$ has cardinality $2^{\aleph_0}$, Claim 14.1 follows.   □

EXERCISE 14.6 (PG). Show that every dense G_δ-subset of the real line has cardinality $2^{\aleph_0}$.

A careful analysis of the proof of Claim 14.1 yields an even stronger theorem.

EXERCISE 14.7 (PG). Show that every uncountable closed subset of a Polish space contains a homeomorphic copy of the Cantor discontinuum defined in Exercise 13.44.

The crux in the proof of Claim 14.1 was the existence of lots of cofinal branches of the full binary tree of height ω. For a cardinal κ, let us call T a κ-*Kurepa tree* if T has height κ, every level of T is of cardinality $< \kappa$, and T has $> \kappa$ cofinal branches. Thus the full binary tree of height ω is an $\aleph_0$-Kurepa tree.

Are there any $\aleph_1$-Kurepa trees? Evidently, the full binary tree of height ω_1 doesn't qualify.

EXERCISE 14.8 (G). Convince yourself that the full binary tree of height κ is a κ-Kurepa tree if and only if κ is a strong limit cardinal, i.e., $2^\lambda < \kappa$ for all $\lambda < \kappa$.

From now on we will call $\aleph_1$-Kurepa trees simply *Kurepa trees*. The following statement is known as the *Kurepa Hypothesis*.

(KH) *There exists a Kurepa tree.*

Theorem 22.8 shows that KH is relatively consistent with ZFC. On the other hand, if the existence of an inaccessible cardinal is consistent with ZFC, then so is the negation of KH.

In many applications of trees it is crucial to know whether a given tree has at least one cofinal branch. The tree in Example 14.6 has no cofinal branches, but rather large levels (of size $|\gamma|$). On the other hand, the slender trees of Examples 14.2 and 14.4 have the property that every branch is cofinal. It is natural to conjecture that trees with relatively small levels will always contain cofinal branches. Let us explore this hunch.

LEMMA 14.2 (König). *Let $\langle T, \leq_T \rangle$ be a tree of height ω such that $T(n)$ is finite for all $n \in \omega$. Then T contains a cofinal branch.*

PROOF. We construct recursively a sequence $(t_n)_{n<\omega}$ of elements of T so that for all $n < \omega$:

(i) $\quad t_n <_T t_{n+1}$;
(ii) $\quad t_n \in T(n)$;
(iii) $\quad |\{s \in T : t_n <_T s\}| = \aleph_0$.

Let t_0 be any element of $T(0)$ such that $|\{s \in T : t_0 <_T s\}| = \aleph_0$.

Suppose t_n has already been constructed in such a way that (ii) and (iii) hold. Since $|T(n+1)| < \aleph_0$, and since for every s with $t_n <_T s$ there exists $t \in T(n+1)$ with $t \leq_T s$, we can find $t \in T(n+1)$ with $t_n <_T t$ such that $|\{s \in T : t <_T s\}| = \aleph_0$. We choose such a t as our t_{n+1}.

Clearly, the set $B = \{t_n : n < \omega\}$ is a cofinal branch of T. □

In the proof of Lemma 14.2, we used the fact that if an infinite set is partitioned into finitely many subsets, then at least one of the sets in the partition must be infinite. This fact is a special case of the following lemma:

LEMMA 14.3. *Let κ be an infinite cardinal, suppose $|X| = \kappa$, and let $\mathcal{Y}$ be a family of cardinality $< cf(\kappa)$ such that $X = \bigcup \mathcal{Y}$. Then at least one of the sets $Y \in \mathcal{Y}$ has cardinality κ.*

EXERCISE 14.9 (G). Prove Lemma 14.3.

Applications of König's Lemma are sometimes called *compactness arguments*. We have already encountered the word "compactness" in two other contexts: as a property of topological spaces and as the name of a theorem in mathematical logic. Does this exemplify the notorious lack of mathematician's creativeness in choosing their vocabulary,[2] or is the same phenomenon popping up in combinatorial, topological, and logical disguise?

The attentive reader will have noticed the word "choose" in the penultimate line of the proof of Lemma 14.2. This word indicates that the Axiom of Choice

---

[2]Did you ever notice how fond mathematicians are of the adjectives *normal, regular,* and *nice*? Quite predictably, set theorists have their normal trees, but no mature, shady, or majestic ones.

(more precisely, the Principle of Dependent Choices) has been used in our proof. Not as a matter of convenience, but out of necessity: König's Lemma is not provable in ZF.

EXERCISE 14.10 (G). Show that the theory ZF + König's Lemma entails that $\prod_{n \in \omega} X_n \neq \emptyset$ for every indexed family $\{X_n : n \in \omega\}$ of nonempty finite sets.

EXERCISE 14.11 (G). If T is as in Lemma 14.2, then each level $T(n)$ is finite and can be wellordered without the help of the Axiom of Choice. Why can't we prove Lemma 14.2 in ZF by always letting t_n be the smallest eligible element of $T(n)$?

We are going to show that König's Lemma is equivalent (i.e., equivalent in ZF) to weak versions of Tychonoff's Theorem and the Compactness Theorem. If you are not curious how such logical niceties illuminate the connection between seemingly disparate notions of compactness, you may want to skip ahead and resume reading right after Exercise 14.23.

Consider the following statements:

(WTY) *Every product $\prod_{n \in \omega} X_n$ of finite nonempty Hausdorff spaces X_n is nonempty and compact.*

(WCT) *Let L be a first-order language without functional symbols, and let $(S_n)_{n \in \omega}$ be an increasing sequence of finite sets of quantifier-free sentences of L. If each S_n has a model, then the theory $S = \bigcup_{n \in \omega} S_n$ also has a model.*

THEOREM 14.4 (ZF). *The following statements are equivalent:*

(i)   *König's Lemma;*
(ii) *WTY;*
(iii) *WCT.*

Before proving Theorem 14.4, let us draw your attention to some subtleties in the wording of WTY and WCT. For $k \in \omega \setminus \{0\}$ let $D(k)$ denote the k-element space $\{0, \ldots, k-1\}$ with the discrete topology. Now consider the statements:

(vWTY) *For every sequence $(k_n)_{n \in \omega}$ of positive integers the product $\prod_{n \in \omega} D(k_n)$ is nonempty and compact.*

(vWCT) *Let L be a first-order language without functional symbols, and let $(\varphi_k)_{k \in \omega}$ be a sequence of quantifier-free sentences of L. If for each $n \in \omega$ the theory $\{\varphi_k : k < n\}$ has a model, then the theory $\{\varphi_k : k \in \omega\}$ also has a model.*

It may appear that vWTY says the same as WTY and vWCT says the same as vWCT, but this is not the case.

EXERCISE 14.12 (X). Prove that vWTY and vWCT are theorems of ZF.

Note also that in stating WCT we used the phrase "S has a model" rather than "S is consistent." Recall that a theory S is consistent if no contradiction can be derived from S. But since formal derivations (i.e., proofs) are finite strings of formulas that can contain only finitely many formulas from S, one does not need the Axiom of Choice to show that if a theory is inconsistent, then it has an inconsistent finite fragment. Of course, by the Completeness Theorem, a theory S is consistent if and only if it has a model. But all known proofs of the Completeness

Theorem use the Axiom of Choice. Thus Theorem 14.4 implies that *any* proof of the Completeness Theorem must use the Axiom of Choice.

PROOF OF THEOREM 14.4. Let us start with the easiest part and show that WTY implies König's Lemma. So suppose that $\langle T, \leq_T \rangle$ is a tree of height ω with all levels finite. We can treat each $T(n)$ as a discrete topological space. Let $X = \prod_{n \in \omega} T(n)$. By WTY, this is a nonempty compact space.

For $k \in \omega$, let $U_k = \{x \in X : x(k) \not\leq_T x(k+1)\}$. Each U_k is open in X.

CLAIM 14.5. *For each $n \in \omega$, the set $\bigcup_{k \leq n} U_k$ does not cover X.*

PROOF. Fix $n \in \omega$ and let $t \in T(n+1)$ be such that for all $m > n$ there exists $r \in T(m)$ with $t \leq_T r$. For each $k \leq n+1$ let $t(k)$ be the unique $s \in T(k)$ such that $s \leq_T t$. By WTY, the space $Y = \prod_{k \leq n+1} \{t(k)\} \times \prod_{k > n+1} T(k)$ is nonempty. Let $y \in Y$. Then $y(k) = t(k)$ for all $k \leq n+1$; hence $y(k) \leq_T y(k+1)$ for all $k \leq n$. Thus $y \in X \setminus \bigcup_{k \leq n} U_k$. □

Now WTY applied to X implies that the family $\{U_n : n \in \omega\}$ is not a cover of X. Let $x \in X \setminus \bigcup_{n \in \omega} U_n$. Then $x(n) \leq_T x(n+1)$ for all $n \in \omega$; hence $\{x(n) : n \in \omega\}$ is an infinite branch of T.

Next let us prove that König's Lemma implies WCT. Let L, $(S_n)_{n \in \omega}$ be as in WCT, and assume that each S_n has a model. Let $\{r_i : i \in I\}$ and $\{c_k : k \in K\}$ be the sets of relational and constant symbols of L. Recall that, in general, atomic formulas of a first-order language are of the form $t_0 = t_1$ or $r(t_0, \ldots, t_{n-1})$, where the t_i's are terms and r is a relational symbol. Since our language L does not contain functional symbols, the only terms of L are variables and constant symbols. Moreover, if $\varphi \in S_n$ for some $n \in \omega$, then φ is quantifier-free, and hence φ is built up from atomic formulas with the help of logical connectives ($\neg, \wedge, \vee, \rightarrow, \leftrightarrow$, or, in a more purist view, just $\neg$ and $\wedge$). Such formulas are called *Boolean combinations* of atomic formulas. If $\varphi \in S_n$ for some $n \in \omega$, then φ is also a sentence, i.e., a formula without free variables. Since φ contains no quantifier, every variable occurring in φ would have to be free, which implies that there are no variables in φ. Let us coin the term *tidy formulas* for Boolean combinations of formulas of the form $c_{k_0} = c_{k_1}$ and $r_i(c_{k_0}, \ldots, c_{k_{m-1}})$, where the c_{k_j}'s are constant symbols, and r_i is a relational symbol of arity m. It follows from the above considerations that all formulas appearing in the S_n's are tidy formulas.

CLAIM 14.6. *Let L be as above, and let $\mathfrak{A} = \langle A, (R_i^A)_{i \in I}, (C_k^A)_{k \in K} \rangle$ and $\mathfrak{B} = \langle B, (R_i^B)_{i \in I}, (C_k^B)_{k \in K} \rangle$ be models of L. Assume that φ is a tidy formula of L, and let $I_0 \subseteq I$, $K_0 \subseteq K$ be such that no relational symbol r_i for $i \in I \setminus I_0$ and no constant symbol c_k for $k \in K \setminus K_0$ occurs in φ. If $C_k^A = C_k^B$ for all $k \in K_0$ and $R_i^A \cap \mathcal{C}^{\tau_0(i)} = R_i^B \cap \mathcal{C}^{\tau_0(i)}$ for all $i \in I$, where $\mathcal{C} = \{C_k^A : k \in K_0\}$ and $\tau_0(i)$ denotes the arity of r_i, then $\mathfrak{A} \models \varphi$ if and only if $\mathfrak{B} \models \varphi$.*

Despite its lengthy text, Claim 14.6 should be quite plausible. You may even wonder why we bothered formulating it.

EXERCISE 14.13 (PG). Give an example of a language L without functional symbols and a sentence φ of L that shows that the assumption of tidiness cannot be dropped in Claim 14.6.

EXERCISE 14.14 (PG). Prove Claim 14.6 by induction over the length of φ.

EXERCISE 14.15 (X). Formulate and prove an analogue of Claim 14.6 for sentences φ without quantifiers that may contain not only constant and relational symbols, but functional symbols as well.

COROLLARY 14.7. *If S is a nonempty set of tidy formulas of L and S has any model at all, then S has a model whose size does not exceed the cardinality of the set of constant symbols that occur in the formulas of S.*

PROOF. Let S be as in the assumptions, and let $\mathfrak{A} = \langle A, (R_i^A)_{i \in I}, (C_k^A)_{k \in K} \rangle$ be a model of S. Define $K_0 = \{k \in K : c_k \text{ occurs in some } \varphi \in S\}$, and let $A_1 = \{C_k : k \in K_0\}$. Since S is nonempty, and since every tidy formula must contain at least one constant symbol, $K_0 \neq \emptyset$. Let $k_0 \in K_0$, and let
$$C_k = C_k^A \text{ for all } k \in K_0, \quad C_k = C_{k_0} \text{ for all } k \in K \backslash K_0.$$
By Claim 14.6, the model $\mathfrak{A}_1 = \langle A_1, (R_i^A \cap (A_1)^{\tau_0(i)})_{i \in I}, (C_k)_{k \in K} \rangle$ is as required. □

Now let us assume that König's Lemma holds and prove that the theory $S = \bigcup_{n \in \omega} S_n$ has a model. For each $n \in \omega$, let $K_n = \{k \in K : c_k \text{ occurs in some } \varphi \in S_n\}$ and $I_n = \{i \in I : r_i \text{ occurs in some } \varphi \in S_n\}$. Since $S_n \subseteq S_{n+1}$, we also have that $K_n \subseteq K_{n+1}$ and $I_n \subseteq I_{n+1}$. For technical reasons we will assume that K_n is a proper subset of K_{n+1} for all $n \in \omega$.[3] Let $m_n = |K_n|$. Since the nature of the objects that comprise the universe of a model does not matter, Corollary 14.7 implies that every theory S_n has a model with universe $\{0, 1, \ldots, m_n - 1\}$. Define:
$$\mathcal{M}_n = \{M : M = \langle m_n, (R_i^M)_{i \in I_n}, (C_k^M)_{k \in K_n} \rangle \land M \models S_n\}.$$
Since the sets I_n, K_n are finite, and since for each $n \in \omega$ and $i \in I_n$ there are only finitely many subsets of $m_n^{\tau_0(i)}$ and finitely many functions from K_n into m_n, the sets $\mathcal{M}_n$ are finite. Now let $T = \bigcup_{n \in \omega} \mathcal{M}_n$, and define a relation $\leq_T$ on T by letting $M \leq_T N$ if and only if M is a submodel of N.

EXERCISE 14.16 (G). Show that $T(n) = \mathcal{M}_n$ for all $n \in \omega$. *Hint:* It is here that you will have to use the extra assumption that K_n is a proper subset of K_{n+1}.

Since each $\mathcal{M}_n$ is finite and nonempty, König's Lemma implies that there exists a cofinal branch $B = \{M_n : n \in \omega\}$ of T such that $M_n \in \mathcal{M}_n$ for each $n \in \omega$. Fix such B and define:
$$M = \langle \omega, (R_i^M)_{i \in \bigcup_{n \in \omega} I_n}, (C_k^M)_{k \in \bigcup_{n \in \omega} K_n} \rangle,$$
where $R_i^M = \bigcup \{R_i^{M_n} : i \in I_n\}$ and $C_k^M = C_k^{M_n}$ for some (equivalently: all) n such that $k \in K_n$.

EXERCISE 14.17 (PG). Show that $M \models S$.

It appears that we have proved WCT. But have we really done that? If $I \neq \bigcup_{n \in \omega} I_n$ or $K \neq \bigcup_{n \in \omega} K_n$, then the model M we just constructed is not a model of L, but only of a fragment of L. But do we *need* to construct a model of L? The last sentence of WCT postulates the existence of a model of S, not a "model of L that is also a model of S." So we are done with our proof, aren't we? Well, yes and no. You will probably admit that the phrase "model of L that is also a model

---

[3] This does not lead to loss of generality, since we can expand L to a new language L^+ by adding new constant symbols $\{d_n : n \in \omega\}$, extend each S_n to $S_n^+ = S_n \cup \{d_n = d_n\}$, and prove the existence of a model $\mathfrak{A}^+$ for $\bigcup_{n \in \omega} S_n^+$. The L-reduct of $\mathfrak{A}^+$ will then be the required model of S.

of S" is a bit awkward. Although mathematicians usually express themselves very explicitly, in polite company it is understood that if a paragraph starts with "Let L be a first-order language" then each subsequent use of the word "model" refers to models of L, unless something to the contrary is said. So let us bow to the rules of etiquette and conclude this part of our argument with one last exercise.

EXERCISE 14.18 (PG). Let $L^- \subseteq L$ be two first-order languages, and let S be a set of sentences in L^-. Show that if there exists a nonempty model M of L^- such that $M \models S$, then there exists a model N of L such that $N \models S$. Be careful to avoid using the Axiom of Choice.

Finally, let us assume WCT and derive WTY. Let $(X_n)_{n \in \omega}$ be a sequence of nonempty finite sets. We first show that $\prod_{n \in \omega} X_n \neq \emptyset$. Define $K = \bigcup_{n \in \omega} X_n \times \{n\}$. Let us treat each $\langle x, n \rangle \in K$ as a constant symbol in a first-order language L. The only other nonlogical symbol of L will be a relational symbol r of arity 1. All models considered below will be models of this language L.

To illustrate our line of reasoning, we will first show you a slightly phony version of the argument. While reading it, you should work on the following:

EXERCISE 14.19 (G). Catch us cheating.

For each $n \in \omega$, let S_n^- be the following set of formulas:

$$S_n^- = \{\varphi_n\} \cup \{\neg(r(\langle a, n \rangle) \wedge r(\langle b, n \rangle)) : \langle a, b \rangle \in (X_n)^2 \setminus \{\langle a, a \rangle : a \in X_n\}\},$$

where φ_n is the disjunction of all formulas $r(\langle a, n \rangle)$, i.e., something like $r(\langle a, n \rangle) \vee r(\langle b, n \rangle) \vee \cdots \vee r(\langle z, n \rangle)$ if X_n happens to have 26 elements. (For larger X_n's, you may have to borrow some characters from the Chinese.)

Now let us put $S_n = \bigcup_{k \leq n} S_n^-$. Each S_n obviously has a model; the S_n's are increasing and composed of tidy sentences. Thus WCT applies, and there exists a model $\mathfrak{A} = \langle A, R, (C_k)_{k \in K} \rangle$ such that $\mathfrak{A} \models S$, where $S = \bigcup_{n \in \omega} S_n$. By the choice of S_n^-, for each $n \in \omega$ there exists a unique $a_n \in X_n$ such that $\mathfrak{A} \models R(\langle a_n, n \rangle)$. Let f be the function with domain ω such that $f(n)$ is this unique a_n for each $n \in \omega$. Then $f \in \prod_{n \in \omega} X_n$, hence $\prod_{n \in \omega} X_n \neq \emptyset$.

So, where did we cheat? Of course, we said "obviously" in one place, and we hope this put you in high alert mode, but this wasn't where we cheated. The act of cheating occured when we called φ_n *the* disjunction of all formulas $r(\langle a, n \rangle)$ for $a \in X_n$. To see why this is problematic, recall that formulas are finite strings of symbols. Thus when choosing *a* disjunction of formulas $r(\langle a, n \rangle)$, we automatically choose the order in which the $\langle a, n \rangle$'s occur in this formula. Hence choosing a sequence $(\varphi_n)_{n \in \omega}$ as in our argument entails choosing l.o. relations on all X_n's simultaneously, which in the absence of the Axiom of Choice may be impossible.

Note that the second part of the definition of S_n^- does not suffer from the same defect: Since we have a finite *set* of formulas, and since for every ordered pair $\langle a, b \rangle \in (X_n)^2 \setminus \{\langle a, a \rangle : A \in X_n\}$ the formula $\neg(r(\langle a, n \rangle) \wedge r(\langle b, n \rangle))$ appears in it, no order on X_n can be extracted from this set alone.[4] This observation suggests how we can overcome the problem. Suppose $X_n = \{a, b, \ldots, z\}$ (with Chinese characters inserted if necessary), and $\sigma : X_n \to X_n$ is any permutation of X_n. Let φ_σ

---

[4]Note the use of ordered pairs here. Even mathematically trained authors might happily define a set of formulas like $\{x = y : \{x, y\} \in [\{a, b, c\}]^2\}$ without ever realizing that this definition is ambiguous. But is the latter set equal to $\{a = b, a = c, b = c\}$, or equal to $\{b = a, c = a, c = b\}$?

be the formula:

$$r(\langle\sigma(a),n\rangle) \vee r(\langle\sigma(b),n\rangle) \vee \cdots \vee r(\langle\sigma(z),n\rangle).$$

Now define:

$$S_n^\circ = \{\varphi_\sigma : \sigma \text{ is a permutation of } X_n\}$$
$$\cup \{\neg(r(\langle a,n\rangle) \wedge r(\langle b,n\rangle)) : \langle a,b\rangle \in (X_n)^2 \setminus \{\langle a,a\rangle : a \in X_n\}\},$$

and let $S_n^\bullet = \bigcup_{k \leq n} S_n^\circ$ for all $n \in \omega$.

EXERCISE 14.20 (PG). Convince yourself that the sequence $(S_n^\bullet)_{n\in\omega}$ does not depend on any particular choice of orderings of the X_n's and can be defined without invoking the Axiom of Choice.

EXERCISE 14.21 (G). Use the sequence $(S_n^\bullet)_{n\in\omega}$ and WCT to show that the product $\prod_{n\in\omega} X_n$ is nonempty.

Now consider each X_n as a discrete topological space, and let $X = \prod_{n\in\omega} X_n$ with the product topology. Recall that the family $\{[f] : \exists N \in \omega \, (f \in \prod_{n<N} X_n)\}$, where $[f] = \{F \in X : f \subset F\}$, is a base for the topology on X. Let $\mathcal{U}$ be a family of open subsets of X. For each $n \in \omega$, let $S_n^\bullet$ be defined as above, and let

$$S_n^+ = S_n^\bullet \cup \{\neg(r(\langle f(0),0\rangle) \wedge r(\langle f(1),1\rangle) \wedge \cdots \wedge r(\langle f(n),n\rangle)) :$$
$$f \in \prod_{k \leq n} X_n \wedge \exists U \in \mathcal{U} \, ([f] \subseteq U)\}.$$

EXERCISE 14.22 (PG). (a) Convince yourself that the sequence $(S_n^+)_{n\in\omega}$ has been defined without reference to the Axiom of Choice.

(b) Show that for every $n \in \omega$ the theory S_n^+ has a model if and only if $\bigcup \mathcal{V} \neq X$ for every $\mathcal{V} \in [\mathcal{U}]^{<\aleph_0}$.

(c) Use WCT to deduce that X is compact. □

EXERCISE 14.23 (PG). Having seen the proof of Theorem 14.4, try your luck with Exercise 14.12 again.

If κ is an infinite cardinal, then a κ-*tree* is a tree T of height κ such that $|T(\alpha)| < \kappa$ for every $\alpha < \kappa$. In particular, a κ-Kurepa tree is a κ-tree. A κ-*Aronszajn tree* is a κ-tree without cofinal branches. An $\aleph_1$-Aronszajn tree is simply called an *Aronszajn tree*. An infinite cardinal κ has the *tree property* if there are no κ-Aronszajn trees.

König's Lemma asserts that $\aleph_0$ has the tree property. Is the analogue of Lemma 14.2 true for other infinite cardinals κ? Let us assume T is a κ-tree, and let us try to construct recursively a cofinal branch $(t_\xi)_{\xi<\kappa}$ as in the proof of Lemma 14.2. To get started, we would want to choose $t_0 \in T(0)$ such that $|\{s \in T : t_0 <_T s\}| = \kappa$. Can we do that? Well, there are fewer that κ nodes in $T(0)$, and by Lemma 14.2 ... Wait a minute. What if κ is singular and $|T(0)| \geq cf(\kappa)$?

EXERCISE 14.24 (G). Show that if κ is a singular uncountable cardinal, then there exists a κ-Aronszajn tree. *Hint:* Modify Example 14.6.

O.k., let us assume that κ is regular. Then we can use Lemma 14.3 to construct $t_0 <_T t_1 <_T t_2 <_T \cdots <_T t_n <_T \cdots$ in such a way that $|\{s \in T : t_n <_T s\}| = \kappa$ for all $n \in \omega$. But what can we do at stage ω? For each individual t_n, the set of

$<_T$-successors is large, but this does not imply that there are many (or in fact any) nodes in T that sit above *all* the t_n's. Sure enough, there must be *some* sequence $t'_0 <_T t'_1 <_T t'_2 <_T \cdots <_T t'_n <_T \cdots <_T t'_\omega$ such that $|\{s \in T : t'_\omega <_T s\}| = \kappa$, but we may have missed it. So what can we do? Scrapping our sequence $(t_n)_{n \in \omega}$ and starting all over again with $(t'_n)_{n \in \omega}$ isn't really an option since we may run into exactly the same problem at levels $\omega + \omega, \ldots, \omega^{\cdot 2}, \ldots, \omega^{\cdot \omega}, \ldots$. So, perhaps there are at least some uncountable regular cardinals κ without the tree property? The next theorem shows that $\aleph_1$ is one of them.

THEOREM 14.8. *There exists an Aronszajn tree.*

PROOF. Let $E = \{e \in {}^{<\omega_1}\omega : e \text{ is an injection and } |\omega \backslash rng(e)| = \aleph_0\}$. On E, we define a binary relation $\subseteq^*$ as follows:

$e_0 \subseteq^* e_1$ iff $dom(e_0) \subseteq dom(e_1)$ and $|\{\xi \in dom(e_0) : e_0(\xi) \neq e_1(\xi)\}| < \aleph_0$.

Note that $\subseteq^*$ is a partial pre-order relation, i.e., reflexive and transitive. Hence, the relation $=^*$ defined by:

$e_0 =^* e_1$ if and only if $e_0 \subseteq^* e_1$ and $e_1 \subseteq^* e_0$

is an equivalence relation on E.

We construct recursively a transfinite sequence $(e_\xi)_{\xi < \omega_1}$ of elements of E such that $dom(e_\xi) = \xi$ and $e_\xi \subseteq^* e_\eta$ for all $\xi < \eta < \omega_1$.

To get started, let $e_0 = \emptyset$.

Given e_ξ, let $e_{\xi+1} = e_\xi \cup \{\langle \xi, \min(\omega \backslash rng(e_\xi)) \rangle\}$.

If $0 < \delta < \omega_1$ is a limit ordinal and e_ξ is defined for all $\xi < \delta$, we choose an increasing sequence $(\xi_n)_{n \in \omega}$ of ordinals less than δ such that $\delta = \sup\{\xi_n : n \in \omega\}$ and use the following lemma to find a suitable e_δ.

LEMMA 14.9. *Suppose $(e_n)_{n \in \omega}$ is a sequence of functions in E such that $e_n \subseteq^* e_{n+1}$ for all $n \in \omega$. Then there exists an $e \in E$ such that $dom(e) = \bigcup_{n \in \omega} dom(e_n)$ and $e_n \subseteq^* e$ for all $n \in \omega$.*

PROOF. Let $(e_n)_{n \in \omega}$ be as in the assumptions. By recursion over ω we construct functions $f_n \in E$ and a set $\{a_n : n \in \omega\} \subseteq \omega$ as follows:

$f_0 = e_0; \; a_0 = \min(\omega \backslash rng(f_0))$.

Given f_n such that $f_n =^* e_n$ and $\{a_0, \ldots, a_n\}$, let

$B_n = \{k \in dom(e_{n+1}) \backslash dom(e_n) : e_{n+1}(k) \in rng(f_n) \cup \{a_0, \ldots, a_n\}\}$.

EXERCISE 14.25 (G). Show that B_n is a finite set.

Let $dom(f_{n+1}) = dom(e_{n+1})$. For $k \in dom(e_{n+1}) \backslash (dom(e_n) \cup B_n)$, let $f_{n+1}(k)$ be equal to $e_{n+1}(k)$, and define $f_{n+1} \restriction B_n$ in such a way that f_{n+1} is a one-to-one function with range disjoint from the set $\{a_0, \ldots, a_n\}$. Since $\omega \backslash rng(e_{n+1})$ is infinite, we have plenty of room to define $f_{n+1} \restriction B_n$ in this way. Finally, let a_{n+1} be the smallest element of the set $(\omega \backslash rng(f_{n+1})) \cup \{a_0, \ldots a_n\}$. Note that $f_{n+1} =^* e_{n+1}$; and hence we are ready for the next step of the construction.

Now let $e = \bigcup_{n \in \omega} f_n$. It is not hard to see that this is a one-to-one function from $\bigcup_{n \in \omega} dom(e_n)$ into ω with $rng(e) \cap \{a_n : n \in \omega\} = \emptyset$. Hence $e \in E$; and a straightforward verification shows that $e_n \subseteq^* e$ for all $n \in \omega$. $\square$

Given a sequence $(e_\xi)_{\xi<\omega_1}$ as above, let $T = \{e \in E : \exists \xi < \omega_1 (e =^* e_\xi)\}$. We show that $\langle T, \subseteq \rangle$ is an Aronszajn tree. First note that if $e \in T$, then $\hat{e} = \{e\lceil\eta : \eta \in dom(e)\}$. This set is wellordered by inclusion and has order type $dom(e)$. Hence T is a tree, and $ht(e) = dom(e)$ for all $e \in T$. It follows that $T(\xi) = \{e \in E : e =^* e_\xi\}$ for $\xi < \omega_1$, and $T(\xi) = \emptyset$ for $\xi \geq \omega_1$. Thus $ht(T) = \omega_1$.

EXERCISE 14.26 (G). Show that every level of T is countable.

It remains to show that T has no uncountable branches. Note that if $(e^\eta)_{\eta<\delta}$ is an increasing chain in T, then $(rng(e^\eta))_{\eta<\delta}$ is a sequence of subsets of ω that is wellordered by inclusion. The following exercise concludes the proof of Theorem 14.8.

EXERCISE 14.27 (G). (a) Show that if $\delta \in \mathbf{ON}$ and $(A_\eta)_{\eta<\delta}$ is a strictly increasing sequence of subsets of ω, then $\delta < \omega_1$.

(b) It is important that the sequence in (a) is wellordered. Give an example of a family $(A_r)_{r \in \mathbb{R}}$ of subsets of ω such that A_r is a proper subset of A_s whenever $r < s$. □

Can we use a similar technique as in the proof of Theorem 14.8 to construct an $\aleph_2$-Aronszajn tree? To explore this question, let $E_1 = \{e \in {}^{<\omega_2}\omega_1 : e$ is an injection and $|\omega_1 \setminus rng(e)| = \aleph_1\}$. On E_1, we define a binary relation $\preceq^*$ as follows:

$$e_0 \preceq^* e_1 \text{ iff } dom(e_0) \subseteq dom(e_1) \text{ and } |\{\xi \in dom(e_0) : e_0(\xi) \neq e_1(\xi)\}| < \aleph_1.$$

Consider the equivalence relation $\approx^*$ defined by:

$$e_0 \approx^* e_1 \text{ iff } e_0 \subseteq^* e_1 \text{ and } e_1 \subseteq^* e_0.$$

We would like to construct a sequence $(e_\xi)_{\xi<\omega_1}$ of elements of E_1 such that

(a) $dom(e_\xi) = \xi$ for all $\xi < \omega_2$;
(b) $e_\xi \preceq^* e_\eta$ for all $\xi < \eta < \omega_2$.

Given a sequence $(e_\xi)_{\xi<\omega_2}$ that satisfies (a) and (b), we'd like to define $T = \{e \in E_1 : \exists \xi < \omega_2 (e \approx^* e_\xi)\}$. The same argument as in the proof of Theorem 14.8 shows that $\langle T, \subseteq \rangle$ is a tree of height ω_2 without cofinal branches. For $\xi < \omega_2$, the ξ-th level $T(\xi)$ of this tree is equal to $\{e \in E_1 : e \approx^* e_\xi\}$. It is not hard to see that $|T(\xi)| = \aleph_1^{\aleph_0}$ if $\omega \leq \xi < \omega_2$. Thus, if CH holds, $\langle T, \subseteq \rangle$ will turn out to be an $\aleph_2$-Aronszajn tree.

But can we construct a sequence $(e_\xi)_{\xi<\omega_2}$ that satisfies conditions (a) and (b)? Let us try to proceed as in the proof of Theorem 14.8. We let $e_0 = \emptyset$. Given e_ξ, one can construct $e_{\xi+1}$ exactly as in the proof of Theorem 14.8. But limit stages pose a new challenge. To illustrate the problem, suppose $\delta > 0$ is a limit ordinal of countable cofinality, and $(\xi_n)_{n\in\omega}$ is an increasing sequence of ordinals with $\lim_{n\in\omega} \xi_n = \delta$. For simplicity, suppose that $e_{\xi_n} \subseteq e_{\xi_{n+1}}$ for all $n \in \omega$.

EXERCISE 14.28 (G). Convince yourself that if $\bigcup_{n\in\omega} rng(e_{\xi_n}) = \omega_1$, then there is no $e_\delta \in E_1$ such that $e_{\xi_n} \preceq^* e_\delta$ for all $n \in \omega$.

How can we prevent the situation described in Exercise 14.28 from ever arising? Evidently, the requirement that $|\omega_1 \setminus rng(e_\xi)| = \aleph_1$ is not sufficient. We can make the requirement more stringent by fixing a proper ideal $\mathcal{I}$ of subsets of ω_1 and considering the following subset of E_1:

$$E_2 = \{e \in {}^{<\omega_2}\omega_1 : e \text{ is an injection and } rng(e) \in \mathcal{I}\}.$$

What properties should $\mathcal{I}$ have so that our construction will work? First of all, we will want $\mathcal{I}$ to contain some uncountable sets. Let us require that $\mathcal{I}$ has the following property:

(i) $\forall A \in \mathcal{I} \exists B \in [\omega_1]^{\aleph_1} (A \cap B = \emptyset \wedge A \cup B \in \mathcal{I})$.

Second, in order to get our construction past limit stages of countable cofinality, we will want $\mathcal{I}$ to be countably complete; i.e., we want $\mathcal{I}$ to satisfy the following:

(ii) If $\mathcal{A} \in [\mathcal{I}]^{\aleph_0}$, then $\bigcup \mathcal{A} \in \mathcal{I}$.

Note that if $\mathcal{I}$ satisfies (ii), $\delta < \omega_2$ is a limit ordinal of countable cofinality, $(\xi_n)_{n \in \omega}$ is an increasing sequence of ordinals cofinal in δ, and for each $n \in \omega$ we have constructed $e_{\xi_n} \in E_2$ such that $e_{\xi_n} \subseteq e_{\xi_{n+1}}$ for all $n \in \omega$, then we can simply define $e_\delta = \bigcup_{n \in \omega} e_{\xi_n}$.

It remains to consider limit stages δ of cofinality ω_1. Let us assume that $(\xi_\alpha)_{\alpha < \omega_1}$ is an increasing sequence of ordinals cofinal in δ, and the e_{ξ_α}'s have already been chosen from the set E_2. For the sake of simplicity, let us assume that $e_{\xi_\alpha} \subseteq e_{\xi_\beta}$ for all $\alpha < \beta < \omega_1$. If there exists $e_\delta \in E_2$ such that $e_{\xi_\alpha} \preceq^* e_\delta$, there must exist $B \in \mathcal{I}$ with $|rng(e_{\xi_\alpha}) \setminus B| < \aleph_1$ for all $\alpha < \omega_1$. In other words, we need to require the following property of $\mathcal{I}$:

(iii) If $\mathcal{A} \in [\mathcal{I}]^{\aleph_1}$, then there exists $B \in \mathcal{I}$ such that $|A \setminus B| < \aleph_1$ for all $A \in \mathcal{A}$.

It turns out that this property also suffices. Given B as above, use (i) to find $C \in [\omega_1]^{\aleph_1}$ that is disjoint from B and still a member of the ideal $\mathcal{I}$. Now define recursively e'_{ξ_α} as follows: If $e_{\xi_\alpha}(\eta) \in B$, then let $e'_{\xi_\alpha}(\eta) = e_{\xi_\alpha}(\eta)$. If $e_{\xi_\alpha}(\eta) \notin B$, then let $e'_{\xi_\alpha}(\eta)$ be the smallest ordinal in C that has not yet been used in the construction. Finally, let $e_\delta = \bigcup_{\alpha \in \omega_1} e'_{\xi_\alpha}$.

EXERCISE 14.29 (G). Convince yourself that this construction yields a sequence $(e_\xi)_{\xi < \omega_1}$ of elements of E_2 that satisfies conditions (a) and (b).

But does there exist an ideal $\mathcal{I}$ that satisfies conditions (i)–(iii)? In Chapter 21 (Lemma 21.11) we will prove the following:

LEMMA 14.10. *Let κ be a regular uncountable cardinal. Then there exists an ideal NS_κ on κ with the following properties:*

(i) $\forall A \in NS_\kappa \exists B \in [\kappa]^\kappa (A \cap B = \emptyset \wedge A \cup B \in NS_\kappa)$;
(ii) *If $\mathcal{A} \in [NS_\kappa]^{<\kappa}$, then $\bigcup \mathcal{A} \in NS_\kappa$;*
(iii) *If $\mathcal{A} \in [NS_\kappa]^\kappa$, then $\exists B \in NS_\kappa \forall A \in \mathcal{A} (|A \setminus B| < \kappa)$.*

EXERCISE 14.30 (G). Use Lemma 14.10 to prove the following:

THEOREM 14.11. *If κ is a regular infinite cardinal such that $2^{<\kappa} = \kappa$, then there exists a κ^+-Aronszajn tree.*

If GCH holds, then $2^{<\kappa} = \kappa$ for every infinite cardinal κ. Hence GCH implies the existence of κ^+-Aronszajn trees for every regular infinite cardinal κ. It follows that if GCH holds, then every uncountable cardinal with the tree property is either the successor of a singular cardinal, or strongly inaccessible. Strongly inaccessible cardinals with the tree property are called *weakly compact cardinals*. In Chapters 15 and 24 we will give some alternative characterizations of weakly compact cardinals.

There are other constructions of Aronszajn trees than the one used in the proof of Theorem 14.8. Let us sketch one of them. Let $\mathbb{P}$ be the set of all nonempty compact subsets of the rationals $\mathbb{Q}$, and let $\leq_e$ denote end-extension, i.e.,

$$K \leq_e L \text{ if and only if } K \subseteq L \text{ and } \forall p \in K \, \forall q \in L \backslash K \, (p \leq q).$$

EXERCISE 14.31 (G). Convince yourself that $\langle \mathbb{P}, \leq_e \rangle$ is a p.o. without uncountable wellordered chains.

So if $T \subseteq \mathbb{P}$ and $\langle T, \leq_e \rangle$ is an ω_1-tree, then $\langle T, \leq_e \rangle$ is automatically an Aronszajn tree. But how to construct an ω_1-subtree of $\mathbb{P}$? By transfinite recursion of course, that is, level by level. It is easy to construct $T(0)$, $T(1)$, $T(2)$,... in such a way that $\langle T_{(\omega)}, \leq_e \rangle$ is a splitting tree of height ω with countable levels. Then $T_{(\omega)}$ has $2^{\aleph_0}$ cofinal branches. Since $T(\omega)$ is supposed to be countable, we will put nodes in $T(\omega)$ only on top of countably many of these branches. Note that a branch in $T_{(\omega)}$ will remain a branch in T if and only if it has no extension in $T(\omega)$. But are there any cofinal branches of $T_{(\omega)}$ that *can* be extended? Not necessarily. Note that an increasing chain $K_0 \leq_e K_1 \leq_e K_2 \leq_e \cdots$ is unbounded in $\langle \mathbb{P}, \leq_e \rangle$ if and only if $\lim_{n \to \infty} \max(K_n) = \infty$. Thus, had we chosen for example the $T(n)$'s in such a way that $\max(K) \geq n$ for all $n \in \omega$ and $K \in T(n)$, then none of the branches in $T_{(\omega)}$ could be extended. The trick is to give ourselves the necessary flexibility by requiring that the following condition $(*)_\alpha$ holds for every $\alpha < \omega_1$:

$(*)_\alpha \; \forall \xi < \eta < \alpha \, \forall K \in T(\xi) \, \forall q \in \mathbb{Q} \, (q > \max K \to \exists L \in T(\eta) \, K \leq_e L \wedge \max L \leq q).$

EXERCISE 14.32 (PG). (a) Suppose that $\delta \in \omega_1 \cap \mathbf{LIM}$ and $\langle T_{(\delta)}, \leq_e \rangle$ is constructed as above. Show that if $(*)_\delta$ holds, then at least some cofinal branches of $T_{(\delta)}$ can be extended.
  (b) Construct a $T \subseteq \mathbb{P}$ such that $\langle T, \leq_e \rangle$ is an ω_1-tree.

An *antichain* in a tree $\langle T, \leq_T \rangle$ is a subset $A \subseteq T$ of pairwise *incomparable* elements of T. This notion of "antichain" is different from the one in Chapter 13, but there is a close connection between the two concepts.

EXERCISE 14.33 (G). Show that if T is a tree and $A \subseteq T$, then A consists of pairwise incomparable elements if and only if it consists of elements that are pairwise incompatible in the dual p.o. $\langle T, \leq_T^* \rangle$.

EXAMPLE 14.7. Every level $T(\alpha)$ of a tree T is an antichain in T.
  The set $\{n_1, q_1, t_1, y_1\}$ is an antichain in the tree of Example 14.1.
  If σ_n denotes the function with domain $n+1$ that takes values $\sigma_n(n) = 1$ and $\sigma_n(k) = 0$ for all $k < n$, then $\{\sigma_n : n \in \omega\}$ is an antichain in the full binary tree of height ω.

EXERCISE 14.34 (G). Show that if T is a splitting tree and T has a branch of cardinality κ, then T also has an antichain of cardinality κ.

Trees of height ω_1 without uncountable chains or antichains are called *Suslin trees*.[5] The *Suslin Hypothesis*, abbreviated SH, is the following statement:
(SH)   There are no Suslin trees.

Note that the Kurepa Hypothesis asserts that there *are* Kurepa trees, while the Suslin Hypothesis asserts that there *are no* Suslin trees. Was Suslin a nihilist?

Obviously, Suslin trees are Aronszajn trees. The converse is not so obvious, since the absence of uncountable antichains of the form $T(\alpha)$ does not automatically imply that a tree T has no antichains whatsoever. In fact, the assertion that every

---

[5] Read: "Soosleen." One often sees the spelling "Souslin."

Aronszajn tree is Suslin is simply false. To see this, consider the tree $\langle T, \leq_e \rangle$ constructed in Exercise 14.32. Each element of T is a nonempty compact subset of $\mathbb{Q}$ and hence has a maximum element. Let $f : T \to \mathbb{Q}$ be the function that assigns to each $t \in T$ its maximum element. Then f satisfies the following condition:

(∗) $\quad \forall s, t \in T \, (s <_T t \to f(s) <_T f(t))$.

A function $f : T \to \mathbb{Q}$ that satisfies (∗) will be called a $\mathbb{Q}$-*embedding of T*, and a tree T will be called $\mathbb{Q}$-*embeddable* if there exists a $\mathbb{Q}$-embedding of T. An Aronszajn tree T is *special* if T is a union of countably many antichains. Note that if $f : T \to \mathbb{Q}$ is a $\mathbb{Q}$-embedding of T, then $f^{-1}\{q\}$ is an antichain for every $q \in \mathbb{Q}$. Thus $\mathbb{Q}$-embeddable Aronszajn trees are special. In particular, the tree constructed in Exercise 14.32 is special. Since Aronszajn trees have $\aleph_1$ nodes, it follows from Lemma 14.3 that each special Aronszajn tree has an uncountable antichain. Thus special Aronszajn trees are never Suslin.

Is the Aronszajn tree constructed in the proof of Theorem 14.8 also special? Yes, but we will have to use a different argument to prove it.

So let $\langle T_0, \subseteq \rangle$ be an Aronszajn tree as in the proof of Theorem 14.8, and let $S = \bigcup_{\alpha < \omega_1} T_0(\alpha + 1)$. Clearly, $\langle S, \subseteq \rangle$ is still an Aronszajn tree. If $s \in S$, then $s : \alpha + 1 \overset{1:1}{\to} \omega$ for some $\alpha \in \omega_1$. Define a function $\pi : S \to \omega$ by $\pi(s) = s(\max \, dom(s))$.

EXERCISE 14.35 (G). Convince yourself that π is a *specializing function* for S, i.e., show that $\pi^{-1}\{n\}$ is an antichain in S for every $n \in \omega$.

We have shown that S is special. How about T_0 itself?

EXERCISE 14.36 (G). Let $h : T_0 \to S$ be a function that assigns to each $t \in T_0(\alpha)$ a node $h(t) \in T_0(\alpha+1)$ with $t \subseteq h(t)$. Let π be as above, and define $\pi_1 = \pi \circ h$. Show that π_1 is a specializing function for T.

We have shown that T_0 is special. But is T_0 also $\mathbb{Q}$-embeddable? Yes.

LEMMA 14.12. *Let T be an Aronszajn tree. Then the following are equivalent:*
(i) $\quad T$ *is special;*
(ii) $\quad T$ *is $\mathbb{Q}$-embeddable.*

PROOF. We have already seen that $\mathbb{Q}$-embeddable Aronszajn trees are special. To prove the converse, let us assume that $\langle T, \leq_T \rangle$ is a special Aronszajn tree, and let $\pi : T \to \omega$ be a specializing function. Define a function f with domain T by:

$$f(t) = \sum_{\{n:\, \exists s \, (s \leq_T t \wedge \pi(s) = n)\}} 2^{-n}.$$

EXERCISE 14.37 (G). Show that f is a $\mathbb{Q}$-embedding of T. □

In Chapter 19 we will show that Martin's Axiom (a statement relatively consistent with ZFC) implies that every Aronszajn tree is special, which in turn implies the Suslin Hypothesis. In Chapter 22 we will prove that the existence of a Suslin tree, i.e., the negation of the Suslin Hypothesis, is also relatively consistent with ZFC.

CLAIM 14.13. *Let κ be a regular infinite cardinal, and let T be a κ-tree without cofinal branches. Then there exists a splitting κ-subtree S of T.*

PROOF. Let κ, T be as in the assumption, and define:
$$S = \{s \in T : |\{t \in T : s <_T t\}| = \kappa\}.$$

EXERCISE 14.38 (G). Show that S is a splitting κ-subtree of T. □

Clearly, $\aleph_1$-subtrees of Aronszajn trees are Aronszajn, and $\aleph_1$-subtrees of Suslin trees are Suslin. It follows that every Aronszajn tree contains a splitting Aronszajn subtree, and every Suslin tree contains a splitting Suslin subtree.

Recall from Chapter 2 that if $\langle A, \preceq_A \rangle$ and $\langle B, \preceq_B \rangle$ are p.o.'s, then the *simple product* $\langle A, \preceq_A \rangle \otimes^s \langle B, \preceq_B \rangle$ is the p.o. $\langle A \times B, \preceq_s \rangle$, where

$$\langle a, b \rangle \preceq_s \langle c, d \rangle \quad \text{if and only if} \quad a \preceq_A b \wedge c \preceq_B d.$$

Suslin trees, if they exist, provide examples of c.c.c. partial orders whose simple product does not satisfy the c.c.c.: By Exercise 14.33, if $\langle T, \leq_T \rangle$ is a Suslin tree, then $\langle T, \leq_T^* \rangle$ satisfies the c.c.c. On the other hand, we have the following:

LEMMA 14.14. *If $\langle T, \leq_T \rangle$ is an uncountable splitting tree, then $\langle T, \leq_T \rangle \otimes^s \langle T, \leq_T \rangle$ does not satisfy the c.c.c.*

PROOF. Let T be as in the assumption. For every $t \in T$, choose $t_\ell, t_r \in T$ such that $t <_T t_\ell$, $t <_T t_r$, $t_\ell \neq t_r$, $ht(t_\ell) = ht(t_r)$, and if $ht(t) < \beta < ht(t_\ell)$, then there exists exactly one $u \in T(\beta)$ such that $t <_T u$. Let $A = \{\langle t_\ell, t_r \rangle : t \in T\}$.

EXERCISE 14.39 (PG). Show that A is an antichain in $\langle T, \leq_T \rangle \otimes^s \langle T, \leq_T \rangle$. □

We have proved the following result:

THEOREM 14.15. *If SH fails, then there exists a c.c.c. p.o. $\langle \mathbb{P}, \preceq \rangle$ such that $\langle \mathbb{P}, \preceq \rangle \otimes^s \langle \mathbb{P}, \preceq \rangle$ does not satisfy the c.c.c. In particular, if $\langle T, \preceq_T \rangle$ is any Suslin tree, then the p.o. $\langle T, \preceq_T^* \rangle$ is as above.*

Trees can be used to construct various fancy linear orders. We conclude this chapter with a sample of such applications.

Let $\langle T, \leq_T \rangle$ be a tree. If $t \in T(\alpha)$ and $\beta \leq \alpha$, then we let $\hat{t}(\beta)$ denote the unique $s \in T(\beta)$ such that $s \leq_T t$. If $s, t \in T$ are incomparable in T, then we define

$$\Delta(s,t) = \min\{\beta : \hat{s}(\beta) \neq \hat{t}(\beta)\}.$$

Now let $\preceq$ be a l.o. relation on T. We define a relation $\leq_\ell$ on T, called the *lexicographical ordering on T induced by $\preceq$*, as follows:

$s \leq_\ell t$ iff $s \leq_T t$ or s and t are incomparable in T and $\hat{s}(\Delta(s,t)) \prec \hat{t}(\Delta(s,t))$.

EXERCISE 14.40 (G). (a) Convince yourself that for every l.o. relation $\preceq$ on T the lexicographical ordering on T induced by $\preceq$ is a l.o. relation.
(b) Consider the tree of Example 14.1 and let $\preceq$ be the alphabetical order of its nodes. Find the lexicographical ordering of the tree induced by $\preceq$.

Not surprisingly, the structure of the l.o. $\langle T, \leq_\ell \rangle$ is closely related to the structure of the tree $\langle T, \leq_T \rangle$. For example, every $\leq_T$-chain is also a $\leq_\ell$-chain of the same order type. If some levels of $\langle T, \leq_T \rangle$ are uncountable, then $\langle T, \leq_\ell \rangle$ may have subsets of order type ω_1 even if $\langle T, \leq_T \rangle$ does not have uncountable branches. However, no such chains can be introduced if $\langle T, \leq_T \rangle$ is an Aronszajn tree.

LEMMA 14.16. *Suppose $\langle T, \leq_T \rangle$ is an Aronszajn tree, $\preceq$ is a l.o. relation on T, and $\leq_\ell$ is the lexicographical ordering of T induced by $\preceq$. Then $\langle T, \leq_\ell \rangle$ has no subsets of order type ω_1 or ω_1^*.*

PROOF. Let $T, \leq_T$, and $\leq_\ell$ be as in the assumptions. We will prove that there are no chains of order type ω_1 in $\langle T, \leq_\ell \rangle$; the same argument works for chains of order type ω_1^*.

So suppose towards a contradiction that $\langle t_\xi : \xi < \omega_1 \rangle$ is such that $t_\xi <_\ell t_\eta$ for $\xi < \eta < \omega_1$. For $\alpha \in \omega_1$ define:
$$S(\alpha) = \{s \in T(\alpha) : |\{\xi < \omega_1 : s <_\ell t_\xi\}| = \aleph_1\}.$$

EXERCISE 14.41 (G). Show that $S(\alpha) \neq \emptyset$ for all $\alpha < \omega_1$.

Now we show that $|S(\alpha)| = 1$ for all $\alpha < \omega_1$. Suppose not, and let α_0 be the least α such that $|S(\alpha)| > 1$. Let r, s be two different elements of $S(\alpha_0)$. Pick $\xi < \omega_1$ such that $r <_\ell t_\xi$. Since there are uncountably many $\eta < \omega_1$ with $s <_\ell t_\eta$, we can find $\eta > \xi$ such that $s <_\ell t_\eta$. By the choice of α_0, we must have $r = \hat{t}_\xi(\Delta(t_\xi, t_\eta))$, $s = \hat{t}_\eta(\Delta(t_\xi, t_\eta))$. Since $t_\xi <_\ell t_\eta$, the definition of $\leq_\ell$ implies that $r \prec s$. But the same argument can be made with the roles of r and s reversed, in which case it yields the conclusion that $s \prec r$. This is impossible though, since the relation $\prec$ is asymmetric, and the contradiction we have reached shows that $|S(\alpha)| = 1$ for each $\alpha < \omega_1$.

To wrap up the argument, let $s(\alpha)$ denote the unique element of $S(\alpha)$. A moment's reflection shows that $s(\alpha) <_T s(\beta)$ whenever $\alpha < \beta$. Hence, the set $\{s(\alpha) : \alpha < \omega_1\}$ is an uncountable branch of $\langle T, \leq_T \rangle$, which contradicts the assumption that T is an Aronszajn tree. □

We have shown that if $\leq_\ell$ is a lexicographical ordering of an Aronszajn tree T, then $\langle T, \leq_\ell \rangle$ is an uncountable l.o. without uncountable wellordered chains. There is nothing fancy about that: The real line has the same property.

EXERCISE 14.42 (G). Convince yourself that $\langle \mathbb{R}, \leq \rangle$ contains no chains of order type ω_1 or ω_1^*.

But the l.o. $\langle T, \leq_\ell \rangle$ is radically different from subsets of the real line with the natural order. Recall that a set $D \subseteq L$ is *dense* in a l.o. $\langle L, \preceq \rangle$ if
$$\forall a, b \in L \, (a \prec b \rightarrow \exists d \in D \, (a \preceq d \wedge d \preceq b)).$$
The real line doesn't contain uncountable wellordered subsets because it has a countable dense subset. Compare this with the following:

LEMMA 14.17. *Suppose $\langle T, \leq_T \rangle$ is an Aronszajn tree, $\preceq$ is a l.o. relation on T, and $\leq_\ell$ is the lexicographical ordering of T induced by $\preceq$. If S is an uncountable subset of T, then $\langle S, \leq_\ell \rangle$ does not contain a countable dense subset.*

PROOF. Let $S, T, \leq_T$, and $\leq_\ell$ be as in the assumption, and let D be a countable subset of S. Choose $\alpha < \omega_1$ such that $D \subseteq T_{(\alpha)}$. Since $T_{(\alpha+1)}$ is countable, we can pick $s, t \in S \setminus T_{(\alpha+1)}$ such that $s <_\ell t$ and $\hat{s}(\alpha) = \hat{t}(\alpha)$.

EXERCISE 14.43 (G). Convince yourself that if $d \in D$ and s, t are as above, then $d \leq_\ell s$ if and only if $d \leq_\ell t$. Conclude that there is no $d \in D$ with $s \leq_\ell d$ and $d \leq_\ell t$. □

A linear order $\langle A, \preceq_A \rangle$ is called an *Aronszajn line* if

(1) A is uncountable;
(2) $\langle A, \preceq_A \rangle$ doesn't contain chains of order type ω_1 or ω_1^*; and
(3) A doesn't contain an uncountable subset B such that $\langle B, \preceq_A \rangle$ is isomorphic to a subset of the real line with the natural order.

Lemmas 14.16 and 14.17 entail the following:

COROLLARY 14.18. *Suppose $\langle T, \leq_T \rangle$ is an Aronszajn tree, $\preceq$ is a l.o. relation on T, and $\leq_\ell$ is the lexicographical ordering of T induced by $\preceq$. Then $\langle T, \leq_\ell \rangle$ is an Aronszajn line.*

Let κ be an infinite cardinal. We say that a l.o. $\langle A, \preceq \rangle$ satisfies the *κ-chain condition* (abbreviated κ-c.c.) if and only if the order topology induced on A by $\preceq$ satisfies the κ-chain condition. In other words, $\langle A, \preceq \rangle$ satisfies the κ-c.c. if and only if there is no family $\{(a_\xi, b_\xi) : \xi < \kappa\}$ of pairwise disjoint nonempty open intervals in $\langle A, \preceq \rangle$.

LEMMA 14.19. *Let κ be an infinite cardinal and let $\langle T, \leq_T \rangle$ be a splitting tree. If there exists a l.o. relation $\preceq$ on T such that the lexicographical order $\langle T, \leq_\ell \rangle$ induced by $\preceq$ satisfies the κ-c.c., then $\langle T, \leq_T^* \rangle$ also satisfies the κ-c.c.*

PROOF. We prove the contrapositive. Assume $\langle T, \leq_T \rangle$ is a splitting tree such that $\langle T, \leq_T^* \rangle$ does not satisfy the κ-c.c. By Exercise 14.33, there exists an antichain $\{r_\xi : \xi < \kappa\}$ of size κ in T. Since T is splitting, we may choose for every $\xi < \kappa$ nodes $s_\xi, t_\xi, u_\xi \in T$ such that $r_\xi <_T s_\xi, t_\xi, u_\xi$, and $s_\xi <_\ell t_\xi <_\ell u_\xi$. Now the following exercise shows that $\langle T, \leq_\ell \rangle$ does not satisfy the κ-c.c.

EXERCISE 14.44 (G). Show that $\{(s_\xi, u_\xi) : \xi < \kappa\}$ is a family of pairwise disjoint nonempty intervals in $\langle T, \leq_\ell \rangle$. □

Is the converse of Lemma 14.19 true? Not for every splitting tree. If s, t are nodes of a tree T, then we say that t is an *immediate successor of s* if $s <_T t$ and $ht(t) = ht(s) + 1$.

CLAIM 14.20. *Let κ be an infinite cardinal, and let $\langle T, \leq_T \rangle$ be a splitting tree of height κ such that every $t \in T$ has only finitely many immediate successors. If $\leq_\ell$ is any lexicographical ordering of T, then $\langle T, \leq_\ell \rangle$ does not satisfy the κ-c.c.*

PROOF. Let $T, \leq_\ell$ be as in the assumption and let $\preceq$ be a l.o. relation on T that induces $\leq_\ell$. For each $\alpha \in \kappa$ pick a node $r_\alpha \in T(\alpha)$. Since T is splitting, for each r_α the set of its immediate successors is nonempty. By assumption, this set is finite. Hence, each r_α has a $\preceq$-smallest immediate successor s_α. Similarly, each s_α has a $\preceq$-smallest immediate successor t_α.

EXERCISE 14.45 (G). Show that if $r, s, t \in T$ are such that s is the $\preceq$-smallest immediate successor of r and t is the $\preceq$-smallest immediate successor of s, then $(r, t) = \{u \in T : r <_\ell u <_\ell t\} = \{s\}$.

Now Exercise 14.45 implies that the family $\{(r_\alpha, t_\alpha) : \alpha \in \kappa\}$ is a family of nonempty, pairwise disjoint intervals in $\langle T, \leq_\ell \rangle$. Thus $\langle T, \leq_\ell \rangle$ does not satisfy the κ-c.c. □

Let us call a tree T *bushy* if every node of T has infinitely many immediate successors. For bushy trees the converse of Lemma 14.19 is true.

LEMMA 14.21. *Let κ be an infinite cardinal, and let $\langle T, \leq_T \rangle$ be a bushy tree such that $\langle T, \leq_T^* \rangle$ has the κ-c.c. Suppose $\preceq$ is a l.o. relation on T such that for every $t \in T$ the set of immediate successors of t has no $\preceq$-minimal element, and let $\leq_\ell$ be the lexicographical ordering of T induced by $\preceq$. Then $\langle T, \leq_\ell \rangle$ has the κ-c.c.*

PROOF. Again we prove the contrapositive. Let $\langle T, \leq_T \rangle$ be a bushy tree, and let $\preceq, \leq_\ell$ be as in the assumptions. Suppose $\{(a_\xi, b_\xi) : \xi < \kappa\}$ is a family of nonempty pairwise disjoint intervals in $\langle T, \leq_\ell \rangle$. For each ξ, choose $t_\xi \in (a_\xi, b_\xi)$ such that $a_\xi <_T t_\xi$ and $t_\xi \not\leq_T b_\xi$. A suitable t_ξ can always be found: If $a_\xi \not<_T b_\xi$, any immediate successor t_ξ of a_ξ will do the job. If $a_\xi <_T b_\xi$, let s_ξ be the immediate successor of a_ξ that satisfies $s_\xi \leq_T b_\xi$. By the assumption on $\preceq$, there exists an immediate successor t_ξ of a_ξ such that $t_\xi \prec s_\xi$. Clearly, $t_\xi \in (a_\xi, b_\xi)$ and $t_\xi \not\leq_T b_\xi$.

The following exercise shows that the set $\{t_\xi : \xi < \kappa\}$ is an antichain in $\langle T, \leq_T^* \rangle$.

EXERCISE 14.46 (G). Suppose $(a_\xi, b_\xi), (a_\eta, b_\eta)$ are disjoint $\leq_\ell$-intervals, $t_\xi \in (a_\xi, b_\xi)$, $t_\eta \in (a_\eta, b_\eta)$, $t_\xi \not\leq_T b_\xi$, and $t_\eta \not\leq_T b_\eta$. Then t_ξ and t_η are incomparable with respect to $\leq_T$. □

In order to deduce the converse of Lemma 14.19 from Lemma 14.21 we still need to show that for every bushy tree a l.o. relation $\preceq$ as in the assumptions of Lemma 14.21 exists.

EXERCISE 14.47 (G). (a) Show that if T is a bushy tree, then there exists a l.o. relation $\preceq$ on T such that for every t in T the set of immediate successors of t has no $\preceq$-minimal element.

(b) Show that if T is bushy, $\preceq$ is as in point (a), and $\leq_T$ is the lexicographical ordering induced by $\preceq$, then $\langle T, \leq_T \rangle$ is a dense linear order.

A *Suslin line* is a linear order $\langle L, \leq_L \rangle$ that satisfies the c.c.c. but has no countable dense subset.[6] From Lemma 14.17, Lemma 14.21, and Exercise 14.47 we can deduce the following:

COROLLARY 14.22. *If $\langle T, \leq_T \rangle$ is a bushy Suslin tree, then there exists a lexicographical ordering $\leq_\ell$ of T such that $\langle T, \leq_\ell \rangle$ is a densely ordered Suslin line.*

Thus, if bushy Suslin trees exist, then so do Suslin lines. We have already mentioned that the existence of Suslin trees is relatively consistent with, but not provable in, ZFC. Can consistent examples of Suslin trees be bushy?

THEOREM 14.23. *Let $\langle T, \leq_T \rangle$ be an Aronszajn tree. Then there exists $S \subseteq T$ such that $\langle S, \leq_T \rangle$ is a bushy Aronszajn tree. If T is Suslin, then so is S.*

PROOF. The idea of the proof is deceptively simple: Let $\langle T, \leq_T \rangle$ be an Aronszajn tree. Construct recursively an increasing transfinite sequence $\langle \alpha_\xi : \xi < \omega_1 \rangle$ of countable ordinals such that every node $t \in T(\alpha_\xi)$ has infinitely many successors at level $\alpha_{\xi+1}$. Let $S = \bigcup_{\xi < \omega_1} T(\alpha_\xi)$. Then $\langle S, \leq_T \rangle$ is an ω_1-subtree of $\langle T, \leq_T \rangle$, and thus is an Aronszajn tree. If T is Suslin, then so is S. Moreover, the set of immediate successors in S of a node $t \in T(\alpha_\xi)$ is the set of successors of t at level $T(\alpha_{\xi+1})$. Thus S is bushy.

O.k., if a sequence $\langle \alpha_\xi : \xi < \omega_1 \rangle$ as above can be found, then the lemma is proved. But the problem is: Given α_ξ, how can we find $\alpha_{\xi+1} > \alpha_\xi$ such that every node in $T(\alpha_\xi)$ has infinitely many successors at level $\alpha_{\xi+1}$?

---

[6]Not all Suslin lines are Aronszajn lines. It is true though that Suslin lines do not contain subsets of order type ω_1 or ω_1^*.

Maybe Claim 14.13 helps. It implies that we may without loss of generality assume that $\langle T, \leq_T \rangle$ is splitting. If T is splitting and α_ξ is given, we can use the fact that $T(\alpha_\xi)$ is countable to find a countable ordinal $\alpha_{\xi,1} > \alpha_\xi$ such that for every $s \in T(\alpha_\xi)$ there exist $\leq_T$-incomparable nodes $t_0(s), t_1(s) \in T_{(\alpha_{\xi,1})}$ such that $s <_T t_0(s)$ and $s <_T t_1(s)$. Similarly, we can find a countable ordinal $\alpha_{\xi,2} > \alpha_{\xi,1}$ such that for every $s \in T(\alpha_\xi)$ there are pairwise incomparable nodes $t_{00}(s), t_{01}(s), t_{10}(s), t_{11}(s) \in T_{(\alpha_{\xi,2})}$ such that $t_0(s) <_T t_{00}(s), t_{01}(s)$ and $t_1(s) <_T t_{10}(s), t_{11}(s)$. Continuing like this, we can recursively construct an increasing sequence of countable ordinals $\langle \alpha_{\xi,n} : n \in \omega \setminus \{0\} \rangle$ such that each $s \in T(\alpha_\xi)$ has at least 2^n pairwise incomparable successors in $T_{(\alpha_{\xi,n})}$. Now let $\alpha_{\xi+1} = \sup\{\alpha_{\xi,n} : n \in \omega \setminus \{0\}\}$. Then for every $m \in \omega$ and $s \in T(\alpha_\xi)$, there is a set of m pairwise incomparable successors of s in $T_{(\alpha_{\xi+1})}$. If $r, r' \in T_{(\alpha_{\xi+1})}$ are $\leq_T$-incomparable and $t, t' \in T(\alpha_{\xi+1})$ are such that $r <_T t, r' <_T t$, then $t \neq t'$. It follows that each $s \in T(\alpha_\xi)$ has infinitely many successors at level $\alpha_{\xi+1}$.

Or does it? How do we know that a given $s \in T(\alpha_\xi)$ has *any* successors at level $\alpha_{\xi+1}$? The bad news is that *a priori* we don't. It is possible that there are $\alpha < \beta < \omega_1$ and $s \in T(\alpha)$ such that s has no successor whatsoever in $T(\beta)$.[7] To get around the problem, we need a new concept. Let $\langle R, \leq_R \rangle$ be a tree of height δ, let $\alpha < \delta$, and let $s \in R(\alpha)$. We say that s is *high* if for every β with $\alpha < \beta < \delta$ there exists $t \in R(\beta)$ such that $s <_R t$. The tree R is *tall* if each node of R is high.

If we could assume that the tree T in the assumption of Theorem 14.23 is tall, then the above reasoning would go through. The following lemma helps.

LEMMA 14.24. *Let $\langle R, \leq_R \rangle$ be an Aronszajn tree. Then there exists a tall Aronszajn subtree T of R.*

EXERCISE 14.48 (G). Derive Theorem 14.23 from Lemma 14.24.

PROOF OF LEMMA 14.24. We need a claim.

CLAIM 14.25. *Let $\langle R, \leq_R \rangle$ be an Aronszajn tree, and let $r \in R$. If r is high, then for each $\beta < \omega_1$ with $\beta > ht(r)$ there exists a high $s \in R(\beta)$ such that $r <_R s$.*

PROOF. Assume not and let R, r be witnesses. Let β be a countable ordinal such that $ht(r) < \beta$ and no $s \in R(\beta)$ with $r <_R s$ is high. Let $X = \{s \in R(\beta) : r <_R s\}$. For each $s \in X$ let γ_s be a countable ordinal $> \beta$ such that there is no $t \in R(\gamma_s)$ with $s <_R t$. Let $\gamma = \sup\{\gamma_s : s \in X\}$. Then there is no $t \in R(\gamma)$ with $r <_R t$, which contradicts the assumption that r is high. □

Now let $T = \{r \in R : r$ is high$\}$. Claim 14.25 implies that T is as required. □□

COROLLARY 14.26. *If there exists a Suslin tree, then there exists a densely ordered Suslin line.*

The technique of the proof of Lemma 14.24 can be used to obtain a more general result.

EXERCISE 14.49 (G). Let $\kappa \geq \omega$. Show that every κ^+-Aronszajn tree contains a tall κ^+-Aronszajn subtree.

While we are at it, let us mention an interesting consequence of Claim 14.25.

---

[7]This is the reason why in the previous paragraph we took the precaution to write $T_{(\alpha_{\xi,n})}$ instead of $T(\alpha_{\xi,n})$.

THEOREM 14.27. *There are no Aronszajn trees with all levels finite.*

PROOF. Suppose towards a contradiction that $\langle T, \leq_T \rangle$ is an Aronszajn tree with all levels finite. By Claim 14.25, the set S of all high nodes of T forms a tall Aronszajn subtree with all levels finite, and we may without loss of generality assume that T is tall to begin with. The latter assumption implies that if $\alpha < \beta < \omega_1$, then $|T(\alpha)| \leq |T(\beta)|$. It follows that there exist $\alpha_0 < \omega_1$ and $n \in \omega$ such that $|T(\beta)| = n$ for all countable ordinals $\beta \geq \alpha_0$. Pick $s \in T(\alpha_0)$. The choice of α_0 implies that the tree does not split above level α_0, and thus the set $\{t \in T : t \leq_T s \vee s \leq_T t\}$ is an uncountable branch of T, which contradicts the assumption that T is an Aronszajn tree. □

If $\langle L, \leq_L \rangle$ is a dense, Dedekind complete linear order with smallest and largest elements, then L will be called a *linearly ordered continuum*. This name suggests a link to topology: Indeed, if $\langle L, \leq_L \rangle$ is a linearly ordered continuum, then the order topology induced by $\leq_L$ is connected and compact.

FACT 14.28. *Every dense linear order can be order-isomorphically embedded as a dense subset of a linearly ordered continuum.*

The proof of Fact 14.28 is an easy EXERCISE, but it is not particularly relevant to the material of this chapter. Much more important is the following:

EXERCISE 14.50 (PG). Suppose $\langle L^+, \preceq \rangle$ is a dense linear order, L is a dense subset of L^+, and $\langle L, \preceq \rangle$ is a Suslin line. Then $\langle L^+, \preceq \rangle$ is also a Suslin line.

The following question is known as the *Suslin Problem*:

(SQ)   Let $\langle L, \leq_L \rangle$ *be a linearly ordered continuum
with no uncountable family of pairwise disjoint open intervals.
Is* $\langle L, \leq_L \rangle$ *necessarily isomorphic to the unit interval* $[0,1]$?

By Fact 14.28 and Exercise 14.50, if a densely ordered Suslin line exists, then the answer to the Suslin problem is negative. Thus it follows from Corollary 14.26 that the negation of the Suslin Hypothesis implies a negative answer to the Suslin Problem.[8] The connection also works the other way round.

FACT 14.29. *Suppose $\langle L, \leq_L \rangle$ is a linearly ordered continuum with no uncountable family of pairwise disjoint open intervals that is not isomorphic to the unit interval $[0,1]$. Then $\langle L, \leq_L \rangle$ is a Suslin line.*

Fact 14.29 is a well-known theorem of general topology. Let us skip its proof and devote the rest of this chapter to the proof of the following:

THEOREM 14.30. *Suppose there exists a Suslin line. Then there exists a Suslin tree.*

PROOF. It will be easier to work with dense Suslin lines; so let us first show that doing so does not lead to loss of generality.

EXERCISE 14.51 (PG). Prove that if there exists a Suslin line, then there exists a dense one. *Hint:* Let $\langle L, \leq_L \rangle$ be a Suslin line, and consider the following equivalence relation $\sim$ on L: $l \sim l'$ if and only if $|(l, l')| < \aleph_1$. Define a relation $\preceq$ on $L/\sim$ by: $l/\sim \preceq l'/\sim$ if and only if $l \leq_L l'$. Show that $\preceq$ is well-defined and that $\langle L/\sim, \preceq \rangle$ is a dense Suslin line.

---

[8]The strange negativity of SH starts to look much less puzzling, doesn't it?

So let $\langle L, \leq_L \rangle$ be a dense Suslin line, and let $\mathcal{I}(L)$ be the set of nonempty open intervals of $\langle L, \leq_L \rangle$. We are going to construct $T \subseteq \mathcal{I}(L)$ such that $\langle T, \subseteq \rangle$ is a Suslin tree.

By recursion over $\alpha < \omega_1$ we construct families $T(\alpha)$ of pairwise disjoint elements of $\mathcal{I}(L)$ as follows:

Let $T(0)$ be any maximal family of pairwise disjoint elements of $\mathcal{I}(L)$.

Suppose $T(\alpha)$ has been constructed. For each interval $(a,b) \in T(\alpha)$ we choose a family $I_{(a,b)} \subseteq \mathcal{I}(L)$ of pairwise disjoint intervals, maximal with respect to the property that (c,d) is a proper subset of (a,b) for each $(c,d) \in I_{(a,b)}$. Note that since $\langle L, \leq_L \rangle$ is dense, $|I_{(a,b)}| > 1$ for each $(a,b) \in T(\alpha)$. Then we let $T(\alpha+1) = \bigcup \{I_{(a,b)} : (a,b) \in T(\alpha)\}$.

At limit stages $\delta > 0$, if $T(\alpha)$ has already been constructed for all $\alpha < \delta$, let $T(\delta)$ be a family I of pairwise disjoint elements of $\mathcal{I}(L)$, maximal with respect to the property that if $(c,d) \in I$, then $\forall \alpha < \delta \exists (a,b) \in T(\alpha) \, ((c,d) \subseteq (a,b))$.

Finally, let $T = \bigcup_{\alpha < \omega_1} T(\alpha)$.

EXERCISE 14.52 (G). (a) Convince yourself that $\langle T, \subseteq \rangle$ is a tree whose α-th level is equal to $T(\alpha)$ for each $\alpha < \omega_1$.

(b) Show that if $(a,b), (c,d) \in T$ are incomparable with respect to $\subseteq$, then $(a,b) \cap (c,d) = \emptyset$.

It follows immediately from the assumption on $\langle L, \leq_L \rangle$ and point (b) of the last exercise that $\langle T, \subseteq \rangle$ does not contain uncountable antichains. Moreover, it follows from the remark made in the construction of successor stages that T is splitting. Hence, by Exercise 14.34, T does not contain uncountable chains either. It is somewhat less obvious that T has height ω_1.

So let α_0 is the smallest ordinal α such that $T(\alpha) = \emptyset$. Clearly, α_0 can be neither a successor ordinal nor 0. Thus, α_0 must be a limit ordinal $\geq \omega$. Consider the set

$$E = \{c \in L : \exists \beta < \alpha_0 \exists (a,b) \in T(\beta) \, (c = a \vee c = b)\}.$$

EXERCISE 14.53 (G). Show that E is dense in $\langle L, \leq_L \rangle$, and conclude that $\alpha_0 = \omega_1$. □

Mathographical Remarks

If you are looking for a source to learn more about trees and their applications, we recommend the article *Trees and linearly ordered sets*, by Stevo Todorčević in the Handbook of Set-Theoretic Topology, K. Kunen and J. E. Vaughan, eds., North-Holland, 1984, 235–293.

More information about equivalences of König's Lemma and weak versions of the Axiom of Choice can be found in the article by J. Truss, *Some cases of König's Lemma*, in: Set Theory and Hierarchy Theory. A Memorial Tribute to Andrzej Mostowski, Bierutowice, Poland, 1975, edited by W. Marek, M. Srebrny, and A. Zarach, Lecture Notes in Mathematics 537, Springer Verlag, 1976.

CHAPTER 15

A Little Ramsey Theory

Here is a bit of common sense: *If more than n pigeons occupy n pigeonholes, then at least one pigeonhole must be occupied by at least two pigeons.* Lemma 14.3 is a mathematical counterpart of this folk wisdom. Let us reformulate it here.

THEOREM 15.1 (Pigeonhole Principle). *Assume κ is an infinite cardinal, A is a set of cardinality κ, and $A = \bigcup_{i<\sigma} A_i$, where $\sigma < cf(\kappa)$. Then there exists $i < \sigma$ such that $|A_i| = \kappa$.*

The name *Pigeonhole Principle* is used in the literature also for the common-sense statement at the beginning of this chapter and several of its modifications.

EXERCISE 15.1 (G). Formulate and prove the strongest "pigeonhole principle" for finite cardinals you can think of.

The Pigeonhole Principle is our first example of a *partition theorem*. In order to give you an idea what, in general, a partition theorem is, we need to introduce some notation. Let X be a set, ρ a cardinal. Recall that $[X]^\rho$ denotes the family of all subsets of X of cardinality ρ. An indexed family $\mathcal{P} = \{P_i : i \in I\}$ is a *partition* of $[X]^\rho$ if

$$\bigcup\{P_i : i \in I\} = [X]^\rho \text{ and } P_i \cap P_k = \emptyset \text{ for } i, k \in I, i \neq k.$$

Let $f_\mathcal{P} : [X]^\rho \to I$ be the (unique) function f with $f^{-1}\{i\} = P_i$ for all $i \in I$. We call $f_\mathcal{P}$ the *canonical coloring associated with* $\mathcal{P}$. Conversely, there is a canonical partition of $[X]^\rho$ associated with every coloring $f : [X]^\rho \to \sigma$. A set Y is *homogeneous* for the partition $\mathcal{P}$ (as well as for the canonical coloring associated with it), if there is an $i_0 \in I$ such that $[Y]^\rho \subseteq P_{i_0}$.

Let κ, λ, ρ, σ be cardinals. The symbol[1]

$$\kappa \to (\lambda)^\rho_\sigma$$

stands for the following statement:

Whenever X is a set of cardinality κ, I is a set of cardinality σ, and $\mathcal{P} = \{P_i : i \in I\}$ is a partition of $[X]^\rho$, there exists $Y \subseteq X$ with $|Y| = \lambda$ that is homogeneous for $\mathcal{P}$.[2]

Theorem 15.1 can be expressed in this notation as follows:

THEOREM 15.2. *For all infinite cardinals κ and for all $\sigma < cf(\kappa)$, the relation $\kappa \to (\kappa)^1_\sigma$ holds.*

---

[1] Read: κ arrows λ super ρ sub σ.
[2] Of course, in order to verify that $\kappa \to (\lambda)^\rho_\sigma$ holds, it suffices to consider partitions $\{P_i : i < \sigma\}$ of $[\kappa]^\rho$.

EXERCISE 15.2 (G). Let $\kappa, \lambda, \rho, \sigma, \kappa', \lambda', \rho', \sigma'$ be cardinals such that $\kappa' \geq \kappa$, $\lambda' \leq \lambda$, $\rho' \leq \rho$, and $\sigma' \leq \sigma$. Show that if $\kappa \to (\lambda)_\sigma^\rho$, then also $\kappa' \to (\lambda')_{\sigma'}^{\rho'}$.

Any true statement of the form $\kappa \to (\lambda)_\sigma^\rho$ is a partition theorem, but as we shall see later in this chapter, not all partition theorems are of this kind. As one might expect, partition theorems for $\rho > 1$ are more difficult to prove than Theorem 15.2. If you want to get your feet wet before reading on, we recommend the following two exercises.

EXERCISE 15.3 (PG). Show that $6 \to (3)_2^2$. *Hint:* You may find it easier to prove this theorem if you give it the following interpretation: *Among every group of six people, there are three that are mutual acquaintances, or there are three that are mutual strangers.*

In order to indicate that a statement $\kappa \to (\lambda)_\sigma^\rho$ is false, we write $\kappa \not\to (\lambda)_\sigma^\rho$.

EXERCISE 15.4 (G). By constructing a suitable partition, show that $5 \not\to (3)_2^2$.

The study of partition theorems is called *partition calculus* or *Ramsey Theory* in honor of the author of the following result.

THEOREM 15.3 (Ramsey's Theorem). *For all positive natural numbers k, ℓ:*
$$\omega \to (\omega)_\ell^k.$$

PROOF. By induction over k.

For $k = 1$, this is just a special case of Theorem 15.2.

Now assume the theorem has been proved for k, i.e., assume that $\omega \to (\omega)_\ell^k$ for all $\ell \in \omega$. Let $\mathcal{P} = \{P_i : i < \ell\}$ be a partition of $[\omega]^{k+1}$, and let f be the canonical coloring associated with $\mathcal{P}$. One of the most important tricks in proving partition theorems is reducing the superscript ρ by defining so-called *induced colorings*. The proof of Ramsey's Theorem gives a beautiful illustration of this technique. We shall define a strictly increasing sequence $(n_i)_{i<\omega}$ of positive natural numbers and a decreasing sequence $(A_i)_{i<\omega}$ of infinite subsets of ω so that for all $i < \omega$ the following hold:

(i) $\forall n \in A_i\, (n_i < n)$;
(ii) $\forall a, b \in [A_i]^k\, (f(\{n_i\} \cup a) = f(\{n_i\} \cup b))$.

Let
$$n_0 = 0.$$
For $a \in [\omega \setminus \{0\}]^k$ let $f^{(0)}(a) = f(\{0\} \cup a)$. Note that $f^{(0)}$ is a coloring of all k-element subsets of $\omega \setminus \{0\}$. Thus we have reduced the superscript by one, and we can apply the inductive assumption to find an infinite $A_0 \subseteq \omega \setminus \{0\}$ that is homogeneous for the induced coloring $f^{(0)}$. Note that A_0 satisfies (i) and (ii).

Now assume that n_i, A_i have been constructed for $i < j$. We put
$$n_j = \min A_{j-1}.$$
For $a \in [A_{j-1} \setminus \{n_j\}]^k$ let $f^{(j)}(a) = f(\{n_j\} \cup a)$. Use the inductive assumption to find an infinite subset A_j of $A_{j-1} \setminus \{n_j\}$ that is homogeneous for the induced coloring $f^{(j)}$.

Now consider the set
$$K = \{n_i : i < \omega\}.$$

It follows from the construction of the n_i's and the A_i's that:

If $n \in K$; $a, b \in [K \setminus (n+1)]^k$, then $f(\{n\} \cup a) = f(\{n\} \cup b)$.

Let $g : K \to \omega$ be the function that assigns to each $n \in K$ the $i < \ell$ for which
$$\forall a \in [K \setminus (n+1)]^k \, (f(\{n\} \cup a) = i).$$
Now choose $i_0 < \ell$ such that $Y = g^{-1}(i_0)$ is infinite. We show that Y is homogeneous for f. To this end, let $a \in [Y]^{k+l}$ and $m = \min a$. We have
$$f(a) = f(\{m\} \cup (a \setminus \{m\})) = g(m) = i_0,$$
independently of the choice of a. □

The next theorem is a version of Ramsey's Theorem for finite cardinals.

THEOREM 15.4. *Let m, k, ℓ be positive natural numbers. There exists a natural number n such that*
$$n \to (m)^k_\ell.$$

PROOF. Suppose towards a contradiction that $m, k, \ell \in \omega \setminus \{0\}$ are such that $n \not\to (m)^k_\ell$ for every $n \in \omega$. This means that for every $n \in \omega$ there exists a coloring $f : [n]^k \to \ell$ without a homogeneous subset of size m. Let $F = \{f : $ there exists $n \in \omega$ such that $f \in {}^{[n]^k}\ell$ and there is no homogeneous set of size m for $f\}$, and consider the p.o. $\langle F, \subseteq \rangle$.

EXERCISE 15.5 (G). Show that:
(a) $\langle F, \subseteq \rangle$ is a tree.
(b) The r-th level of the tree $\langle F, \subseteq \rangle$ consists of all functions $f \in F$ with domain $[k+r-1]^k$.

Our assumption implies that every level of the tree $\langle F, \subseteq \rangle$ is nonempty. On the other hand, since there are only finitely many functions from $[r+k-1]^k$ into ℓ, each level of the tree is finite. Thus König's Lemma (Lemma 14.2) implies that the tree $\langle F, \subseteq \rangle$ has an infinite branch.

EXERCISE 15.6 (G). Suppose B is an infinite branch of $\langle F, \subseteq \rangle$, and let $g = \bigcup B$. Show that:
(a) g is a function from $[\omega]^k$ into ℓ;
(b) $g \restriction [n]^k \in F$ for all $n \in \omega$.

If g is as in Exercise 15.6, then Ramsey's Theorem implies that there exists an infinite subset A of ω that is homogeneous for g. Let $a_0 < \cdots < a_{m-1}$ be the first m elements of A, and let $n = a_{m-1} + 1$. Then $\{a_0, \ldots, a_{m-1}\}$ is a homogeneous set of size m for $g \restriction [n]^k$, which contradicts Exercise 15.6(b) and the definition of the set F. □

The technique of the proof of Theorem 15.4 allows us to obtain even stronger partition theorems for finite sets. A subset $A \subseteq \omega$ is called *relatively large* if $|A| \geq \min A$. We write $n \xrightarrow{*} (m)^k_\ell$ if for every coloring $f : [n]^k \to \ell$ there exists a relatively large homogeneous set of size at least m.

THEOREM 15.5 (Paris-Harrington). *For all $m, k, \ell \in \omega \setminus \{0\}$ there exists $n \in \omega$ such that $n \xrightarrow{*} (m)^k_\ell$.*

EXERCISE 15.7 (G). Prove Theorem 15.5.

The remarkable feature of Theorems 15.4 and 15.5 is that we used a result about infinite sets to prove a property of natural numbers. Could we obtain the same results without employing the notion of an infinite set?

THEOREM 15.6. *For all $m \in \omega\setminus\{0\}$, $4^m \to (m+1)_2^2$.*

PROOF. We need a lemma.

LEMMA 15.7. *Let $m \in \omega\setminus\{0\}$, and let $f : [4^m]^2 \to 2$ be a coloring. Then there exist a tree $\langle T, \leq_T \rangle$ of height $2m$ and an injection $h : T \to 4^m$ such that $f(\{h(r), h(s)\}) = f(\{h(r), h(t)\})$ for all $r, s, t \in T$ with $r <_T s <_T t$.*

Before we prove the lemma, let us show how it implies Theorem 15.6. Let $B \subseteq T$ be a branch of T of length $2m$, and let $A = \{h(t) : t \in B\}$. Since h is injective, we may define a l.o. relation $\preceq$ on A as follows: $a \preceq b$ if and only if there exist $s, t \in B$ with $a = h(s)$, $b = h(t)$, and $s \leq_T t$. Let $\langle a_0, \ldots a_{2m-1}\rangle$ be the enumeration of A in $\prec$-increasing order. Now define an induced coloring $f_* : \{a_0, \ldots a_{2m-2}\} \to 2$ by:

$$f_*(a_j) = f(\{a_j, a_{j+1}\}).$$

By the choice of T and A we have:

(15.1) $\qquad f(\{a_j, a_k\}) = f_*(a_j)$ for all $j < k < 2m$.

Since the domain of f_* has $2m - 1$ elements, there exists $i_0 \in \{0, 1\}$ such that $|f_*^{-1}\{i_0\}| \geq m$. Now it follows from 15.1 that $f_*^{-1}\{i_0\} \cup \{a_{2m-1}\}$ is a homogeneous set of size at least $m+1$ for f.

It remains to prove Lemma 15.7.

PROOF OF LEMMA 15.7. The elements of T will be functions $t \in {}^{<2m}2$; the p.o. relation $\leq_T$ will be inclusion.

With each $t \in T$ we will associate $h(t) \in 4^m$ and $C_t^+, C_t \subseteq 4^m$ as follows: $h(\emptyset) = 0$, $C_\emptyset^+ = 4^m$, $C_\emptyset = 4^m\setminus\{0\}$. Given $t \in T \cap {}^{<2m-1}2$, $C_t \subseteq 4^m$, and $i \in \{0, 1\}$, we let $C_{t^\frown i}^+ = \{a \in 4^m : f(\{h(t), a\}) = i\} \cap C_t$. We put $t^\frown i$ into T if and only if $C_{t^\frown i}^+ \neq \emptyset$. In this case we let $h(t^\frown i) = \min C_{t^\frown i}^+$, and we define $C_{t^\frown i} = C_{t^\frown i}^+ \setminus \{h(t^\frown i)\}$.

Note that this construction ensures that h is an injection. Moreover, note that if $r <_T s <_T t$, then $h(s), h(t) \in C_{r^\frown s(|r|)}^+$, and hence $f(\{h(r), h(s)\}) = f(\{h(r), h(t)\}) = s(|r|)$, as required.

It remains to show that T has height $2m$, i.e., that $t \in T$ for at least one $t \in {}^{2m-1}2$. Since $t \in T$ whenever $C_t^+ \neq \emptyset$, the following exercise gives us what we need.

EXERCISE 15.8 (G). Let $j < 2m$. Show that

$$\bigcup \{C_t^+ : |t| = j\} = 4^m \setminus \{h(s) : s \in T \cap {}^{<j}2\}.$$

Conclude that $C_t^+ \neq \emptyset$ for at least one $t \in {}^j 2$. $\qquad\square\square$

The proof of Theorem 15.6 doesn't use infinite sets. Moreover, the theorem gives an upper bound for the number n in the first nontrivial instance of Theorem 15.4. Let us introduce some notation to express the last observation more concisely. For $m, k, \ell \in \omega\setminus\{0\}$, let $R(m, k, \ell)$ be the smallest natural number n

such that $n \to (m)_\ell^k$ holds.[3] Similarly, define $R^*(m,k,\ell)$ as the smallest natural number n such that $n \overset{*}{\to} (m)_\ell^k$ holds. Theorems 15.4 and 15.5 tell us that the numbers $R(m,k,\ell)$ and $R^*(m,k,\ell)$ are well defined, but they give us no idea about their magnitude. Theorem 15.6 asserts that $R(m+1,2,2) \leq 4^m$. Exercises 15.3 and 15.4 combined imply that $R(3,2,2) = 6$.

EXERCISE 15.9 (PG). Show that $R(m+1,2,\ell) \leq \ell^{\ell m}$ for all $m, \ell \in \omega \setminus \{0,1\}$.

EXERCISE 15.10 (R). Modify the proof of Theorem 15.6 to find upper bounds for $R(m,k,\ell)$ for arbitrary positive natural numbers m, k, ℓ. Be sure not to use infinite sets in your proof.

It seems unreasonable to expect that we will ever be able to find explicit formulas for $R(m,k,\ell)$; the best one can hope for is to pinpoint the rate of growth of this function. At the time of this writing, even the following question is open:

Does there exist a constant c so that $\lim_{m \to \infty}(R(m,2,2))^{1/m} = c$? If so, what is the value of c?

It is known that c, if it exists, must be between $\sqrt{2}$ and 4. The upper bound follows from Theorem 15.6. The lower bound can be obtained by using random colorings. We include the proof of the next theorem to illustrate this interesting technique.

THEOREM 15.8. $R(m,2,2) \geq 2^{(m-1)/2}$ for all $m \geq 2$.

PROOF. Let m be as above, and let $n < 2^{(m-1)/2}$. We show that there is a partition $\mathcal{P}$ of $[n]^2$ into two subsets without a homogeneous subset of size m. Rather than constructing such a partition, we merely show its existence.

Let Ω be the set of all partitions $\mathcal{P}$ of $[n]^2$ into two subsets. We consider Ω as a probability space with the uniform probability distribution. One gets this probability distribution by performing the following experiment: For each $a \in [n]^2$, toss a fair coin. If it comes up heads, put a into P_0; if it comes up tails, put a into P_1. Let $\langle A_i : i < \frac{n!}{(n-m)!m!} \rangle$ be an enumeration of all m-element subsets of n. For each i, define a random variable ξ_i on Ω by: $\xi_i(\mathcal{P}) = 1$ if A_i is homogeneous for $\mathcal{P}$, and $\xi_i(\mathcal{P}) = 0$ otherwise. Note that $E(\xi_i)$, the expected value of ξ_i, is equal to the probability that A_i is homogeneous for $\mathcal{P}$. Thus

$$E(\xi_i) = \frac{2}{2^{(m-1)m/2}}.$$

Let $\xi = \sum_i \xi_i$. Recall that the expected value $E(\xi)$ is the sum of the expected values $E(\xi_i)$. The number of m-element subsets of n can be crudely estimated as follows:[4]

$$\frac{n!}{(n-m)!m!} < \frac{n^m}{m!} \leq \frac{n^m}{2} < \frac{2^{(m-1)m/2}}{2}.$$

The last of these inequalities follows from the assumption that $n < 2^{(m-1)/2}$. Multiplying this estimate by $E(\xi_i)$ we get:

$$E(\xi) < 1.$$

Since ξ takes only natural numbers as values, it follows that ξ must take the value 0 for some partition $\mathcal{P}$. Note that $\xi(\mathcal{P})$ counts the number of homogeneous subsets

---
[3]The numbers $R(m,k,\ell)$ are often called *Ramsey numbers*.
[4]More skillful estimates give the result $R(m,2,2) \geq \frac{m}{e+o(1)} 2^{(m-1)/2}$.

of size m of $\mathcal{P}$. We conclude that for some $\mathcal{P} \in \Omega$, no homogeneous subset of size m exists. □

So far, we have given no estimate of the numbers $R^*(m, k, \ell)$. For good reason:

THEOREM 15.9 (Ketonen-Solovay). *$R^*(m+1, m, m)$ defines a function from $\omega \setminus \{0\}$ into ω that grows faster than any function from $\omega \setminus \{0\}$ into ω that is provably total in Peano Arithmetic.*

Let us dissect the meaning of the above theorem. Peano Arithmetic (abbreviated PA) was mentioned in Chapter 6; it is an axiomatization of number theory. Hereditarily finite sets can be coded as natural numbers, and it has been shown that the properties of hereditarily finite sets provable in PA are exactly the properties provable in the theory one obtains from ZFC by replacing the Axiom of Infinity by its negation. One can thus think of the theorems provable in PA as precisely the theorems about hereditarily finite sets that are provable without employing the notion of an infinite set. Note that Theorem 15.5 is a theorem about hereditarily finite sets: All colorings and homogeneous sets mentioned in it are hereditarily finite. It is not hard to write down a crude algorithm for computing $R^*(m+1, m, m)$ without reference to infinite sets.

EXERCISE 15.11 (G). Describe an algorithm for computing $R^*(m+1, m, m)$. Assume that you don't have constraints upon computing time and memory.

By Theorem 15.5, the function $R^*(m+1, m, m)$ is total, i.e., the algorithm you described in Exercise 15.11 eventually yields an output for every input m. By the Ketonen-Solovay Theorem, this fact cannot be proved in Peano Arithmetic. In other words, *every* proof of Theorem 15.5 must use infinite sets in one way or another.[5]

After this excursion into the realm of finite combinatorics, let us return to set theory proper and study some sufficient conditions for the existence of *uncountable* homogeneous sets. The first questions that come to a set theorist's mind after seeing Ramsey's Theorem are: *Does $\omega_1 \to (\omega_1)_2^2$? Or, more modestly, is there any cardinal κ such that $\kappa \to (\omega_1)_2^2$?*

Let α and β be ordinals. Recall that the *lexicographical order* $\leq_l$ on $^\beta\alpha$ is defined as follows: Let $f, g \in {}^\beta\alpha$. Then $f <_l g$ if and only if there exists $\gamma < \beta$ such that $f(\gamma) < g(\gamma)$ and $f(\delta) = g(\delta)$ for all $\delta < \gamma$. By Exercise 2.35, $<_l$ is a strict l.o. relation.

LEMMA 15.10. *In $\langle {}^\kappa 2, \leq_l \rangle$ there are no strictly increasing or strictly decreasing transfinite sequences of length κ^+.*

PROOF. We show that there is no increasing sequence of length κ^+ in $\langle {}^\kappa 2, \leq_l \rangle$. The proof for decreasing sequences is similar.

Assume towards a contradiction that $(f_\alpha)_{\alpha < \kappa^+}$ is a squence in $^\kappa 2$ with $f_\alpha <_l f_\beta$ for $\alpha < \beta < \kappa^+$. For $\alpha < \beta < \kappa^+$ define:

$$\Delta(\alpha, \beta) = min\{\delta : f_\alpha(\delta) \neq f_\beta(\delta)\};$$

and

$$d(\alpha) = min\{\Delta(\alpha, \beta) : \alpha < \beta < \kappa^+\}.$$

---

[5] A similar phenomenon was discussed in the Mathographical Remark at the end of Chapter 10.

Note that d is a function from κ^+ into κ.

Now the following claim yields a contradiction.

CLAIM 15.11. *The function d is nondecreasing and not eventually constant.*

EXERCISE 15.12 (G). Prove Claim 15.11. □

THEOREM 15.12. $2^\kappa \not\to (\kappa^+)^2_2$ *for each infinite cardinal κ.*

PROOF. Let $(f_\alpha)_{\alpha < 2^\kappa}$ be an enumeration of $^\kappa 2$, and let $\leq_l$ be the lexicographical order on $^\kappa 2$. We define a partition $\mathcal{P} = \{P_0, P_1\}$ as follows:

$$P_0 = \{\{\alpha, \beta\} : f_\alpha <_l f_\beta \text{ and } \alpha < \beta\};$$
$$P_1 = [2^\kappa]^2 \setminus P_0.$$

Thus P_0 is the set of pairs where the lexicographical order agrees with the enumeration, and P_1 is the set of pairs where the enumeration reverses the lexicographical order. If Y is homogeneous for $\mathcal{P}$, then Y yields a monotonic sequence of length $|Y|$ in $^\kappa 2$. By Lemma 15.10, $|Y| < \kappa^+$. □

By Exercise 15.2, Theorem 15.12 implies in particular that $\omega_1 \not\to (\omega_1)^2_2$. This gives a negative answer to our first question. The answer to the second question is positive. In order to state the relevant theorem in its full generality, we need to introduce some notation.

Let κ be an arbitrary cardinal. By recursion over $n \in \omega$ we define:

(15.2)
$$\exp_0(\kappa) = \kappa;$$
$$\exp_{n+1}(\kappa) = 2^{\exp_n(\kappa)}.$$

THEOREM 15.13 (Erdős-Rado). $(\exp_n(\kappa))^+ \to (\kappa^+)^{n+1}_\kappa$ *for every infinite cardinal κ and each $n \in \omega$.*

Note that for $n = 0$ the Erdős-Rado Theorem simply says that $\kappa^+ \to (\kappa^+)^1_\kappa$, which is a special case of the Pigeonhole Principle. For $n = 1$, the theorem says that $(2^\kappa)^+ \to (\kappa^+)^2_\kappa$, which in view of Theorem 15.12 is the best possible result for superscript 2, even if the subscript κ is replaced by 2. For $n = 2$, the theorem says that $(2^{2^\kappa})^+ \to (\kappa^+)^3_\kappa$, and for larger n's we get higher stacks of exponents on the left hand side.

PROOF OF THEOREM 15.13. By induction over n.

As we already mentioned, for $n = 0$ the theorem is a special case of the Pigeonhole Principle.

So assume $n > 0$, let κ be an infinite cardinal, and assume that $(\exp_{n-1}(\kappa))^+ \to (\kappa^+)^n_\kappa$. To streamline notation, let $\lambda = \exp_{n-1}(\kappa)$, $\mu = (\exp_n(\kappa))^+$. Fix a coloring $f : [\mu]^{n+1} \to \kappa$. We have the following analogue of Lemma 15.7:

LEMMA 15.14. *There exist a tree $\langle T, \leq_T \rangle$ of height $\lambda^+ + 1$ and an injection $h : T \to \mu$ such that for all $s, t \in T$ with $s <_T t$ and all $a \in [\hat{s}]^n$ the equality $f(h[a] \cup \{h(s)\}) = f(h[a] \cup \{h(t)\})$ holds.*

Before proving Lemma 15.14, let us show how it implies Theorem 15.13. Given T, h as in the lemma, fix any t at the top level of T, and let $A = \{h(s) : s \in \hat{t}\}$. Define an induced coloring $f_* : [A]^n \to \kappa$ by:

$$f_*(a) = f(a \cup \{h(t)\}).$$

Since $|A| = \lambda^+$, the inductive assumption implies that f_* has a homogeneous set $D \subseteq A$ of size κ^+.

EXERCISE 15.13 (G). Show that if D is homogeneous for f_*, then $D \cup \{h(t)\}$ is homogeneous for f.

EXERCISE 15.14 (PG). Assume the tree in Lemma 15.14 is only of height λ^+, but otherwise the lemma is unchanged. Could we still derive Theorem 15.13 by defining f_* as in the proof of Theorem 15.6 (i.e., without taking advantage of the existence of a top level of T)?

PROOF OF LEMMA 15.14. This argument is a modification of the proof of Lemma 15.7. The elements of T will be functions $t : [\alpha]^n \to \kappa$, where $\alpha \leq \lambda^+$. The p.o. relation $\leq_T$ will be inclusion. With every $t \in T$ we will associate $h(t) \in \mu$ and $C_t^+, C_t \subseteq \mu$ such that $C_t^+ \neq \emptyset$, $h(t) = \min C_t^+$, and $C_t = C_t^+ \setminus \{h(t)\}$.

Let us now describe the recursive construction of the t's and C_t^+'s. If $\alpha = 0$, then the only function with domain $[\alpha]^n$ is $\emptyset$. We make this the root of the tree[6] and let $C_\emptyset^+ = \mu$.

Now suppose $\alpha = \beta + 1 \leq \lambda^+$, $t : [\alpha]^n \to \kappa$, and $s = t\upharpoonright[\beta]^n \in T$. We let $C_t^+ = \{\xi \in C_s : f(\{h(t\upharpoonright[\gamma]^n) : \gamma \in a\} \cup \{\xi\}) = t(a) \text{ for all } a \in [\alpha]^n\}$, and put t into T if and only if $C_t^+ \neq \emptyset$.

At limit stages α, if $t : [\alpha]^n \to \kappa$ and $t\upharpoonright[\beta]^n$ is defined for all $\beta < \alpha$, we let $C_t^+ = \bigcap \{C_{t\upharpoonright[\beta]^n} : \beta < \alpha\}$ and put t into T if and only if $C_t^+ \neq \emptyset$.

EXERCISE 15.15. (a) Show that the function $h : T \to \kappa$ is injective.
(b) Assume $s, t \in T$ are such that $s <_T t$ and $a \in [\hat{s}]^n$. Show that
$$f(a \cup \{h(s)\}) = f(a \cup \{h(t)\}).$$
(c) Show that if $\alpha \leq \lambda^+$, then
$$\bigcup \{C_t^+ : t \in {}^{[\alpha]^n}\kappa\} = \mu \setminus \bigcup_{\beta < \alpha} \{h(t) : t \in T \cap {}^{[\beta]^n}\kappa\}.$$

Since $\lambda \geq \kappa$ and $\mu = (2^\lambda)^+ = (\kappa^\lambda)^+ = (\lambda^+ \cdot \kappa^\lambda)^+$ and since for all $\alpha \leq \lambda^+$ the set $\bigcup_{\beta < \alpha} {}^{[\beta]^n}\kappa$ has cardinality at most $\lambda^+ \cdot \kappa^\lambda$, it follows that for each $\alpha \leq \lambda^+$ there exists some $t \in {}^{[\alpha]^n}\kappa$ with $C_t^+ \neq \emptyset$. Thus, our construction implies that the tree $\langle T, \leq_T \rangle$ has height $\lambda^+ + 1$. □□

EXERCISE 15.16 (PG). If you passed on Exercise 15.10 when you first saw it, return to it now and give it another try.

Can Theorem 15.13 be strengthened to $\exp_n(\kappa) \to (\kappa^+)_\kappa^{n+1}$ or to $(\exp_n(\kappa))^+ \to (\kappa^+)_{\kappa^+}^{n+1}$? Theorem 15.12 rules out the first possibility for $n = 1$. Concerning the second possiblity, the following results show limitations for $n = 0$ and $n = 1$.

EXERCISE 15.17 (G). Show that $\kappa^+ \not\to (2)_{\kappa^+}^1$.

---

[6] The attentive reader will have noticed that for every $\alpha < n$ the only function with domain $[\alpha]^n$ is the empty set. Our successor step stipulates that at the first level of T we put the function with domain $[1]^n$, and if $n > 1$, then we would put the same node at several levels of the tree. The standard way to overcome this difficulty is to add some decorations to the first n nodes of the tree, but this would only obscure the notation. So we will pretend that the empty functions with domains $[0]^n, \ldots, [n-1]^n$ are actually different sets and ask the reader's kind forgiveness for a white lie.

THEOREM 15.15. $2^\kappa \not\to (3)^2_\kappa$ for each infinite cardinal κ.

PROOF. Fix κ. For $f, g \in {}^\kappa 2$ let $\Delta(f, g) = \min\{\delta : f(\delta) \neq g(\delta)\}$. Define:
$$P_\alpha = \{\{f, g\} : f, g \in {}^\kappa 2 \text{ and } \Delta(f, g) = \alpha\}.$$

The family $\{P_\alpha : \alpha < \kappa\}$ is a partition of a set of cardinality 2^κ into κ pieces for which there is no homogeneous set of cardinality 3. $\square$

What about $n > 1$? The following result, which is called the *Negative Stepping-Up Lemma*, gives an answer.

LEMMA 15.16. *Suppose $n \geq 2$, and κ, λ, σ are cardinals such that κ, λ are infinite, $\sigma \geq 2, \lambda$ is regular, and $\kappa \not\to (\lambda)^n_\sigma$. Then $2^\kappa \not\to (\lambda)^{n+1}_\sigma$.*

PROOF. We prove how to step up from superscript 2 to superscript 3 in the case when $\sigma \geq \omega$. For the complete proof of Lemma 15.16, as well as some variations on it, we refer the reader to Lemma 24.1 of [EHMR] (see Mathographical Remarks).

So let $F : [\kappa]^2 \to \sigma$ be a coloring without a homogeneous subset of size λ. For $\{f, g\} \in [{}^\kappa 2]^2$ let
$$\Delta(f, g) = \min\{\alpha : f(\alpha) \neq g(\alpha)\}.$$

EXERCISE 15.18 (G). Show that if $\{f, g, h\} \in [{}^\kappa 2]^3$ and $f <_l g <_l h$ (where $<_l$ stands for the lexicographical ordering), then $\Delta(f, g) \neq \Delta(g, h)$ and, moreover, $\{\Delta(f, g), \Delta(f, h), \Delta(g, h)\} \in [\kappa]^2$.

If $\{f, g, h\} \in [{}^\kappa 2]^3$ and $f <_l g <_l h$, let
$$br(f, g, h) = \begin{cases} 0, & \text{if } \Delta(f, g) < \Delta(g, h); \\ 1, & \text{if } \Delta(f, g) > \Delta(g, h). \end{cases}$$

The function br characterizes the branching pattern of the triple $\{f, g, h\}$. Now we can define a function $G : [{}^\kappa 2]^3 \to \sigma \times \{0, 1\}$ as follows:
$$G(\{f, g, h\}) = \langle F(\{\Delta(f, g), \Delta(f, h), \Delta(g, h)\}), br(f, g, h)\rangle.$$

Since we are proving the lemma for the case where σ is an infinite cardinal, we have $\sigma \cdot 2 = \sigma$, and it suffices to show that G does not have a homogeneous set of size λ. Let
$$A = \{\Delta(f, g) : \{f, g\} \in [B]^2\}.$$

We will derive a contradiction by showing that $|A| = \lambda$ and that A is homogeneous for F.

Case 1: $br(f, g, h) = 0$ for all $\{f, g, h\} \in [B]^3$.

In this case, for each $f \in B$ there exists an ordinal $\alpha_f \in \kappa$ such that $\Delta(f, g) = \alpha_f$ for all $g \in B$ such that $f <_l g$. For all but the largest element of B (in the lexigrahical order), the ordinal α_f is uniquely determined. So let us assume without generality that B does not have a largest element in the lexicographical order. Moreover, if $\{f, g\} \in [B]^2$, then $\alpha_f \neq \alpha_g$, and it follows that $|A| = |B| = \lambda$. Now consider $\{\alpha, \beta\} \in [A]^2$. Pick $f_0, f_1, f_2, f_3 \in B$ such that $\Delta(f_0, f_1) = \alpha$ and $\Delta(f_2, f_3) = \beta$. We may without loss of generality assume that f_0 is the lexicographically smallest of these elements. Then $\alpha = \Delta(f_0, f_1) = \alpha_{f_0} = \Delta(f_0, f_2) = \Delta(f_0, f_3)$. It follows that $G(f_0, f_2, f_3) = \langle F(\{\alpha, \beta\}), 0\rangle$, and since B was assumed homogeneous for G, the first coordinate of this function value does not depend on the choice of $\{\alpha,\beta\}$. In other words, A is a homogeneous set for F and we have reached a contradiction.

Case 2: $br(f,g,h) = 1$ for all $\{f,g,h\} \in [B]^3$.

We leave this case to the reader.

EXERCISE 15.19. Finish the above proof for Case 2. □

COROLLARY 15.17. *Let $n \in \omega$ and let κ be an infinite cardinal. Then $\exp_n(\kappa) \not\to (\kappa^+)_2^{n+1}$. Moreover, if GCH holds, then $(\exp_n(\kappa))^+ \not\to (\kappa^+)_{\kappa^+}^{n+1}$.*

EXERCISE 15.20 (G). Prove Corollary 15.17. *Hint:* Apply Lemma 15.16 to Theorem 15.12 and Exercise 15.17.

We have seen that $\omega_1 \not\to (\omega_1)_2^2$. Naturally, the question arises whether there is *any* uncountable cardinal κ such that $\kappa \to (\kappa)_2^2$.

DEFINITION 15.18. A cardinal κ is *weakly compact* if κ is uncountable and $\kappa \to (\kappa)_2^2$.

EXERCISE 15.21 (G). Show that if κ is weakly compact, then $2^\lambda < \kappa$ for all $\lambda < \kappa$, i.e., κ is a strong limit cardinal.

LEMMA 15.19. *Weakly compact cardinals are regular.*

PROOF. Suppose $cf(\kappa) = \lambda < \kappa$. We show that κ is not weakly compact by constructing a function $f : [\kappa]^2 \to \kappa$ without homogeneous subset of size κ. Let $(\kappa_i)_{i<\lambda}$ be an increasing sequence of cardinals with $\sup\{\kappa_i : i < \lambda\} = \kappa$. Define f by:
$$f(\{\alpha,\beta\}) = 0 \text{ iff } \forall i < \lambda\, (\alpha < \kappa_i \leftrightarrow \beta < \kappa_i).$$
Now suppose $X \subseteq \kappa$ is homogeneous for f.

If $f(\{\alpha,\beta\}) = 1$ for all $\{\alpha,\beta\} \in [X]^2$, then $|X| \leq \lambda < \kappa$.

If $f(\{\alpha,\beta\}) = 0$ for all $\{\alpha,\beta\} \in [X]^2$, then $X \subseteq \kappa_i$ for some $i < \kappa$, and again we have $|X| < \kappa$. □

COROLLARY 15.20. *Every weakly compact cardinal is strongly inaccessible.*

Thus weakly compact cardinals are another example of large cardinals: Their definition generalizes a property of ω, their existence implies the consistency of ZFC,[7] and therefore their existence is not provable in ZFC.

EXERCISE 15.22 (PG). One can show that the smallest strongly inaccessible cardinal is not weakly compact. Use this information to show that the existence of a weakly compact cardinal implies consistency of the theory ZFC + "there exists a strongly inaccessible cardinal." *Hint:* Use the same technique as in the proof of Corollary 12.23.

Their dubious existence notwithstanding, weakly compact cardinals have some neat properties that make them worth studying.

THEOREM 15.21. *If κ is weakly compact, then κ has the tree property.*

PROOF. Assume κ is weakly compact. Let $\langle T, \leq_T \rangle$ be a κ-tree, i.e., a tree of height κ such that every level contains fewer than κ elements. Fix a wellorder $\leq$ of the nodes of T of order type κ, and let $\leq_\ell$ be the lexicographical ordering on T

---

[7] See Chapter 12 for a discussion of the same phenomenon for inaccessible cardinals.

induced by $\leq$, as defined in Chapter 14. Now define a partition $\mathcal{P} = \{P_0, P_1\}$ of $[T]^2$ by putting the pair $\{s,t\}$ into P_0 if and only if the relations $\leq_\ell$ and $\leq$ agree on $\{s,t\}$, i.e., if $s < t$ implies $s <_\ell t$.

Now suppose $X \subseteq T$ is a set of cardinality κ homogeneous for $\mathcal{P}$. Let
$$Y = \{s \in T : |\{t \in X : s <_T t\}| = \kappa\}.$$
Theorem 15.21 is an immediate consequence of the following exercise.

EXERCISE 15.23 (PG). (a) Show that $|Y \cap T(\alpha)| = 1$ for all $\alpha < \kappa$.
(b) Conclude that the set Y is linearly ordered by $\leq_T$. □

If κ is inaccessible, then the converse of Theorem 15.21 also holds.

THEOREM 15.22. *If κ is strongly inaccessible and has the tree property, then $\kappa \to (\kappa)^n_\lambda$ for all $n \in \omega \setminus \{0\}$ and $\lambda < \kappa$.*

PROOF. By induction over n. For $n = 1$, the theorem is a special case of the Pigeonhole Principle.

So assume $n > 1$, κ is as in the assumptions, $\lambda < \kappa$, and let $f : [\kappa]^n \to \lambda$. We have the following analogue of Lemma 15.14

LEMMA 15.23. *There exist a κ-tree $\langle T, \leq_T \rangle$ and an injection $h : T \to \kappa$ such that for all $s, t \in T$ with $s <_T t$ and all $a \in [\hat{s}]^{n-1}$ the equality $f(h[a] \cup \{h(s)\}) = f(h[a] \cup \{h(t)\})$ holds.*

EXERCISE 15.24 (PG). Prove Lemma 15.23.

Now suppose $\langle T, \leq_T \rangle$ is as in Lemma 15.23. Since κ has the tree property, there exists a cofinal branch B of T.[8] Let $A = h[B]$. Since h is an injection and B is cofinal, we have $|A| = \kappa$. Define a wellorder relation $\preceq$ on A as in the proof of Theorem 15.6: $a \preceq b$ if and only if there are $s, t \in B$ such that $s <_T t$, $a = h(s)$, and $b = h(t)$. For $c \in [A]^{n-1}$ define $f_*(c) = f(c \cup \{a(c)\})$, where $a(c)$ is the $\preceq$-minimal element of A above all elements of c. Then $f_* : [A]^{n-1} \to \lambda$, and by the inductive assumption, there is a $D \in [A]^\kappa$ that is homogeneous for the induced coloring f_*. Now the same reasoning as in the proofs of Theorems 15.6 and 15.13 shows that D is also homogeneous for f. □

Let us summarize what we have learned about weakly compact cardinals.

COROLLARY 15.24. *Let κ be an uncountable cardinal. The following are equivalent:*
(a) $\kappa \to (\kappa)^2_2$;
(b) $\kappa \to (\kappa)^n_\lambda$ *for every* $n \in \omega \setminus \{0\}$ *and* $\lambda < \kappa$;
(c) κ *is strongly inaccessible and has the tree property.*

Of course, any one of the properties (a)–(c) can serve as the defining property of a weakly compact cardinal.

Having seen partition theorems $\kappa \to (\lambda)^\rho_\sigma$ for arbitrarily large λ, one naturally wonders whether the superscript ρ could be made infinite. Unfortunately, this is asking for too much.

THEOREM 15.25. *For each infinite cardinal κ, $\kappa \not\to (\omega)^\omega_2$.*

---

[8]Compare this with Exercise 15.14. Of course the right answer was *No*.

PROOF. Let $\preceq$ be an arbitrary wellorder relation on $[\kappa]^{\aleph_0}$. We define a function $f : [\kappa]^{\aleph_0} \to 2$ by:
$$f(X) = 0 \text{ if and only if } \forall Y \in [X]^{\aleph_0} \, (X \preceq Y).$$
We show that f has no infinite homogeneous set. Suppose towards a contradiction that X is an infinite homogeneous set for f. Let X^* be the $\preceq$-minimal element of $[X]^{\aleph_0}$. Then $f(X^*) = 0$, and homogeneity of X implies that $f(Y) = 0$ for all $Y \in [X]^{\aleph_0}$. Consider a sequence $(Y_i)_{i<\omega}$ of denumerable subsets of X such that Y_i is a proper subset of Y_{i+1} for each $i \in \omega$. Since $f(Y_i) = 0$, the choice of f implies that $Y_{i+1} \prec Y_i$ for all $i \in \omega$. Hence $(Y_i)_{i\in\omega}$ is a strictly decreasing sequence in $\langle [\kappa]^{\aleph_0}, \preceq \rangle$, contradicting the assumption that the latter structure is a w.o. □

Since infinite superscripts don't give positive partition relations, we might try our luck with superscripts of the form $< \omega$.

DEFINITION 15.26. Let κ, λ, σ be cardinals. The symbol $\kappa \to (\lambda)_\sigma^{<\omega}$ stands for the statement: *Whenever $\{f_n : n \in \omega \backslash \{0\}\}$ is a family of functions such that $f_n : [\kappa]^n \to \sigma$, there exists a set $A \subseteq \kappa$ of cardinality λ such that A is homogeneous for all f_n's simultaneously.*

Note the word "simultaneously." The relation $\kappa \to (\lambda)_\sigma^{<\omega}$ is much stronger than the statement $\forall n \in \omega \backslash \{0\} \, (\kappa \to (\lambda)_\sigma^n)$.

EXERCISE 15.25 (G). Show that $\omega \not\to (\omega)_2^{<\omega}$.

Let λ be an infinite cardinal. The least κ such that $\kappa \to (\lambda)_2^{<\omega}$ is called the *Erdős cardinal* $\kappa(\lambda)$. If $\kappa \to (\kappa)_2^{<\omega}$, then κ is called a *Ramsey cardinal*. One can show that $\kappa(\omega)$ (if it exists) is larger than the first weakly compact cardinal, and that $\lambda < \mu$ implies $\kappa(\lambda) < \kappa(\mu)$. Thus the first Ramsey cardinal is larger than $\kappa(\omega), \kappa(\omega_1), \ldots, \kappa(\omega_\omega), \ldots$. Corollary 23.18 shows that measurable cardinals are Ramsey.

If $\mathcal{P} = \{P_i : i < \sigma\}$ is a partition of a set $[X]^n$, then we say that $A \subseteq X$ is *homogeneous of color i*, or *i-homogeneous* if $[A]^n \subseteq P_i$, i.e., if the induced coloring takes the value i for all $a \in [A]^n$. Let us now consider the question: How bad can counterexamples to partition relations get? For example, we have seen that $\omega_1 \not\to (\omega_1)_2^2$. It follows from Theorem 15.3 that every partition $\{P_0, P_1\}$ of $[\omega_1]^2$ has an infinite homogeneous set. But could it happen that there is neither an uncountable homogeneous set of color 1 nor an infinite homogeneous set of color 0? In order to express this type of questions succinctly, let us introduce a new symbol.

DEFINITION 15.27. Let $n \in \omega \backslash \{0\}$, $\sigma \in \mathbf{ON}$, κ, λ_i be cardinals for $i < \sigma$. The symbol $\kappa \to (\lambda_i)_{i<\sigma}^n$ abbreviates the statement: *For every partition $\mathcal{P} = \{P_i : i < \sigma\}$ of $[\kappa]^n$ there exist an $i < \sigma$ and a homogeneous set A of color i and cardinality λ_i.*

In this new terminology, the above question about ω_1 can be expressed as follows: *Does $\omega_1 \not\to (\omega_1, \omega)^2$?* The next theorem gives an answer.

THEOREM 15.28 (Erdős-Dushnik-Miller). *Let κ be an infinite cardinal. Then $\kappa \to (\kappa, \omega)^2$.*

PROOF. Let κ be an infinite cardinal, and let $\mathcal{P} = \{P_0, P_1\}$ be a partition of κ. For $\alpha \in \kappa$ we define
$$B(\alpha) = \{\beta < \kappa : \{\alpha, \beta\} \in P_1\}.$$

CLAIM 15.29. *If for every $A \in [\kappa]^\kappa$ there exists $\alpha \in A$ such that $|B(\alpha) \cap A| = \kappa$, then there exists an infinite homogeneous set for $\mathcal{P}$ of color 1.*

PROOF. Let $F : [\kappa]^\kappa \to \kappa$ be a function such that $|B(F(A)) \cap A| = \kappa$ for each $A \in dom(F)$. We define recursively:

$$C_0 = \kappa;$$

$$C_{n+1} = \{\alpha \in C_n : F(C_n) < \alpha \wedge \{F(C_n), \alpha\} \in P_1\}.$$

The set $\{F(C_n) : n \in \omega\}$ is infinite and homogeneous of color 1. □

Now let us return to the proof of Theorem 15.28. We assume that there is no infinite homogeneous set of color 1, and show that there must be a homogeneous set of color 0 and size κ. Claim 15.29 allows us to find $A \in [\kappa]^\kappa$ such that $|B(\alpha) \cap A| < \kappa$ for all $\alpha \in A$. We fix such A for the rest of this argument.

Case 1: κ is regular.

EXERCISE 15.26 (G). Convince yourself that in this case there exists a strictly increasing sequence $(\gamma_\alpha)_{\alpha < \kappa}$ of elements of A such that $\gamma_\alpha > \sup \bigcup_{\beta < \alpha} B(\gamma_\beta)$ for all $\alpha < \kappa$, and derive the theorem from this fact.

Case 2: κ is singular.

This is the more interesting situation. Let $\lambda = cf(\kappa)$, and let $(\kappa_\xi)_{\xi < \lambda}$ be an increasing sequence of cardinals cofinal in κ. We would like to somehow reduce this case to the previous one. So, let us choose the above sequence in such a way that each of the κ_ξ's is regular and, just in case it'll help, $\kappa_0 > \lambda$. Let $\{X_\xi : \xi < \lambda\}$ be a partition of A such that $|X_\xi| = \kappa_\xi$ for all $\xi < \lambda$. By the result of Case 1, for each ξ we may pick $C_\xi \in [X_\xi]^{\kappa_\xi}$ that is homogeneous of color 0 for $\mathcal{P}$.

Let $C = \bigcup_{\xi < \lambda} C_\xi$. This is a set of size κ, and $C \cap X_\xi$ is 0-homogeneous for each ξ. Moreover, $|B(\alpha) \cap C| < \kappa$ for every $\alpha \in C$. However, the sets $B(\alpha)$ may look rather erratic; in particular, it could happen that $B(\alpha) \cap C_\xi \neq \emptyset$ for all $\xi \neq \alpha$. The only thing we know is that for sufficiently large ξ, the set $B(\alpha) \cap C_\xi$ will be of size smaller than that of C_ξ. We need to get as much mileage as we can out of this rudimentary well-behavedness. Unfortunately, the meaning of "sufficiently large" may be different for different α's, even for different α's from the same C_ξ.

Let us try to remedy the latter problem. For $\xi, \eta \in \lambda$ we define:

$$C_{\xi, \eta} = \{\alpha \in C_\xi : |B(\alpha) \cap C| < \kappa_\eta\}.$$

Note that $C_\xi = \bigcup_{\eta < \lambda} C_{\xi, \eta}$. Since κ_ξ is regular and larger than λ,[9] the Pigeonhole Principle implies that there is some $\delta(\xi)$ with $|C_{\xi, \delta(\xi)}| = \kappa_\xi$. Moreover, there is a set $L \in [\lambda]^\lambda$ such that

$$\forall \xi, \eta \in L \, (\xi < \eta \to \delta(\xi) \leq \eta).$$

Note that the set $\bigcup_{\xi \in L} C_{\xi, \delta(\xi)}$ is still of cardinality κ. Now define:

$$E = \bigcup_{\eta \in L} (C_{\eta, \delta(\eta)} \setminus \bigcup \{B(\alpha) : \exists \xi < \eta \, (\xi \in L \wedge \alpha \in C_{\xi, \delta(\xi)})\}).$$

A straightforward verification shows that E has cardinality κ and is 0-homogeneous for $\mathcal{P}$. □

---

[9] Didn't we have the right inkling when we chose κ_0 to be larger than λ?

EXERCISE 15.27 (PG). Let κ be an infinite cardinal, $f : \kappa \to \kappa$.
(a) Show that there exists $X \in [\kappa]^\kappa$ such that $f(\alpha) \leq f(\beta)$ whenever $\alpha, \beta \in X$ are such that $\alpha < \beta$.
(b) Show that if κ is regular, there exists $Y \in [\kappa]^\kappa$ such that $f\lceil Y$ is either constant or strictly increasing.

So far, we have been concerned only with the cardinality of homogeneous sets. But as long as n-tuples of ordinals are being partitioned, it also makes sense to ask whether homogeneous sets of a given order type exist.

DEFINITION 15.30. Let $n \in \omega \backslash \{0\}$, $\alpha, \beta_i, \sigma \in \mathbf{ON}$, where $i < \sigma$. The symbol $\alpha \to (\beta_i)_{i<\sigma}^n$ abbreviates the statement: *For every partition $\mathcal{P} = \{P_i : i < \sigma\}$ of $[\alpha]^n$ there exist $i < \sigma$ and a homogeneous set A of color i and order type β_i.*

The symbols $\alpha \to (\beta)_\sigma^n$ and $\alpha \to (\beta)_\sigma^{<\omega}$ are defined analoguosly. One might think that symbols like $\omega_2 \to (\omega_1)_2^2$ are ambiguous. But since every subset of $\mathbf{ON}$ of cardinality κ has order type at least κ, $\aleph_2 \to (\aleph_1)_2^2$ means exactly the same as $\omega_2 \to (\omega_1)_2^2$.

Some partition results for cardinalities can be generalized to order types. For example, one can define the Erdős cardinal $\kappa(\alpha)$ for each ordinal $\alpha \geq \omega$ and show that $\alpha < \beta$ implies that $\kappa(\alpha) < \kappa(\beta)$.

Other partition relations for cardinals can be somewhat sharpened if one considers the order types of the resulting homogeneous sets. The Erdős-Rado Theorem is a prime example.

EXERCISE 15.28 (PG). By analyzing the proof of Theorem 15.13, show that $(\exp_n(\kappa))^+ \to (\kappa^+ + n)_\kappa^{n+1}$ for every infinite cardinal κ and $n \in \omega \backslash \{0\}$.

As long as κ is regular and uncountable, one can also replace the ω by $\omega + 1$ in the Erdős-Dushnik-Miller Theorem, but this requires a separate proof and will not be done here.

How about the Pigeonhole Principle? If there are only finitely many pigeonholes, nothing exciting happens.

EXERCISE 15.29. (a) Show that if $k \in \omega \backslash \{0\}$, then $\omega \cdot \omega \to (\omega \cdot \omega)_k^1$ and $\omega^{\cdot \omega} \to (\omega^{\cdot \omega})_k^1$.
(b) What is the largest α such that $\omega \cdot 9 \to (\alpha)_4^1$?

However, partitions of uncountable cardinals into countably many pieces can be surprisingly short on homogeneous subsets.

THEOREM 15.31 (Miller-Rado). *Let κ be an infinite cardinal and let $\alpha < \kappa^+$. Then $\alpha \not\to (\kappa^{\cdot \omega})_\omega^1$.*

PROOF. Let κ be an infinite cardinal. We want to show that for every $\alpha < \kappa^+$ there exists a partition $\{A_n(\alpha) : n \in \omega\}$ of α such that

$(*)_\alpha \quad ot(A_n(\alpha)) \leq \kappa^{\cdot n}$ for each $n \in \omega$.

We construct the $A_n(\alpha)$'s by recursion over α.
For $\alpha = 0$, we let $A_n(0) = \emptyset$ for all $n \in \omega$.
Suppose $\alpha = \beta + 1$ and the $A_n(\beta)$'s have been constructed and satisfy $(*)_\beta$. We let $A_0(\alpha) = \{\beta\}$ and $A_{n+1}^\alpha = A_n^\beta$ for all $n \in \omega$. Clearly, $(*)_\alpha$ holds.

Now suppose α is a limit ordinal and $(*)_\beta$ holds for all $\beta < \alpha$. Let $\langle \beta_\xi : \xi < cf(\alpha) \rangle$ be an increasing sequence of ordinals cofinal in α such that $ot([\beta_\xi, \beta_{\xi+1})) = \beta_{\xi+1}$ for all ξ and $\beta_\delta = \bigcup_{\xi < \delta} \beta_\xi$ for all limit ordinals $\delta < cf(\alpha)$. Define:

$$A_0(\alpha) = \emptyset, \qquad A_{n+1}(\alpha) = \bigcup_{\xi < cf(\alpha)} \{\beta_\xi + \gamma : \gamma \in A_n(\beta_{\xi+1})\}.$$

Then $\alpha = \bigcup_{n \in \omega} A_n(\alpha)$, and the $A_n(\alpha)$'s are pairwise disjoint. Moreover,

$$ot(A_{n+1}(\alpha)) \leq \sup\{ot(A_n(\beta_{\xi+1})) : \xi < cf(\alpha)\} \cdot cf(\alpha).$$

By the inductive assumption and since $cf(\alpha) \leq \kappa$, the right hand side is an ordinal not exceeding $\kappa^{\cdot n} \cdot \kappa = \kappa^{\cdot n+1}$. Hence $(*)_\alpha$ holds and we are done. □

Can Ramsey's Theorem also be generalized to "reasonable" countable ordinals? For example, can one show that $\omega^{\cdot 2} \to (\omega^{\cdot 2})^2_2$? Or can one at least prove $\omega^{\cdot 2} \to (\omega^{\cdot 2}, \omega)^2$, in analogy with the Erdős-Dushnik-Miller Theorem?

THEOREM 15.32. $\omega^{\cdot 2} \not\to (\omega + 1, \omega)^2$.

PROOF. Consider $\omega \times \omega$ with the lexicographical ordering. This is a set of order type $\omega^{\cdot 2}$. Define a coloring $f_0 : [\omega \times \omega]^2 \to 2$ by letting $f_0(\{\langle a,b \rangle, \langle c,d \rangle\}) = 1$ if and only if $a < c$ and $b > d$, i.e., if the order on the second coordinate reverses the order on the first coordinate.

EXERCISE 15.30 (G). Show that there are no 0-homogeneous sets for f_0 of order type $\omega + 1$ and every 1-homogeneous set for f_0 is finite. □

EXERCISE 15.31 (PG). (a) Define colorings $f_i : [\omega \times \omega]^2 \to 2$ for $1 \leq i < 6$ as follows: $f_2(\{\langle a,b \rangle, \langle c,d \rangle\}) = 1$ if and only if $a < c < b < d$. Let f_0 be as in the proof of Theorem 15.32, and let f_4 be such that $f_4^{-1}\{1\} = f_0^{-1}\{1\} \cup f_2^{-1}\{1\}$. Now define f_{2j+1} for $j < 3$ by letting $f_{2j+1}^{-1}\{1\} = f_{2j}^{-1}\{1\} \cup \{\{\langle a,c \rangle, \langle b,c \rangle\} : a,b,c \in \omega \wedge a,b < c\}$. Show that for each $i < 6$, there are no 0-homogeneous sets of order type $\omega \cdot 2 + 1$ (in the lexicographical ordering on $\omega \times \omega$) and no infinite 1-homogeneous sets for any of the f_i's.

(b) Let $A \in [\omega]^{\aleph_0}$. Define $A^\Delta = \{\langle n,k \rangle : n,k \in A \wedge n < k\}$. Show that A^Δ is a subset of $\omega \times \omega$ of order type $\omega^{\cdot 2}$ in the lexicographical ordering, and for every $i < 6$, $f_i \restriction [A^\Delta]^2$ has 1-homogeneous sets of arbitrarily large finite size.

The following theorem shows that partitions of $\omega^{\cdot 2}$ cannot get much messier than the ones we just constructed.

THEOREM 15.33 (Specker). *If* $m \in \omega$, *then* $\omega^{\cdot 2} \to (\omega^{\cdot 2}, m)$.

One remarkable thing about Specker's Theorem is that it doesn't generalize to $\omega^{\cdot k}$ for $k > 2$; in fact, one can show that $\omega^{\cdot k} \not\to (\omega^{\cdot k}, 3)$ whenever $3 \leq k < \omega$. For $k = \omega$, the situation changes again; by a result of Chang, $\omega^{\cdot \omega} \to (\omega^{\cdot \omega}, m)$ for all $m \in \omega$.

We will prove Specker's Theorem by showing that, in a sense, the functions f_i constructed in Exercise 15.31 are the *only* counterexamples witnessing that $\omega^{\cdot 2} \not\to (\omega^{\cdot 2}, \omega)$. More precisely, we will prove the following:

LEMMA 15.34. *Suppose* $g : [\omega \times \omega]^2 \to 2$ *is a coloring without 0-homogeneous set of order type* $\omega^{\cdot 2}$ *(in the lexicographical order on* $\omega \times \omega$*) and without infinite 1-homogeneous set. Then there exist* $A \in [\omega]^{\aleph_0}$ *and* $i < 6$ *such that* $g \restriction [A^\Delta]^2 = f_i \restriction [A^\Delta]^2$.

PROOF. Let g be as in the assumptions. The idea of the proof is to whittle down the domain of g so that what remains will resemble more and more closely one of the f_i's.

As a first step, consider the coloring $h_1 : [\omega]^3 \to 2$ defined by:
$$h_1(\{a,b,c\}) = g(\langle a,b\rangle, \langle a,c\rangle) \text{ for } a < b < c < \omega.$$
Let $C \in [\omega]^{\aleph_0}$ be homogeneous for h_1. It follows from the definition of h_1 that $g\lceil [C^\Delta]^2$ is homogeneous on all vertical sections, and since g has no infinite 1-homogeneous sets, all these vertical sections must be 0-homogeneous.

Now consider colorings $h_2 : [C]^4 \to 2$ and $h_3 : [C]^3 \to 2$ defined by
$$h_2(\{a,b,c,d\}) = g(\langle a,b\rangle, \langle c,d\rangle) \text{ for } a < b < c < d < \omega,$$
$$h_3(\{a,b,c\}) = g(\langle a,b\rangle, \langle b,c\rangle) \text{ for } a < b < c < \omega.$$
Applying Ramsey's Theorem twice, we find $B \in [C]^{\aleph_0}$ that is a homogeneous set for both h_2 and h_3.

EXERCISE 15.32 (G). Show that B must be homogeneous of color 0 for both h_2 and h_3.

We need to consider three more colorings h_4, h_5, h_6, where $h_4 : [B]^3 \to 2$ is defined by
$$h_4(\{a,b,c\}) = g(\langle a,c\rangle, \langle b,c\rangle) \text{ for } a < b < c < \omega,$$
and $h_5, h_6 : [B]^4 \to 2$ are defined as follows:
$$h_5(\{a,b,c,d\}) = g(\langle a,d\rangle, \langle b,c\rangle) \text{ for } a < b < c < d < \omega,$$
$$h_6(\{a,b,c,d\}) = g(\langle a,c\rangle, \langle b,d\rangle) \text{ for } a < b < c < d < \omega.$$
Applying Ramsey's theorem three more times, we find a set $A \in [B]^{\aleph_0}$ such that A is homogeneous for $h_4, h_5,$ and h_6 simultaneously.

EXERCISE 15.33 (PG). (a) Show that A is 1-homogeneous for at least one of the functions h_5, h_6.

(b) Show that A is 1-homogeneous for all three functions h_4, h_5, h_6 if and only if $g\lceil [A^\Delta]^2 = f_5 \lceil [A^\Delta]^2$.

(c) Formulate and prove analogues of (b) for the functions f_i where $i < 5$. □

The example we constructed to show that $2^{\aleph_0} \not\to (\omega_1)_2^2$ was a rather unnatural concoction. One might ask whether it is possible to show that every "nice" partition of $[\mathbb{R}]^2$ into two subsets has an uncountable homogeneous set. To make mathematical sense of this question, we have to specify what "nice" means. One possible interpretation would be "topologically uncomplicated." Suppose X is a topological space, and $\mathcal{P} = \{P_0, P_1\}$ is a partition of $[X]^2$. One can identify P_0 with the set $P_0^{sym} = \{\langle x,y\rangle : \{x,y\} \in P_0\}$. We say that $\mathcal{P}$ is an *open partition* if P_0^{sym} is an open subset of $X \times X$ with the product topology.[10] The following statement is known as the *Open Coloring Axiom* (abbreviated OCA):

(OCA) *If X is a separable metrizable space and*
  $\mathcal{P} = \{P_0, P_1\}$ *is an open coloring of $[X]^2$, then*
  • *either there exists an uncountable 0-homogeneous subset of X,*
  • *or X is a union of countably many 1-homogeneous sets.*

---

[10]Note that $\mathcal{P}$ is open if and only if P_1^{sym} is a closed subspace of $X \times X \setminus \{\langle x,x\rangle : x \in X\}$.

Note that, in particular, OCA implies that no open partition of a set of pairs of reals witnesses the negative partition relation $2^{\aleph_0} \not\to (\omega_1)_2^2$.

The Open Coloring Axiom is not a theorem of ZFC. However, Stevo Todorčević has shown that if ZFC is a consistent theory, then so is the theory ZFC + OCA. Thus, every statement provable in ZFC + OCA is relatively consistent with ZFC.

Let us conclude this chapter with an interesting consequence of OCA.

THEOREM 15.35. *Suppose* OCA *holds and f is a function from an uncountable subset of the reals into the reals. Then there exists an uncountable subset $D \subseteq dom(f)$ such that $f\lceil D$ is weakly monotonic. In fact, if there is no uncountable $D \subseteq dom(f)$ such that $f\lceil D$ is strictly increasing, then there exists a sequence $(f_n)_{n \in \omega}$ of nonincreasing functions such that $f = \bigcup_{n \in \omega} f_n$.*

PROOF. It suffices to prove the second statement, since the first one follows from it. Let f be as in the assumptions, and let $X = \{\langle x, f(x)\rangle : x \in dom(f)\}$ be the graph of f. Then X is a subspace of $\mathbb{R}^2$; hence X is separable metrizable. Let

$$P_0 = \{\{\langle x_0, f(x_0)\rangle, \langle x_1, f(x_1)\rangle\} \in [X]^2 : x_0 < x_1 \wedge f(x_0) < f(x_1)\};$$
$$P_1 = [X]^2 \backslash P_0.$$

EXERCISE 15.34 (PG). (a) Convince yourself that $\{P_0, P_1\}$ is an open partition of $[X]^2$.
(b) Derive Theorem 15.35. □

Mathographical Remarks

There are several good sources for learning more about Ramsey theory. Neil H. Williams' text *Combinatorial Set Theory*, North-Holland, 1977, contains a fairly extensive treatment of partition calculus for infinite cardinals and ordinals. The standard reference for partition theorems for infinite cardinals nowadays is [EHMR], i.e., the encyclopaedical treatise *Combinatorial Set Theory: Partition Relations for Cardinals*, by P. Erdős, A. Hajnal, A. Máté, and R. Rado, North-Holland, 1984. *Mathematics of Ramsey Theory*, edited by J. Nešetřil and V. Rödl, Springer Verlag, 1990, is a collection of survey articles covering various topics in Ramsey theory and related fields. If you want to study finite Ramsey theory, you may be interested in the text *Ramsey Theory*, by R. L. Graham, B. L. Rothschild, and J. H. Spencer; John Wiley, 1990 (2nd edition), or in the shorter *Rudiments of Ramsey Theory*, by Ronald L. Graham, published by the AMS in 1981.

CHAPTER 16

The Δ-System Lemma

A family A is called a Δ-*system* if there is a set r such that $a \cap b = r$ whenever $a, b \in A$ and $a \neq b$. If $|A| \geq 2$, the set r is uniquely determined by A and is called its *root* by some authors, and its *kernel* by others.

As a warm-up for this chapter, consider the following.

THEOREM 16.1. *Every uncountable family of finite sets contains an uncountable Δ-system.*

PROOF. Let $n > 0$ be a positive natural number, and consider the following statement:

(Δ_n)   Every uncountable $B \subseteq [\omega_1]^n$ contains an uncountable Δ-system.

CLAIM 16.2. (Δ_n) *holds for every natural number $n > 0$.*

EXERCISE 16.1 (G). Derive Theorem 16.1 from Claim 16.2.

PROOF OF CLAIM 16.2. By induction over $n > 0$. To get started, note that every B as in (Δ_1) is a Δ-system with root $\emptyset$. Now suppose (Δ_n) holds, and let $B = \{b_\xi : \xi < \omega_1\} \subseteq [\omega_1]^{n+1}$. In order to show that B contains an uncountable Δ-system, we distinguish two cases.

Case 1: For each $\alpha \in \omega_1$, the set $\{b \in B : \alpha \in b\}$ is countable.

In this case, for each countable $C \subset \omega_1$ the set $\{\xi < \omega_1 : b_\xi \cap \bigcup_{\eta \in C} b_\eta \neq \emptyset\}$ is countable. We define recursively a function $h : \omega_1 \to \omega_1$ as follows:

$$h(0) = 0; \quad h(\xi) = min\{\eta : b_\eta \cap \bigcup_{\zeta < \xi} b_{h(\zeta)} = \emptyset\}.$$

The family $A = \{b_{h(\xi)} : \xi < \omega_1\}$ is a Δ-system with kernel $\emptyset$.

Case 2: There is an $\alpha \in \omega_1$ such that $|\{b \in B : \alpha \in b\}| = \aleph_1$.

Fix such α. Let $C = \{b \in B : \alpha \in b\}$ and $C' = \{b \setminus \{\alpha\} : b \in C\}$. By ($\Delta_n$), there is an uncountable Δ-system $A' \subseteq C'$. Let r denote the root of A', and let $A = \{a \cup \{\alpha\} : a \in A'\}$. This is a Δ-system with root $r \cup \{\alpha\}$. □□

EXERCISE 16.2 (R). Let B be an uncountable family of finite subsets of ω_1. Show that there exists a countable subset $N \subset \omega_1$ such that every $b \in B$ which is not a subset of N is a member of some uncountable Δ-system $A \subseteq B$ with kernel $r \subseteq N$.[1]

Theorem 16.1 can be generalized as follows:

THEOREM 16.3 (Δ-System Lemma). *Let κ and λ be infinite cardinals such that λ is regular and the inequality $\nu^{<\kappa} < \lambda$ holds for all $\nu < \lambda$. If B is a set*

---

[1] A solution to Exercise 16.2 will be given in Chapter 24.

of cardinality at least λ such that $|b| < \kappa$ for all $b \in B$, then there exists a Δ-system $A \subseteq B$ with $|A| = \lambda$.

PROOF. Let κ, λ, B be as in the assumption.

EXERCISE 16.3 (G). Show that the assumptions imply the inequality $\kappa < \lambda$.

Without loss of generality we may assume that $|B| = \lambda$. Since every element of B has cardinality less than λ, we may also restrict ourselves to the case where $\bigcup B \subseteq \lambda$.

Each $b \in B$ is a set of ordinals of order type less than κ. By regularity of λ, and since $\kappa < \lambda$, there exist an ordinal $\rho < \kappa$ and a subset $C \subseteq B$ of cardinality λ such that every $c \in C$ has order type ρ. Fix such ρ and C.

EXERCISE 16.4 (G). Show that $\bigcup C$ is cofinal in λ.

Enumerate $C = \{c_\xi : \xi < \lambda\}$. For $\zeta < \rho$ and $c \in C$, let $c(\zeta)$ denote the ζ-th element of c.

EXERCISE 16.5 (G). Show that there exists a $\zeta < \rho$ such that $\{c(\zeta) : c \in C\}$ is unbounded in λ.

Let ζ_0 be the smallest ζ as in Exercise 16.5, and define:

$$\alpha_0 = sup\{c(\eta) + 1 : c \in C \wedge \eta < \zeta_0\}.$$

Then $\alpha_0 < \lambda$ and $c(\eta) < \alpha_0$ for all $c \in C$, $\eta < \zeta_0$.

We recursively define a function $h : \lambda \to \lambda$ by $h(0) = \min\{\xi : c_\xi(\zeta_0) > \alpha_0\}$, $h(\alpha) = \min\{\xi : c_\xi(\zeta_0) > \bigcup\{c_{h(\gamma)}(\delta) : \gamma < \alpha \wedge \delta < \rho\}\}$. Note that h is strictly increasing. Let $D = \{c_{h(\alpha)} : \alpha < \lambda\}$.

EXERCISE 16.6 (G). Convince yourself that the intersection of any two different elements of D is a subset of α_0.

Since $|\alpha_0|^{<\kappa} < \lambda$, there exist a subset $E \subseteq D$ of cardinality λ and a set $r \subseteq \alpha_0$ such that $\forall e \in E \, (e \cap \alpha_0 = r)$. By Exercise 16.6, the set E is a Δ-system with kernel r. □

EXERCISE 16.7 (G). Assume CH holds, and let $S = \{s_\alpha : \alpha < \omega_2\}$ be a family of $\aleph_2$ countable sets. Show that there exists a Δ-system $A \subseteq S$ such that $|A| = \aleph_2$.

EXERCISE 16.8 (R). Show that in Theorem 16.3, the assumption that λ is regular cannot be replaced by "$cf(\lambda) > \kappa$." However, if we also slightly weaken the assertion, then a modification of the proof of Theorem 16.3 yields a correct and meaningful theorem. Find this theorem and prove it.

We conclude this section with an application to topology. Recall that in general, the product of finitely many c.c.c. spaces does not have the c.c.c. (compare Theorem 14.15). The following theorem asserts that if the product of a family of spaces does not have the c.c.c., then the c.c.c. fails already for the product of a finite subfamily.

THEOREM 16.4. *Let $\{\langle X_i, \tau_i \rangle : i \in I\}$ be a family of topological spaces such that for every finite $F \subseteq I$ the product $\prod_{i \in F} X_i$ has the c.c.c. Then $\prod_{i \in I} X_i$ also satisfies the c.c.c.*

PROOF. Assume towards a contradiction that $\{U_\alpha : \alpha < \omega_1\}$ is a family of pairwise disjoint, nonempty open subsets of $\prod_{i \in I} X_i$. By shrinking the U_α's if necessary, we may without loss of generality assume that they are basic open sets. This means that $U_\alpha = \prod_{i \in I} V_i^\alpha$, where V_i^α is a nonempty open subset of X_i for each $i \in I$, and $V_i^\alpha \neq X_i$ only if $i \in b_\alpha$, where b_α is a finite subset of I.

EXERCISE 16.9 (PG). Fix $\alpha < \beta < \omega_1$ and let $F = b_\alpha \cap b_\beta$. Moreover, let π be the projection of $\prod_{i \in I} X_i$ onto $\prod_{i \in F} X_i$. Show that $\pi[U_\alpha]$ and $\pi[U_\beta]$ are disjoint open subsets of $\prod_{i \in F} X_i$. Show also that $F \neq \emptyset$.

By Theorem 16.1, there exist a finite $F \subset \omega_1$ and an uncountable $A \subseteq \omega_1$ such that the set $\{b_\alpha : \alpha \in A\}$ is a Δ-system with kernel F. By Exercise 16.9, $F \neq \emptyset$ and the family $\{\pi[U_\alpha] : \alpha \in A\}$ witnesses that $\prod_{i \in F} X_i$ does not satisfy the c.c.c. This contradicts one of our assumptions. $\square$

CHAPTER 17

Applications of the Continuum Hypothesis

In this chapter we illustrate how the Continuum Hypothesis can be used in mathematical proofs. In Section 17.1 we concentrate on the impact of CH on the structure of the ideals of meager and null sets of reals. In Section 17.2 we present some typical applications of CH to a variety of problems.

17.1. Applications to Lebesgue measure and Baire category

Let $A \subseteq \mathbb{R} \times \mathbb{R}$. The *vertical section* of A at x is the set $A_x = \{y \in \mathbb{R} : \langle x, y \rangle \in A\}$, and the *horizontal section* of A at y is the set $A^y = \{x \in \mathbb{R} : \langle x, y \rangle \in A\}$.

THEOREM 17.1 (CH). *There exists $A \subseteq [0,1] \times [0,1]$ such that each horizontal section of A is countable and each vertical section of A contains all but countably many reals.*

PROOF. Assume CH. Let $\preceq \, \subseteq [0,1] \times [0,1]$ be a binary relation on $[0,1]$ such that $\langle [0,1], \preceq \rangle$ is a w.o. of order type ω_1.

EXERCISE 17.1 (G). Show that $A = \preceq$ is as required in Theorem 17.1. □

EXERCISE 17.2 (PG). Show that the converse of Theorem 17.1 is also true.

EXERCISE 17.3 (PG). (a) Show that the set A of Theorem 17.1 is not Lebesgue measurable. *Hint:* Let $\chi_A : [0,1] \times [0,1] \to \mathbb{R}$ be defined by:

$$\chi_A(a,b) = \begin{cases} 1, & \text{if } \langle a,b \rangle \in A; \\ 0, & \text{otherwise.} \end{cases} \tag{17.1}$$

If A were measurable, then χ_A would be Lebesgue-integrable, and by Fubini's Theorem,

$$\int_0^1 \int_0^1 \chi_A \, dx \, dy = \int_0^1 \int_0^1 \chi_A \, dy \, dx$$

(b) Conclude that if $\langle [0,1], \preceq \rangle$ is a w.o. of order type ω_1, then $\preceq$ is not a Borel subset of $[0,1]^2$.

A subset L of $\mathbb{R}$ is called a *Luzin set* if L is uncountable, but for every nowhere dense subset K of $\mathbb{R}$ the intersection $K \cap L$ is countable.

THEOREM 17.2 (CH). *There exists a Luzin set.*

PROOF. Since a set is nowhere dense in a topological space if and only if its closure is nowhere dense, we can equivalently define a Luzin set as an uncountable set of reals that has countable intersection with every closed nowhere dense set of reals. It follows from Claim 11.6 that there are exactly $2^{\aleph_0}$ nowhere dense closed subsets of $\mathbb{R}$. Assume CH holds, and let $\langle K_\xi : \xi < \omega_1 \rangle$ be an enumeration of all closed nowhere dense subsets of $\mathbb{R}$. Construct $L = \{x_\eta : \eta < \omega_1\}$ in such a way that

$x_\eta \in \mathbb{R} \setminus (\bigcup_{\xi < \eta} K_\xi \cup \{x_\xi : \xi < \eta\})$ for all $\eta < \omega_1$. This is possible since by the Baire Category Theorem (Theorem 26.5) the real line is not a union of countably many nowhere dense subsets. Clearly, L is uncountable. Moreover, if K is a nowhere dense subset of $\mathbb{R}$, then $cl(K) = K_\xi$ for some $\xi < \omega_1$, and $L \cap K \subseteq \{x_\eta : \eta \leq \xi\}$. It follows that L is a Luzin set. □

A subset S of $\mathbb{R}$ is called a *Sierpiński set*, if S is uncountable, but for every $N \subseteq \mathbb{R}$ of Lebesgue measure zero the intersection $N \cap S$ is countable.

EXERCISE 17.4 (G). Show that CH implies the existence of a Sierpiński set.

Note that one gets an equivalent definition of a Luzin set if one replaces the phrase "of Lebesgue measure zero" in the definition of a Sierpiński set by the phrase "of first Baire category." It quite often happens that if φ is a statement about subsets of the real line and φ^* is the statement obtained from φ by replacing every occurrence of "Lebesgue measure zero" by "first Baire category" and vice versa, then a proof of φ can be translated mutatis mutandis into a proof of φ^*. Apparently, the notions of Lebesgue measure zero and first Baire category are "dual" to each other. Is there a way to give a precise meaning to this apparent duality? If CH holds, then the next theorem explains the observed phenomenon. Since repeating the phrases "first Baire category" and "Lebesgue measure zero" gets tiring after a while, let us speak of *meager* and *null* sets from now on. The family of meager sets of reals will be denoted by $\mathcal{M}$, and the family of null sets by $\mathcal{N}$. Both $\mathcal{M}$ and $\mathcal{N}$ are countably complete ideals in $\mathcal{P}(\mathbb{R})$.

THEOREM 17.3 (Erdős-Sierpiński). *Assume CH. Then there exists a bijection* $f : \mathbb{R} \to \mathbb{R}$ *such that for all* $A \subseteq \mathbb{R}$:

(i) $f[A] \in \mathcal{M} \leftrightarrow A \in \mathcal{N}$;
(ii) $f[A] \in \mathcal{N} \leftrightarrow A \in \mathcal{M}$.

EXERCISE 17.5 (G). Convince yourself that if f is as in Theorem 17.3, then $f[L]$ is a Sierpiński set whenever L is a Luzin set, and $f[S]$ is a Luzin set for every Sierpiński set S.

PROOF OF THEOREM 17.3. We need two lemmas.

LEMMA 17.4. *There exists a decomposition of* $\mathbb{R}$ *into disjoint sets* $F_0 \in \mathcal{M}$ *and* $G_0 \in \mathcal{N}$.

PROOF. Choose an enumeration of the set of rationals $\mathbb{Q} = \{q_i : i \in \omega\}$. For every $n \in \omega$, let $U_n = \bigcup_{i \in \omega}(q_i - 2^{-(n+i)}, q_i + 2^{-(n+i)})$. Define $G_0 = \bigcap_{n \in \omega} U_n$ and $F_0 = \mathbb{R} \setminus G_0$. Note that the Lebesgue measure of U_n is at most 2^{-n+2}. Thus G_0 is a null set. Since each U_n is a dense open set of reals, F_0 is meager. □

EXERCISE 17.6 (PG). (a) Show that in the proof of Lemma 17.4, the sequence $(q_i)_{i \in \omega}$ can be chosen in such a way that $[0,1] \setminus U_3$ is homeomorphic with the Cantor set D.

(b) Conclude that the property of being a null set is not a topological property, i.e., is not preserved by homeomorphisms.

Let $\mathcal{I}$ be an ideal of subsets of a set X. We say that $\mathcal{B}$ is a *base* of the ideal $\mathcal{I}$ if $\mathcal{B} \subseteq \mathcal{I}$ and for every $A \in \mathcal{I}$ there exists $B \in \mathcal{B}$ such that $A \subseteq B$.

17.1. APPLICATIONS TO LEBESGUE MEASURE AND BAIRE CATEGORY

LEMMA 17.5 (CH). *There exist bases $\{F_\alpha : \alpha < 2^{\aleph_0}\}$ of $\mathcal{M}$ and $\{G_\alpha : \alpha < 2^{\aleph_0}\}$ of $\mathcal{N}$ such that $F_\alpha \subseteq F_\beta$, $G_\alpha \subseteq G_\beta$, $|F_{\alpha+1} \setminus F_\alpha| = |G_{\alpha+1} \setminus G_\alpha| = 2^{\aleph_0}$ for all $\alpha < \beta < 2^{\aleph_0}$, and $F_\alpha = \bigcup_{\beta<\alpha} F_\beta$, $G_\alpha = \bigcup_{\beta<\alpha} G_\beta$ whenever $\alpha > 0$ is a limit ordinal.*

PROOF. We need the following:

FACT 17.6. *If F is a meager subset of $\mathbb{R}$, then there exists a meager set $F^+ \supset F$ such that $|F^+ \setminus F| = 2^{\aleph_0}$. Similarly, if G is a null set, then there exists a null set $G^+ \supset G$ such that $|G^+ \setminus G| = 2^{\aleph_0}$.*

EXERCISE 17.7 (PG). (a) Prove Fact 17.6.

(b) Show that both the ideal of meager sets and the ideal of null sets have bases of size $2^{\aleph_0}$.

Now let $\{A_\alpha : \alpha < 2^{\aleph_0}\}$ be a base for the ideal of meager sets. Construct F_α by recursion over α such that for all α:

(a) F_α is meager;
(b) $F_{\alpha+1} \supseteq F_\alpha \cup A_\alpha$;
(c) $|F_{\alpha+1} \setminus F_\alpha| = 2^{\aleph_0}$; and
(d) $F_\alpha = \bigcup_{\beta<\alpha} F_\beta$ if α is a countable limit ordinal.

Since the union of countably many (and in particular: two) meager sets is meager, conditions (b) and (d) do not force us to violate condition (a). By Fact 17.6, the same can be said of condition (c). Clearly, conditions (a)–(d) imply that $\{F_\alpha : \alpha < 2^{\aleph_0}\}$ is as required. The construction of $\{G_\alpha : \alpha < 2^{\aleph_0}\}$ is analogous. We have proved Lemma 17.5. □

Wait a minute. Lemma 17.5 was supposed to be a consequence of CH.

EXERCISE 17.8 (G). Where in the proof of Lemma 17.5 did we use CH?

Now we are ready to prove Theorem 17.3. Let $\{F_\alpha : \alpha < 2^{\aleph_0}\}$ and $\{G_\alpha : \alpha < 2^{\aleph_0}\}$ be as in Lemma 17.5. Without loss of generality we may assume that the pair $\langle F_0, G_0 \rangle$ is a decomposition of $\mathbb{R}$ as in Lemma 17.4. For each α, choose a bijection $f_\alpha : F_{\alpha+1} \setminus F_\alpha \to G_{\alpha+1} \setminus G_\alpha$, and let $\tilde{f} = \bigcup_{\alpha < 2^{\aleph_0}} f_\alpha$. Note that $\tilde{f}$ is a bijection with $dom(\tilde{f}) = \mathbb{R} \setminus F_0 = G_0$ and $rng(\tilde{f}) = \mathbb{R} \setminus G_0 = F_0$. Moreover, every meager subset of G_0 is contained in F_β for some $\beta < 2^{\aleph_0}$ and gets mapped by $\tilde{f}$ to a subset of the null set $G_\beta \cap F_0$. Similarly, $\tilde{f}^{-1}$ maps each null set $F_0 \cap G_\beta$ to the meager set $G_0 \cap F_\beta$. Let $f = \tilde{f} \cup \tilde{f}^{-1}$.

EXERCISE 17.9 (G). Verify that f is a bijection that satisfies conditions (i) and (ii) of Theorem 17.3. □

Naturally, if a theorem has been proved with the help of CH, one should try to prove the same theorem from a weaker assumption, or, even better, in ZFC alone. Let us go over the proof of Theorem 17.3 and see what we really need to make it work. The only place where CH was used is the proof of Lemma 17.5. More specifically, we constructed recursively sequences $\langle F_\alpha : \alpha < 2^{\aleph_0} \rangle$ and $\langle G_\alpha : \alpha < 2^{\aleph_0} \rangle$ of meager (respectively null) sets. At limit stages $\alpha < 2^{\aleph_0}$ we defined $F_\alpha = \bigcup_{\beta<\alpha} F_\beta$ and $G_\alpha = \bigcup_{\beta<\alpha} G_\beta$. The Continuum Hypothesis $2^{\aleph_0} = \aleph_1$ implies that all the relevant α's are countable limit ordinals. Since the union of countably many meager sets is meager, and the union of countably many null sets is a null set, this still gave us meager sets F_α and null sets G_α, and we could proceed with the recursive construction.

Do we need the full force of CH for the argument? To explore this question, let us introduce some new concepts. If $\mathcal{I}$ is an ideal, define:

$$add(\mathcal{I}) = \min\{\kappa : \exists \{A_\xi : \xi < \kappa\} \subseteq \mathcal{I} \ (\bigcup_{\xi<\kappa} A_\xi \notin \mathcal{I})\}.$$

The cardinal $add(\mathcal{I})$ is called the *additivity* of the ideal $\mathcal{I}$. Note that $\mathcal{I}$ is countably complete if and only if $add(\mathcal{I}) \geq \aleph_1$.

The numbers $add(\mathcal{M})$ and $add(\mathcal{N})$ are our first examples of *cardinal invariants of the continuum*. Many consequences of the Continuum Hypothesis or Martin's Axiom can be derived from suitable assumptions about cardinal invariants.

EXERCISE 17.10 (G). (a) Convince yourself that CH implies that $add(\mathcal{M}) = add(\mathcal{N}) = 2^{\aleph_0}$.

(b) Show that Theorem 17.3 remains valid if we replace CH by the assumption that $add(\mathcal{M}) = add(\mathcal{N}) = 2^{\aleph_0}$.

In Chapter 19 we will show that the equalities $add(\mathcal{M}) = add(\mathcal{N}) = 2^{\aleph_0}$ are consistent with the negation of the Continuum Hypothesis. It is not possible, however, to prove in ZFC that $add(\mathcal{M}) = add(\mathcal{N}) = 2^{\aleph_0}$, nor even that $add(\mathcal{M}) = add(\mathcal{N})$.

EXERCISE 17.11 (G). Convince yourself that if $add(\mathcal{M}) \neq add(\mathcal{N})$, then there is no bijection $f : \mathbb{R} \to \mathbb{R}$ that satisfies conditions (i) and (ii) of Theorem 17.3.

Let us say that L is a *generalized Luzin set* if $L \subseteq \mathbb{R}$, $|L| = 2^{\aleph_0}$, and $|L \cap F| < 2^{\aleph_0}$ for each meager set F. Similarly, let us say that S is a *generalized Sierpiński set* if $S \subseteq \mathbb{R}$, $|S| = 2^{\aleph_0}$, and $|S \cap G| < 2^{\aleph_0}$ for each null set G.

EXERCISE 17.12 (G). Show that if $add(\mathcal{M}) = add(\mathcal{N}) = 2^{\aleph_0}$, then there exist generalized Luzin sets and generalized Sierpiński sets.

A set $X \subseteq \mathbb{R}$ has *strong measure zero* if for every sequence $(\varepsilon_n)_{n \in \omega}$ of positive reals there exists a sequence $(I_n)_{n \in \omega}$ of open intervals such that I_n has length $\leq \varepsilon_n$ for each $n \in \omega$ and $X \subseteq \bigcup_{n \in \omega} I_n$.

Clearly, every countable set of reals has strong measure zero, and strong measure zero sets are null sets. But uncountable sets of strong measure zero are difficult to come by.

EXERCISE 17.13 (R). (a) Show that no uncountable closed set of reals has strong measure zero.

(b) Deduce from (a) that no uncountable Borel set of reals has strong measure zero.

Presumably motivated by the result of Exercise 17.13, E. Borel made the following conjecture:

THE BOREL CONJECTURE: *Every strong measure zero set of reals is countable.*

The Borel Conjecture turned out to be relatively consistent with ZFC. But it is not a theorem of ZFC.

THEOREM 17.7 (CH). *There exists an uncountable set of reals of strong measure zero.*

Here is a clever way to prove Theorem 17.7:

EXERCISE 17.14 (PG). Show that every Luzin set has strong measure zero.

And here is a more pedestrian approach:

PROOF OF THEOREM 17.7. Assume CH and let $\langle(\varepsilon_n^\eta)_{n\in\omega} : \eta < \omega_1\rangle$ be an enumeration of all sequences of positive reals. We want to construct recursively a sequence $\langle x_\xi : \xi < \omega_1\rangle$ of real numbers and a double sequence $\langle I_n^\eta : n \in \omega, \eta < \omega_1\rangle$ of open intervals such that

(a) I_n^η has length $\leq \varepsilon_n^\eta$ for all $n \in \omega$ and $\eta < \omega_1$;
(b) $\mathbb{Q} \cup \{x_\xi : \xi < \eta\} \subseteq \bigcup_{n\in\omega} I_n^\eta$ for all $\eta < \omega_1$; and
(c) $x_\eta \in (\bigcap_{\xi\leq\eta} \bigcup_{n\in\omega} I_n^\xi) \setminus \{x_\xi : \xi < \eta\}$ for all $\eta < \omega_1$.

The construction can be carried out as follows: At stage η, we are given x_ξ and I_n^ξ for $\xi < \eta$ and $n \in \omega$. Since countable sets have strong measure zero, we can choose intervals I_n^η such that (a) and (b) hold. Note that the presence of $\mathbb{Q}$ in (b) ensures that $\bigcup_{n\in\omega} I_n^\xi$ is a dense set of reals for every $\xi \leq \eta$. Thus the set $\bigcap_{\xi\leq\eta} \bigcup_{n\in\omega} I_n^\xi$ is a dense G_δ-set of reals and hence uncountable (see Theorem 26.5 and Exercise 14.6). So we can pick $x_\eta \in (\bigcap_{\xi\leq\eta} \bigcup_{n\in\omega} I_n^\xi) \setminus \{x_\xi : \xi < \eta\}$ as required in (c).

Now suppose x_ξ and I_n^ξ have been constructed for all $\xi < \omega_1$ and $n \in \omega$ so that (a)–(c) hold. Let $X = \{x_\xi : \xi < \omega_1\}$. By (c), X is uncountable. Moreover, if $\xi < \omega_1$, then (b) implies that $x_\xi \in \bigcup_{n\in\omega} I_n^\eta$ for $\xi < \eta$, and (c) implies that $x_\xi \in \bigcup_{n\in\omega} I_n^\eta$ for $\xi \geq \eta$. Thus $X \subseteq \bigcup_{n\in\omega} I_n^\eta$ for each $\eta < \omega_1$. Since every sequence of positive reals $(\varepsilon_n)_{n\in\omega}$ is listed as $(\varepsilon_n^\eta)_{n\in\omega}$ for some $\eta < \omega_1$, (a) implies that X has strong measure zero. □

Can we replace CH in Theorem 17.7 by an assumption about a suitable cardinal invariant of the continuum? An analysis of the "pedestrian" proof of Theorem 17.7 may shed light on this question. The Continuum Hypothesis was used to enumerate all sequences of positive reals by countable ordinals. This of course cannot be done with the help of anything less than CH. But perhaps we do not need to consider *all* sequences of positive reals? Wouldn't it be enough if for every sequence $(\delta_n)_{n\in\omega}$ of positive reals our enumeration contained a sequence $(\varepsilon_n^\eta)_{n\in\omega}$ such that $\varepsilon_n^\eta \leq \delta_n$ for all $n \in \omega$? Let us define:

$\mathfrak{d}_\varepsilon = \min\{\kappa : \exists\{(\varepsilon_n^\eta)_{n\in\omega} : \eta < \kappa\} \subseteq {}^\omega(\mathbb{R}^+) \forall (\delta_n)_{n\in\omega} \in {}^\omega(\mathbb{R}^+) \exists \eta < \kappa \forall n\, (\varepsilon_n^\eta \leq \delta_n)\}$,

where $\mathbb{R}^+ = (0,\infty)$.

EXERCISE 17.15 (PG). (a) Show that $\mathfrak{d}_\varepsilon > \aleph_0$.
(b) Conclude that $\text{CH} \to \mathfrak{d}_\varepsilon = \aleph_1$.
(c) Show that Theorem 17.7 remains valid if CH is replaced by the assumption that $\mathfrak{d}_\varepsilon = \aleph_1$.

The cardinal invariant $\mathfrak{d}_\varepsilon$ is usually defined in a different way. Let $f, g \in {}^\omega\omega$. We write $f <^* g$ and say that g *eventually dominates* f if $\exists m \forall n > m\, (f(n) < g(n))$. A subset $G \subseteq {}^\omega\omega$ is said to be a *dominating family* if $\forall f \in {}^\omega\omega \exists g \in G\, (f <^* g)$. Define

$$\mathfrak{d} = \min\{|G| : G \subseteq {}^\omega\omega \wedge G \text{ is a dominating family}\}.$$

17. APPLICATIONS OF THE CONTINUUM HYPOTHESIS

The cardinal invariant $\mathfrak{d}$ is called the *dominating number*.

EXERCISE 17.16 (G). Show that $\mathfrak{d} = \mathfrak{d}_\varepsilon$.

We will prove in Chapter 19 that the equality $\mathfrak{d} = \aleph_1$ is not a theorem of ZFC. However, it can be shown that this equality is consistent with the negation of CH. A sequence $\langle f_\alpha : \alpha < \kappa \rangle$ of elements of $^\omega\omega$ is called a κ-*scale* if $f_\alpha <^* f_\beta$ for all $\alpha < \beta < \kappa$ and $\{f_\alpha : \alpha < \kappa\}$ is a dominating family.

EXERCISE 17.17 (PG). Show that $\mathfrak{d} = \aleph_1$ implies the existence of an ω_1-scale.

Let us examine your solution of Exercise 17.17. (You did complete it before reading on, didn't you?) You had to deal with situations where for a given countable family $\{f_n : n \in \omega\} \subseteq {}^\omega\omega$ you wanted to find $g \in {}^\omega\omega$ such that $f_n <^* g$ for all $n \in \omega$. Superficially, this may look like an application of Exercises 17.15(a) and 17.16, but let us look a little closer at the problem. The inequality $\mathfrak{d} > \aleph_0$ only implies that some $g \in {}^\omega\omega$ is not eventually dominated by any of the f_n's, not that there exists $g \in {}^\omega\omega$ which eventually dominates all f_n's. Let us say that a family $F \subseteq {}^\omega\omega$ is *bounded* if there exists $g \in {}^\omega\omega$ such that $f <^* g$ for all $f \in F$, and *unbounded* if no such g exists. Define:

$$\mathfrak{b} = \min\{|F| : F \subseteq {}^\omega\omega \wedge F \text{ is an unbounded family}\}.$$

The cardinal $\mathfrak{b}$ is called the *bounding number*.

EXERCISE 17.18 (PG). (a) Show that $\aleph_0 < \mathfrak{b} \leq \mathfrak{d}$.
(b) Show that there exists a sequence $\langle f_\alpha : \alpha < \mathfrak{b}\rangle$ of functions in $^\omega\omega$ such that $f_\alpha <^* f_\beta$ for all $\alpha < \beta < \mathfrak{b}$ and the family $\{f_\alpha : \alpha < \mathfrak{b}\}$ is unbounded.
(c) Infer from (a) and (b) that $\mathfrak{b}$ is a regular uncountable cardinal.

Does the notion of strong measure zero sets have a "dual" for Baire category? This is far from obvious, since there is no category analogue for the length of an interval. But how about this: Let us call a set $M \subseteq \mathbb{R}$ *perfectly meager*[1] if $M \cap K$ is meager in K whenever K is a perfect (i.e., closed without isolated points) subset of $\mathbb{R}$. Perfectly meager sets share some properties of strong measure zero sets. For example, every countable set of reals is perfectly meager and there are no uncountable perfectly meager Borel subsets of $\mathbb{R}$. Moreover, the equality $\mathfrak{b} = \aleph_1$ allows us to construct an uncountable perfectly meager set. Here is how.

LEMMA 17.8. *Suppose X is a Polish space, $\langle J_\alpha : \alpha < \omega_1\rangle$ is a decreasing sequence of F_σ-subsets of X such that $\bigcap_{\alpha<\omega_1} J_\alpha = \emptyset$, and K is a perfect subset of X. Then there exists $\alpha < \omega_1$ such that $K \cap J_\alpha$ is meager in K.*

PROOF. Suppose $\langle J_\alpha : \alpha < \omega_1\rangle$ is as in the assumption. For each α, let $J_\alpha = \bigcup_{m\in\omega} J_{\alpha,m}$, where the $J_{\alpha,m}$'s are closed in X. Let K be a perfect subset of X. Then K is also a Polish space. In the remainder of this proof, all topological notions (basic open set, meager set, etc.) refer to the topological space K. Since for each $\alpha < \omega_1$, $K \cap J_\alpha$ has the Baire property, either $K \cap J_\alpha$ is meager for some α, or for every $\alpha < \omega_1$ there exist a nonempty basic open set U_α and $n(\alpha) \in \omega$ such that $K \cap J_{\alpha,n(\alpha)} \cap U_\alpha$ is comeager in U_α. Let's assume towards a contradiction that the second alternative holds. Since K is second countable, the Pigeonhole Principle implies that there exist a basic open U and a cofinal subset A of ω_1 such that $U_\alpha = U$ for all $\alpha \in A$. But each comeager subset of U is dense in U. Since

---

[1] Another name for perfectly meager sets is *always first category* sets.

$K \cap J_{\alpha,n(\alpha)}$ is closed, we must have $J_{\alpha,n(\alpha)} \supseteq U$ for each $\alpha \in A$. It follows that $\bigcap_{\alpha \in A} J_\alpha \supseteq U$, and since the sequence $\langle J_\alpha : \alpha < \omega_1 \rangle$ was assumed decreasing, we have $\bigcap_{\alpha < \omega_1} J_\alpha \supseteq U$. This contradicts the assumption that $\bigcap_{\alpha < \omega_1} J_\alpha = \emptyset$. □

LEMMA 17.9. *Suppose X is a Polish space without isolated points, $\langle J_\alpha : \alpha < \omega_1 \rangle$ is a strictly decreasing sequence of F_σ-subsets of X such that $\bigcap_{\alpha < \omega_1} J_\alpha = \emptyset$, and $y_\alpha \in J_\alpha \setminus J_{\alpha+1}$ for all $\alpha < \omega_1$. Then the set $Y = \{y_\alpha : \alpha < \omega_1\}$ is perfectly meager.*

PROOF. Let X, Y, J_α be as in the assumptions and let K be a perfect subset of X. By Lemma 17.8, there is an $\alpha_0 < \omega_1$ such that the set $K \cap \{y_\alpha : \alpha_0 < \alpha < \omega_1\}$ is meager in K. The set $K \cap \{y_\alpha : \alpha \leq \alpha_0\}$ is countable, and hence also meager in K. Thus, the set $K \cap Y$ is the union of two meager sets in K, and hence meager in K. □

LEMMA 17.10. *Let $X = {}^\omega\omega$ with the product topology. If $\mathfrak{b} = \aleph_1$, then there exists a strictly decreasing sequence $\langle J_\alpha : \alpha < \omega_1 \rangle$ of nonempty F_σ-subsets of X such that $\bigcap_{\alpha < \omega_1} J_\alpha = \emptyset$.*

PROOF. Assume $\mathfrak{b} = \aleph_1$. Let $\langle f_\alpha : \alpha < \omega_1 \rangle$ be a sequence as in Exercise 17.18(b). Define:

$$J_\alpha = \{g \in {}^\omega\omega : f_\alpha <^* g\}.$$

EXERCISE 17.19 (G). Convince yourself that the J_α's are as required. *Hint:* To show that J_α is a countable union of closed sets, consider $J_{\alpha,m} = \{g \in {}^\omega\omega : \forall n > m \, (f_\alpha(n) < g(n))\}$. □

Since ${}^\omega\omega$ is homeomorphic to the set of irrationals with the usual topology, it follows from Lemmas 17.8–17.10 that the equality $\mathfrak{b} = \aleph_1$ implies the existence of an uncountable perfectly meager set of reals.

Have we found category counterparts to all properties of strong measure zero sets discussed earlier in this section? How about Exercise 17.14?

EXERCISE 17.20 (PG). Show that every Sierpiński set is perfectly meager.

How about the Borel Conjecture? Is it consistent that every perfectly meager set is countable? In Chapter 20 (Theorem 20.2 and Claim 20.3) we will prove in ZFC the existence of a strictly decreasing sequence of length ω_1 of F_σ-subsets of the Cantor set with empty intersection. Thus, by Lemma 17.9, the existence of uncountable perfectly meager sets can be proved in ZFC. Oops, down the drain goes our "dual" Borel conjecture.

Did we discover a blemish on the face of Mathematica, or did we pick the wrong candidate for the dual notion of "strong measure zero?" To be completely honest with you: Mathematica's beauty is not flawless. Not always does the most aesthetically pleasing among competing conjectures turn out to be true. For example, the inequality $add(\mathcal{N}) \leq add(\mathcal{M})$ is provable in ZFC, but it is consistent that $add(\mathcal{N}) < add(\mathcal{M})$. The Lebesgue measure side of Mathematica's face is not a perfect mirror image of the Baire category side. But such little blemishes only add spice to the intimate knowledge of Mathematica's features. Much of the time, mathematical beauty is a surprisingly reliable indicator of mathematical truth.

In the case discussed here, we simply suspected the wrong kind of duality. This becomes evident if you consider the following notion: A set $X \subseteq \mathbb{R}$ is said to have *universal measure zero* if for every Borel measure μ that vanishes on the singletons[2] there exists a Borel set A such that $X \subseteq A$ and $\mu(A) = 0$. Now *this* appears to be the right measure-theoretic counterpart of the notion of perfectly meager sets, doesn't it? And vice versa, of course.

EXERCISE 17.21 (R). (a) Show that every strong measure zero set has universal measure zero.

(b) Show that no uncountable closed set (and hence no uncountable Borel set) of reals has universal measure zero.

In Chapter 20 we will show in ZFC that there exists a strictly decreasing sequence $\langle J_\alpha : \alpha < \omega_1 \rangle$ of F_σ-subsets of $\mathbb{R}$ such that $\bigcap_{\alpha < \omega_1} J_\alpha = \emptyset$ and, if $y_\alpha \in J_{\alpha+1} \setminus J_\alpha$ for all α, then the set $\{y_\alpha : \alpha < \omega_1\}$ has universal measure zero. Do we need to persuade you further that the notions of strongly meager sets and sets of universal measure zero are dual to each other?

Now back to the drawing board: What is the right category analogue of strong measure zero? The following theorem points in a promising direction. It shows that strong measure zero sets can be characterized by their behavior with respect to meager sets.

THEOREM 17.11 (Galvin-Mycielski-Solovay). *A set $X \subseteq \mathbb{R}$ has strong measure zero iff for every meager set F there exists a real y such that $(y + X) \cap F = \emptyset$.*

The set $y + X$ of Theorem 17.11 is defined as the shift of X by y units, i.e., $y + X = \{y + x : x \in X\}$. Upon seeing Theorem 17.11 it is almost impossible to resist the temptation to call a set $Y \subseteq \mathbb{R}$ *strongly meager* if for every null set G there exists $x \in \mathbb{R}$ such that $(Y + x) \cap G = \emptyset$. Moreover, one is led to the following:

DUAL BOREL CONJECTURE: *Every strongly meager set is countable.*

Speaking of temptations: Here are some good ones for you.

EXERCISE 17.22 (X). Show that no uncountable closed set of reals is strongly meager.

EXERCISE 17.23 (XX). Show that every Sierpiński set is strongly meager.

EXERCISE 17.24 (XXX). Show that the family of strongly meager sets forms an ideal of subsets of $\mathbb{R}$.

The status of the Dual Borel Conjecture is the same as the status of the Borel Conjecture: It is consistent with, but not provable in ZFC. Let us demonstrate the latter:

THEOREM 17.12. *Assume $add(\mathcal{N}) = 2^{\aleph_0}$. Then there exists a strongly meager set of reals of size $2^{\aleph_0}$.*

PROOF. Let $\langle G_\zeta : \zeta < 2^{\aleph_0} \rangle$ be an enumeration of all G_δ-sets of Lebesgue measure zero. We want to construct sequences of reals $\langle x_\xi : \xi < 2^{\aleph_0} \rangle$ and $\langle y_\xi : \xi < 2^{\aleph_0} \rangle$ such that if $X = \{x_\xi : \xi < 2^{\aleph_0}\}$, then

(+) $(y_\zeta + X) \cap G_\zeta = \emptyset$ for all $\zeta < 2^{\aleph_0}$.

---

[2]This is shorthand for saying that μ is a measure defined at least on the σ-field of Borel subsets of $\mathbb{R}$ such that $\mu(\{x\}) = 0$ for each $x \in \mathbb{R}$.

Since $\{G_\zeta : \zeta < 2^{\aleph_0}\}$ is a base for the ideal of null sets, condition (+) implies that X is strongly meager. But how can we construct x_ξ, y_ξ by recursion over $\xi < 2^{\aleph_0}$ and ensure that (+) holds?

At stage η of the construction, we will have chosen x_ξ, y_ξ for $\xi < \eta$, and we may assume that

$(+)_\eta \quad \forall \zeta < \eta \, ((y_\zeta + \{x_\xi : \xi < \eta\}) \cap G_\zeta = \emptyset).$

Consider the set $H_\eta = \bigcup_{\xi < \eta, \zeta \leq \eta} (G_\zeta - x_\xi)$. Since H_η is a union of fewer than $2^{\aleph_0}$ null sets, $add(\mathcal{N}) = 2^{\aleph_0}$ implies that H_η is a null set. So we may choose $y_\eta \in \mathbb{R} \setminus H_\eta$. Note that $y \in G - x$ if and only if $y + x \in G$. Thus our choice of y_η implies:

$(++)_\eta \quad \forall \zeta \leq \eta \, ((y_\zeta + \{x_\xi : \xi < \eta\}) \cap G_\zeta = \emptyset).$

Now let us choose x_η. We have already incurred a lot of obligations: For every $\zeta \leq \eta$ we must make sure that $y_\zeta + x_\eta \notin G_\zeta$. Choosing $x_\eta \in \mathbb{R} \setminus \bigcup_{\zeta \leq \eta} (G_\zeta - y_\zeta)$ allows us to meet all these obligations. Again, $add(\mathcal{N}) = 2^{\aleph_0}$ implies that the set $\mathbb{R} \setminus \bigcup_{\zeta \leq \eta} (G_\zeta - y_\zeta)$ is nonempty, and hence a suitable choice for x_η exists. □

EXERCISE 17.25 (G). Show that if $add(\mathcal{M}) = 2^{\aleph_0}$, then there exists a strong measure zero set of reals of size $2^{\aleph_0}$.

If you have become a true believer in cardinal invariants by now, here is a good exercise for you. For an ideal $\mathcal{I}$ of subsets of $\mathbb{R}$, define:

(17.2) $\qquad cof(\mathcal{I}) = \min\{|\mathcal{B}| : \mathcal{B} \text{ is a base of } \mathcal{I}\},$

$\qquad cov(\mathcal{I}) = \min\{|\mathcal{A}| : \mathcal{A} \subseteq \mathcal{I} \wedge \bigcup \mathcal{A} = \mathbb{R}\}.$

EXERCISE 17.26 (G). Show that if $cof(\mathcal{N}) \leq cov(\mathcal{N})$, then there exists a strongly meager set $Y \subseteq \mathbb{R}$ of cardinality $cof(\mathcal{N})$, and if $cof(\mathcal{M}) \leq cov(\mathcal{M})$, then there exists a strong measure zero set $X \subseteq \mathbb{R}$ of cardinality $cof(\mathcal{M})$.

Those readers who want to learn more about meager sets, null sets, and cardinal invariants connected with these notions will find suggestions for further reading in the Mathographical Remarks at the end of this chapter.

17.2. Miscellaneous applications of CH

Our next example is an application of CH to partition calculus.

THEOREM 17.13 (Erdős-Rado). CH *implies that there exists a partition of* $\omega \times \omega_1$ *into disjoint sets* H *and* K *such that* $\forall a \in [\omega]^{\aleph_0} \forall B \in [\omega_1]^{\aleph_1} \, (a \times B \not\subseteq H \wedge a \times B \not\subseteq K)$.

PROOF. We need a lemma.

LEMMA 17.14 (CH). *There exists a sequence* $\langle b_\alpha : \alpha < \omega_1 \rangle$ *of subsets of* ω *such that* $\forall a \in [\omega]^{\aleph_0} \exists \alpha_a < \omega_1 \forall \alpha \geq \alpha_a \, (|a \cap b_\alpha| = |a \setminus b_\alpha| = \aleph_0)$.

EXERCISE 17.27 (PG). Prove Lemma 17.14.

Let $\langle b_\alpha : \alpha < \omega_1 \rangle$ be a sequence as in Lemma 17.14. Define:
$$H = \{\langle n, \alpha \rangle : \alpha < \omega_1, n \in b_\alpha\},$$
$$K = \{\langle n, \alpha \rangle : \alpha < \omega_1, n \notin b_\alpha\}.$$
Consider $a \in [\omega]^{\aleph_0}, B \in [\omega_1]^{\aleph_1}$. Let α_a be as in Lemma 17.14, and pick $\alpha \in B$ such that $\alpha \geq \alpha_a$. If $m \in a \cap b_\alpha$ and $n \in a \setminus b_\alpha$, then $\langle n, \alpha \rangle \in (a \times B) \setminus H$ and $\langle m, \alpha \rangle \in (a \times B) \setminus K$. □

Can CH in Theorem 17.13 (or, more precisely, in Lemma 17.14) be replaced by an assumption about a suitable cardinal invariant; preferably one that has already been introduced? By using a clever and useful trick, one can show that Theorem 17.13 remains valid if CH is replaced by the assumption $\mathfrak{d} = \aleph_1$.

Let f be a function from ω into ω such that $f(n) > n$ for all $n \in \omega$. For $k \in \omega$, define recursively a function $f^k \in {}^\omega\omega$ as follows: $f^0(n) = n$, $f^{k+1}(n) = f(f^k(n))$ for all $n \in \omega$. By the assumption about f, the sequence $(f^k(0))_{k \in \omega}$ is strictly increasing. Thus we can associate with f a set $b_{(f)} \subseteq \omega$ as follows:
$$b_{(f)} = \bigcup_{k \in \omega} [f^{2k}(0), f^{2k+1}(0)),$$
where $[\ell, m) = \{n \in \omega : \ell \leq n < m\}$.

For every $a \in [\omega]^{\aleph_0}$ let us define a function $g_{(a)} \in {}^\omega\omega$ by:
$$g_{(a)}(n) = \min\{m \in a : m > n\}.$$

EXERCISE 17.28 (G). Let $a \in [\omega]^{\aleph_0}$, and assume that $f \in {}^\omega\omega$ is such that $f(n) > n$ for all $n \in \omega$ and $f_{(a)} <^* f$. Show that $|a \cap b_{(f)}| = |a \setminus b_{(f)}| = \aleph_0$.

Now suppose $\mathfrak{d} = \aleph_1$, and $\langle f_\alpha : \alpha < \omega_1 \rangle$ is an ω_1-scale. Let $b_\alpha = b_{(f_\alpha)}$ for $\alpha < \omega_1$. It follows from Exercise 17.28 that the sequence $\langle b_\alpha : \alpha < \omega_1 \rangle$ is as postulated in Lemma 17.14.

By considering another cardinal invariant of the continuum, one can show that Theorem 17.13 is not provable in ZFC. The *splitting number* $\mathfrak{s}$ is defined as the smallest cardinality of a *splitting family*, where a family $S \subseteq \mathcal{P}(\omega)$ is splitting if for every $a \in [\omega]^{\aleph_0}$ there exists $s \in S$ such that $|a \cap s| = |a \setminus s| = \aleph_0$. Note that the family $\{b_\alpha : \alpha < \omega_1\}$ of Lemma 17.14 is a (rather special) splitting family of cardinality $\aleph_1$. In Chapter 19 (Theorem 19.20) we will show that MA + ¬CH implies that $\mathfrak{s} > \aleph_1$.

THEOREM 17.15. *Assume* $\mathfrak{s} > \aleph_1$. *Then for every partition of* $\omega \times \omega_1$ *into disjoint sets H and K there exist* $a \in [\omega]^{\aleph_0}$ *and* $B \in [\omega_1]^{\aleph_1}$ *such that* $a \times B \subseteq H$ *or* $a \times B \subseteq K$.

PROOF. Let $\omega \times \omega_1 = H \cup K$, where $H \cap K = \emptyset$. For $\alpha < \omega_1$, let $s_\alpha = \{n \in \omega : \langle n, \alpha \rangle \in H\}$. If $\mathfrak{s} > \aleph_1$, then the family $\{s_\alpha : \alpha < \omega_1\}$ is not splitting. This means, we can pick $a_0 \in [\omega]^{\aleph_0}$ such that for all $\alpha < \omega_1$, either $|a_0 \cap s_\alpha| < \aleph_0$ or $|a_0 \setminus s_\alpha| < \aleph_0$.

Case 1: $|\{\alpha < \omega_1 : |a_0 \cap s_\alpha| < \aleph_0\}| = \aleph_1$.

For $n \in \omega$, let $B_n = \{\alpha : a_0 \cap s_\alpha \subseteq n\}$. By the Pigeonhole Principle, $|B_n| = \aleph_1$ for some $n \in \omega$. Pick the smallest such n, and let $B = B_n$, $a = a_0 \setminus n$. Then $a \times B \subseteq K$.

Case 2: Not Case 1. Then $|\{\alpha < \omega_1 : |a_0\setminus s_\alpha| < \aleph_0\}| = \aleph_1$.

Reasoning as in Case 1, we find $a \in [a_0]^{\aleph_0}$ and $B \in [\omega_1]^{\aleph_1}$ such that $a \times B$ is a subset of H. □

In topology, CH has been used to construct many examples of spaces that exhibit various bizarre combinations of topological properties. Let us present here just one of these examples, the so-called *Kunen line*. One of the counter-intuitive properties of the Kunen line: It is not a linearly ordered space.

THEOREM 17.16 (CH). *Let $\langle x_\alpha : \alpha < \omega_1 \rangle$ be an enumeration of $\mathbb{R}$. Then there exists a topology τ on $\mathbb{R}$ that refines the usual metric topology τ_M and is such that:*

(i) $\{x_\alpha : \alpha < \beta\} \in \tau$ for each $\beta < \omega_1$;
(ii) $|cl_{\tau_M}(A)\setminus cl_\tau(A)| \leq \aleph_0$ for every $A \in [\mathbb{R}]^{\aleph_0}$;
(iii) $\langle \mathbb{R}, \tau \rangle$ is zero-dimensional;
(iv) $\langle \mathbb{R}, \tau \rangle$ is first countable;
(v) $\langle \mathbb{R}, \tau \rangle$ is locally compact.

Note that (i)–(iii) imply that the Kunen line is an S-space: Regularity follows from (iii). The family $\{\{x_\alpha : \alpha < \beta\} : \beta < \omega_1\}$ is an open cover without countable subcover; hence the Kunen line is not Lindelöf. If X is a subspace of $\mathbb{R}$ and A is a countable subset of X dense in $\langle X, \tau_M \rangle$, then by a (ii), $X\setminus cl_\tau(A)$ is countable, and it follows that $\langle \mathbb{R}, \tau \rangle$ is hereditarily separable.

Moreover, note that $\langle \mathbb{R}, \tau \rangle$ is scattered: This follows from (i), (v), and Theorem 26.4.

EXERCISE 17.29 (G). How do we know that $\langle \mathbb{R}, \tau \rangle$ is Hausdorff?

PROOF OF THEOREM 17.16. Let us begin by enumerating everything relevant. Let $\langle x_\alpha : \alpha < \omega_1 \rangle$ be as in the assumption. For $\beta < \omega_1$, let X_β denote the set $\{x_\alpha : \alpha < \beta\}$. Fix an enumeration $\langle A_\gamma : \gamma < \omega_1 \rangle$ of $[\mathbb{R}]^{\aleph_0}$. For $\beta < \omega_1$, let $\mathcal{C}_\beta = \{A_\gamma : \gamma < \beta \wedge A_\gamma \subseteq X_\beta \wedge x_\beta \in cl_{\tau_M}(A_\gamma)\}$, and fix a sequence $\langle C_{\beta,n} : n \in \omega \rangle$ such that every element of $\mathcal{C}_\beta$ gets listed infinitely often in it.

By recursion over $\beta < \omega_1$, we will construct a sequence of topologies $\langle \tau_\beta : \beta < \omega_1 \rangle$ such that for all $\beta < \omega_1$:

(a) τ_β is a topology on X_β that refines the usual metric topology $\tau_M \lceil X_\beta$;
(b) If $\gamma < \beta$, then $\tau_\gamma = \tau_\beta \lceil X_\gamma$;
(c) If $\beta \in \omega_1 \cap \mathbf{LIM}$, then $\bigcup_{\gamma < \beta} \tau_\gamma$ is a base for τ_β;
(d) If $A \in \mathcal{C}_\beta$, then $x_\beta \in cl_{\tau_{\beta+1}}(A)$;
(e) τ_β is a zero-dimensional topology on X_β;
(f) τ_β is a first countable topology on X_β;
(g) τ_β is a locally compact topology on X_β.

If we succeed in doing this, then $\bigcup_{\beta < \omega_1} \tau_\beta$ is a base for a topology τ on $\mathbb{R}$. It is not hard to see that (a)–(c) imply that τ is a refinement of τ_M that satisfies (i), and, moreover, that $\tau \lceil X_\beta = \tau_\beta$ for each $\beta < \omega_1$. This in turn implies that the whole space $\langle \mathbb{R}, \tau \rangle$ has such "local" properties as zero-dimensionality, first countability, and local compactness if and only if each X_β has these properties. Thus (e) implies (iii), (f) implies (iv), and (g) implies (v).

EXERCISE 17.30 (G). Convince yourself that (d) implies that if $A \in \mathcal{C}_\beta$, then $x_\beta \in cl_\tau(A)$. Infer that $\langle \mathbb{R}, \tau \rangle$ satisfies (ii).

It remains to construct the sequence $\langle \tau_\beta : \beta < \omega_1 \rangle$. For $\beta < \omega$, the space X_β is finite, and the metric topology τ_M on X_β has all the desired properties. At limit stages β, it suffices to observe that $\bigcup_{\gamma < \beta} \tau_\gamma$ is a base for a topology, and choose τ_β as dictated by (c). Repeating the argument preceding Exercise 17.30, one can easily show that if (a)–(g) hold up to stage β, then this topology τ_β will also have all the required properties.

Now assume τ_β has been constructed and satisfies conditions (a)–(c) and (e)–(f). We show how to define $\tau_{\beta+1}$. The basic neighborhoods of x_α for $\alpha < \beta$ have already been determined at previous stages. Now we need to choose a countable base for x_β so that (d) holds. Since $x_\beta \in cl_{\tau_M} C_{\beta,n}$ for each $n \in \omega$, we can choose a sequence $(y_n)_{n \in \omega}$ of reals in X_β such that $y_n \in C_{n,\beta}$ for all $n \in \omega$ and the sequence $(|y_n - x_\beta|)_{n \in \omega}$ of distances between x_β and y_n is strictly decreasing and converges to zero. We are going to choose a decreasing sequence $(V_{\beta,k})_{k \in \omega}$ of subsets of $X_{\beta+1}$ such that $\{y_n : n \geq k\} \subseteq V_{\beta,k}$, $V_{\beta,k} \cap X_\beta$ is open for all $k \in \omega$, and $\bigcap_{k \in \omega} V_{\beta,k} = \{x_\beta\}$. Then we will declare $\tau_\beta \cup \{V_{\beta,k} : k \in \omega\}$ a base for $\tau_{\beta+1}$.

EXERCISE 17.31 (G). Convince yourself that if $\tau_{\beta+1}$ is defined as indicated above, then (d) holds at stage β, and conditions (a)–(c) and (f) will be satisfied at stage $\beta + 1$.

To ensure that $\tau_{\beta+1}$ is zero-dimensional and locally compact, some extra care in choosing the $V_{\beta,k}$'s is needed. Let the y_n's be as above, and choose a sequence $(I_n)_{n \in \omega}$ of pairwise disjoint closed intervals such that $y_n \in int(I_n)$ for all $n \in \omega$. For each $n \in \omega$, pick a compact neighborhood $U_{\beta,n} \in \tau_\beta$ of y_n such that $U_{\beta,n} \subseteq I_n$ and let $V_{\beta,k} = \{x_\beta\} \cup \bigcup_{n \geq k} U_{\beta,n}$.

EXERCISE 17.32 (G). Convince yourself that these $V_{\beta,k}$'s satisfy the specifications described above, and show that the resulting space $\langle X_{\beta+1}, \tau_{\beta+1} \rangle$ is locally compact.

To show that $\langle X_{\beta+1}, \tau_{\beta+1} \rangle$ is zero-dimensional, it suffices to prove that each of the sets $V_{\beta,k}$ is closed. The rest follows from the inductive assumption. So suppose $x \in X_{\beta+1} \setminus V_{\beta,k}$ for some $k \in \omega$. Then $x \neq x_\beta$. Since $\lim_{n \to \infty} y_n = x_\beta$ and the intervals I_n do not overlap, there exist $\ell \in \omega$ and $W \in \tau_M$ such that $W \cap \bigcup_{n > \ell} I_n = \emptyset$ and $x \in W$. In other words,

$$W \cap V_{\beta,k} \subset U_{\beta,k} \cup U_{\beta,k+1} \cup \cdots \cup U_{\beta,\ell}.$$

The right hand side of the last equality is compact and hence closed in τ_β. Since $W \in \tau_\beta$, there exists a τ_β-neighborhood $W_0 \subseteq W$ of x such that $W_0 \cap V_{\beta,k} = \emptyset$. □

The applications of CH we have seen so far all relied on transfinite recursion. Let us give a couple of examples with a different flavor.

Let $\mathcal{A}$ be a family of subsets of a fixed set X. We say that $\mathcal{A}$ is an *almost disjoint family*[3] if $|A| = |X|$ and $|A \cap B| < |X|$ for all $\{A, B\} \in [\mathcal{A}]^2$. While the size of every family of pairwise disjoint subsets of X is bounded by $|X|$, almost disjoint families can be somewhat larger.

---

[3]Some authors would insist on using the phrase "family of pairwise almost disjoint sets of cardinality $|X|$ each." The phrase "almost disjoint" is also being used in the literature as a shorthand for "$|A \cap B| < \min\{|A|, |B|\}$."

THEOREM 17.17. *Let κ be a regular infinite cardinal. Then there exists an almost disjoint family $\mathcal{A}$ of subsets of κ such that $|\mathcal{A}| = \kappa^+$.*

PROOF. Let κ be as in the assumption. By recursion over $\alpha < \kappa^+$ we will construct sets $A_\alpha \in [\kappa]^\kappa$ such that $|A_\alpha \cap A_\beta| < \kappa$ for all $\alpha < \beta < \kappa$.

To get started, let $f : \kappa \to \kappa \times \kappa$ be a bijection and define $A_\alpha = f^{-1}\{\langle \alpha, \beta \rangle : \beta < \kappa\}$ for $\alpha < \kappa$.

Now suppose $\kappa \leq \gamma < \kappa^+$, and $\{B_\alpha : \alpha < \gamma\}$ has been constructed and forms an almost disjoint family. Fix a bijection $g : \kappa \to \gamma$. For $\beta < \kappa$, define recursively $\delta_\beta = \min\{A_\beta \setminus \bigcup_{\alpha < \beta}(B_{g(\alpha)} \cup \{\delta_\alpha\})\}$. Since $|A_\beta \cap (B_{g(\alpha)} \cup \{\delta_\alpha\})| < \kappa$ for each $\alpha < \beta$, and since $\beta < \kappa = cf(\kappa)$, the set $A_\alpha \cap \bigcup_{\alpha < \beta}(B_{g(\alpha)} \cup \{\delta_\alpha\})$ has cardinality less than κ, and hence δ_β is well defined. We let $B_\gamma = \{\delta_\beta : \beta < \kappa\}$. Then $|B_\gamma| = \kappa$ and for each $\beta < \kappa$, the set $B_\gamma \cap B_{g(\beta)}$ is contained in the set $\{\delta_\alpha : \alpha \leq \beta\}$, and hence is of cardinality less than κ. Since the range of the function g is γ, the family $\{B_\alpha : \alpha \leq \gamma\}$ is almost disjoint. □

If $\kappa = \omega$, then Theorem 17.17 can be improved.

THEOREM 17.18. *There exists an almost disjoint family of subsets of ω of size $2^{\aleph_0}$.*

PROOF. Fix a bijection $f : {}^{<\omega}2 \to \omega$. For $a \subseteq \omega$, let $\chi_a : \omega \to 2$ denote the characteristic function of a. Define:

$$C(a) = \{f(\chi_a \restriction n) : n \in \omega\}.$$

EXERCISE 17.33 (G). (a) Show that if $a \neq b$, then $|C(a) \cap C(b)| < \aleph_0$.

(b) Conclude that the family $\mathcal{C} = \{C(a) : a \in \mathcal{P}(\omega)\}$ is an almost disjoint family of size $2^{\aleph_0}$. □

It is not possible to prove in ZFC that there exists an almost disjoint family of size $2^{\aleph_1}$ of subsets of ω_1. However, CH does imply the existence of such a family.

THEOREM 17.19 (CH). *There exists an almost disjoint family of size $2^{\aleph_1}$ of subsets of ω_1.*

PROOF. If CH holds, the set $\bigcup_{\alpha < \omega_1} {}^\alpha 2$ has cardinality $\aleph_1$.

EXERCISE 17.34 (G). Take a close look at the proof of Theorem 17.18 and then prove Theorem 17.19. □

EXERCISE 17.35 (G). Find and prove a generalization of Theorem 17.19 to arbitrary infinite cardinals.

Let κ be an infinite cardinal. A family $\mathcal{A} \subseteq \mathcal{P}(\kappa)$ is called *independent* if for all pairs of disjoint $F, G \in [\mathcal{A}]^{<\aleph_0}$ we have:

$$C_{F,G} = \bigcap_{A \in F} A \cap \bigcap_{A \in G}(\kappa \setminus A) \neq \emptyset.$$

(We assume that $\bigcap \emptyset = \kappa$.) We say that $\mathcal{A}$ is *strongly independent* if $|C_{F,G}| = \kappa$ for every pair $\langle F, G \rangle$ as above.

THEOREM 17.20 (Fichtenholz-Kantorovitch-Hausdorff). *Let κ be an infinite cardinal. Then there exists a strongly independent family $\mathcal{A} \subseteq \mathcal{P}(\kappa)$ such that $|\mathcal{A}| = 2^\kappa$.*

PROOF. Let κ be an infinite cardinal, and let $K = [\kappa]^{<\aleph_0} \times [[\kappa]^{<\aleph_0}]^{<\aleph_0}$. Then $|K| = \kappa$, and we can prove Theorem 17.20 by constructing a strongly independent family $\mathcal{B} \subseteq \mathcal{P}(K)$ of size 2^κ. The desired family $\mathcal{A}$ can then be obtained by fixing a bijection $f : K \to \kappa$ and defining $\mathcal{A} = \{f[B] : B \in \mathcal{B}\}$. For $A \in \mathcal{P}(\kappa)$ let

$$B_A = \{\langle s, T\rangle \in K : s \cap A \in T\}.$$

Let $\mathcal{B} = \{B_A : A \in \mathcal{P}(\kappa)\}$. It is not hard to see that if $A_0 \neq A_1$, then $B_{A_0} \neq B_{A_1}$. Thus $|\mathcal{B}| = 2^\kappa$.

It remains to show that $\mathcal{B}$ is strongly independent. Let $\langle F, G\rangle$ be a pair of finite disjoint subsets of $\mathcal{B}$. Enumerate $F \cup G$ as $\{B_{A_i} : i < n\}$. For $i < j < n$, fix $\xi_{i,j} \in A_i \triangle A_j$. Let

$$s = \{\xi_{i,j} : i < j < n\}, \qquad T = \{s \cap A_i : B_{A_i} \in F\}.$$

CLAIM 17.21. *Let s, T be as above. If $R \in [[\kappa]^{<\aleph_0}]^{<\aleph_0}$ is such that $T \subseteq R$ and $s \cap A_j \notin R$ whenever $B_{A_j} \in G$, then*

$$\langle s, R\rangle \in \bigcap_{B_{A_i} \in F} B_{A_i} \cap \bigcap_{B_{A_j} \in G} (K \setminus B_{A_j}).$$

EXERCISE 17.36 (G). Prove Claim 17.21.

Since the set of R's that satisfy the assumption of Claim 17.21 has cardinality κ, it follows that $\mathcal{B}$ is strongly independent. □

The notions of independence and strong independence can be generalized as follows: Let $\lambda \leq \kappa$ be infinite cardinals. A family $\mathcal{A} \subseteq \mathcal{P}(\kappa)$ will be called λ-*independent* if for all pairs of disjoint $F, G \in [\mathcal{A}]^{<\lambda}$ we have:

$$C_{F,G} = \bigcap_{A \in F} A \cap \bigcap_{A \in G} (\kappa \setminus A) \neq \emptyset.$$

We say that $\mathcal{A}$ is *strongly λ-independent* if $|C_{F,G}| = \kappa$ for every pair $\langle F, G\rangle$ as above.

Of course, (strong) $\aleph_0$-independence is the same thing as (strong) independence.

THEOREM 17.22 (CH). *There exists a strongly $\aleph_1$-independent family $\mathcal{A} \subseteq \mathcal{P}(\omega_1)$ such that $|\mathcal{A}| = 2^{\aleph_1}$.*

PROOF. Let $K = [\omega_1]^{\leq \aleph_0} \times [[\omega_1]^{\leq \aleph_0}]^{\leq \aleph_0}$. If CH holds, then $|K| = \aleph_1$, and we can prove Theorem 17.22 by constructing a strongly independent family $\mathcal{B} \subseteq \mathcal{P}(K)$ of size $2^{\aleph_1}$.

Sounds familiar? So you know already what comes next.

EXERCISE 17.37 (G). Finish the proof of Theorem 17.22. □

So far, we have not given you an explicit example of an application of the Generalized Continuum Hypothesis. Let us include one for good measure.

THEOREM 17.23 (GCH). *For every infinite regular cardinal κ there exists a strongly κ-independent family $\mathcal{A} \subseteq \mathcal{P}(\kappa)$ such that $|\mathcal{A}| = 2^\kappa$.*

EXERCISE 17.38 (G). Prove Theorem 17.23.

Mathographical Remarks

If you would like to see more examples of applications of CH, we recommend W. Sierpiński's classical monograph *Hypothèse du Continu*, Monografie Matematyczne, Warsaw, 1934. This is a collection of 82 different consequences of CH and their proofs.

The inequality $add(\mathcal{N}) \leq add(\mathcal{M})$ was proved independently by T. Bartoszyński in *Additivity of measure implies additivity of category*, Transactions of the American Mathematical Society 281 (1984), 209–213, and by J. Raisonnier and J. Stern in *Mesurabilité et propriété de Baire*, Comptes Rendus, Série I Mathématique 296 (1983), 323–326. Consistency of the Borel Conjecture was proved by R. Laver in the paper *On the consistency of Borel's conjecture*, Acta Mathematica 137 (1976), 151–169. Consistency of the Dual Borel Conjecture was proved by T. Carlson. The proof is published in his paper *Strong measure zero and strongly meager sets*, Proceedings of the American Mathematical Society 118 (1993), 577–586. A solution of Exercise 17.22 can also be looked up in this paper (Theorem 5.11). Exercise 17.23 was a famous open problem for many years. It was finally shown by J. Pawlikowski in 1992 that Sierpiński sets are indeed strongly meager. By a recent result of Bartoszyński and Shelah, it is consistent that the strongly meager sets do not form an ideal. So you should not despair if you were unable to solve Exercise 17.24.

If you want to learn more about measure and category, you may be interested in one of the following items: The book *Measure and Category*, by J. Oxtoby, Springer Verlag, 1971, gives a good elementary introduction to the subject. If you are interested in all sorts of exotic subsets of the real line, we recommend A. W. Miller's article *Special subsets of the real line*, In: Handbook of Set-Theoretic Topology, K. Kunen and J. E. Vaughan, eds., North-Holland, 1984, 201–235. Among other things, this article contains a proof of Theorem 17.11 and a solution of Exercise 17.14. If you are familiar with forcing and curious to learn more about the ideals $\mathcal{M}$ and $\mathcal{N}$, the monograph *Set Theory. On the Structure of the Real Line*, by T. Bartoszyński and H. Judah, A K Peters, 1995, will give you all the information you could possibly want. For example, Corollary 8.1.4 of that book provides a solution of Exercise 17.13(a).

CHAPTER 18

From the Rasiowa-Sikorski Lemma to Martin's Axiom

Martin's Axiom is a powerful tool that has found many applications both inside and outside of set theory. It is not a theorem of ZFC, but it is relativly consistent with ZFC. It can be thought of as a generalization of the Rasiowa-Sikorski Lemma, which is provable in ZFC. This section is organized as follows: First we state and prove the Rasiowa-Sikorski Lemma. Then we show how one can transform, step by step, a proof involving a recursive construction into one using the Rasiowa-Sikorski Lemma. Next we introduce Martin's Axiom and show how it can be applied to derive a more general result from the transformed argument. Finally, we discuss variants of the c.c.c. as well as some consistent and some inconsistent modifications of Martin's Axiom.

Let $\langle \mathbb{P}, \leq \rangle$ be a p.o., and let $\mathcal{D}$ be a family of dense[1] subsets of $\mathbb{P}$. A filter $\mathbb{G}$ in $\mathbb{P}$ is *$\mathcal{D}$-generic* if $D \cap \mathbb{G} \neq \emptyset$ for all $D \in \mathcal{D}$.

LEMMA 18.1 (Rasiowa-Sikorski). *Let $\langle \mathbb{P}, \leq \rangle$ be a p.o., and let $\mathcal{D}$ be a countable family of dense subsets of $\mathbb{P}$. Then there exists a $\mathcal{D}$-generic filter in $\mathbb{P}$.*

PROOF. Let $\mathcal{D} = \{D_i : i < \omega\}$. We construct recursively a sequence $(p_i)_{i<\omega}$ of elements of $\mathbb{P}$ with the following properties:

(i) $p_j \leq p_i$ for all $i < j < \omega$;
(ii) For each $i < \omega$ there exists $q_i \in D_i$ with $p_i \leq q_i$.

We let p_0 be an arbitrary element of D_0. Assume that $0 < n < \omega$ and that the p_i's have already been constructed for $i < n$. Since D_n is dense in $\mathbb{P}$, there exists $p \in D_n$ with $p \leq p_{n-1}$. We choose some such p as our p_n. The filter generated by $\{p_i : i < \omega\}$ is as required. □

Recursive constructions as in the proof of Lemma 18.1 occur quite frequently in mathematical arguments. In many cases it would be possible to replace such constructions by applications of the Rasiowa-Sikorski Lemma. This is not commonly done though, for at least two reasons, one sound and one regrettable. The sound reason is that since the proof of Lemma 18.1 is so simple, a reference to the Lemma would not usually decrease the overall length of an argument. In fact, as we shall see in a moment, this approach may actually result in a longer proof. The other reason is that many mathematicians find it easier to work with recursive constructions than with dense subsets of p.o.'s. This is unfortunate, because recursive constructions have a more limited scope of application. For example, it is not possible to use a procedure as in the proof of Lemma 18.1 when one is trying

---

[1]This notion and other relevant concepts pertaining to p.o.'s were defined in Chapter 13.

to construct a countable object that satisfies uncountably many conditions. An approach using partial orders may still work in such cases, provided there exists a filter that meets all the uncountably many dense sets involved. Martin's Axiom postulates the existence of such filters under certain circumstances. In order to take full advantage of the opportunities this creates, one needs to be at ease with expressing properties of an object constructed from a filter in terms of dense subsets of a relevant p.o. The aim of the next example is to demonstrate, step by step, how a recursive construction can be translated into an argument involving partial orders.

An almost disjoint family A of infinite subsets of ω is *maximal* (or, a *mad family*) if for each infinite $b \subseteq \omega$ there exists $a \in A$ such that $|a \cap b| = \aleph_0$. There are finite mad families; e.g., $\{\omega\}$ is one. The following theorem shows that all infinite mad families are uncountable.

THEOREM 18.2. *Let A be a denumerable almost disjoint family of infinite subsets of ω. Then there exists an infinite $d \subseteq \omega$ such that $|d \cap a| < \aleph_0$ for all $a \in A$.*

This theorem can be proved by a recursive construction as follows:

PROOF 1. Let $A = \{a_i : i < \omega\}$. We construct recursively a sequence $(x_i)_{i<\omega}$ of elements of ω so that
$$x_i \in \omega \setminus (\bigcup_{k<i} a_k \cup \{x_k : k < i\})$$
for each $i \in \omega$. Since the set on the right hand side contains all but finitely many elements of the infinite set a_i, it is infinite, and thus in particular nonempty. Therefore, at each stage of the recursive construction we have a valid choice for x_i. Now let $d = \{x_i : i < \omega\}$. This d is as required, since $d \cap a_i \subseteq \{x_0, \ldots, x_i\}$ for each $i \in \omega$. □

Let us analyze the above reasoning. Our aim was to construct an object d that satisfies for each $i \in \omega$ the following condition:

(S^i) $d \cap a_i \subseteq \{x_0, \ldots, x_i\}$.

We are able to formulate the conditions to be met by d in this way because both the x_i's and the a_i's can be enumerated by the same set ω. But suppose we were trying to prove the following:

STATEMENT 18.3. *Let A be an almost disjoint family of size $\aleph_1$ of infinite subsets of ω. Then there exists an infinite $d \subseteq \omega$ such that $|d \cap a| < \aleph_0$ for all $a \in A$.*

In this case, the index set for the x_i's would still be ω, but for the a_i's we need $\aleph_1$ indices, so we cannot meaningfully express all the conditions on d in the form (S^i). But note that neither the order of the x_i's nor the order of the a_i's really matters in the proof of Theorem 18.2. We could rewrite it as follows:

PROOF 2. Let $\{F_i : i \in \omega\}$ be an increasing sequence of finite subsets of A such that $\bigcup_{i \in \omega} F_i = A$. We construct recursively a sequence $(s_i)_{i<\omega}$ of finite subsets of ω so that
$$s_i \subseteq s_{i+1} \land |s_i| \geq i \land s_{i+1} \setminus s_i \subseteq \omega \setminus \bigcup F_i$$
for each $i \in \omega$. Since the set on the right hand side contains all but finitely many elements of any $a \in A \setminus F_i$, it is infinite, and hence nonempty. Therefore, the rules

18. FROM THE RASIOWA-SIKORSKI LEMMA TO MARTIN'S AXIOM

of the game always allow us to put an extra element into s_{i+1} if need be. Now let $d = \bigcup_{i \in \omega} s_i$. This d is as required, since $d \cap a \subseteq s_i$ for each $a \in F_i$, the sets s_i are finite, and $A = \bigcup_{i \in \omega} F_i$. □

Of course, the difference between Proof 1 and Proof 2 is only notational, but the latter argument allows us to reformulate the requirements on d in a manner independent of their numbering: Let $d = \bigcup_{i \in \omega} s_i$. Then the following requirements should be met for each $j \in \omega$ and $a \in A$:

(B_j) There is some s_i with $|s_i| \geq j$;
(W^a) There is some F_i with $a \in F_i$.

Now you may suspect that one should be able to eliminate the subscripts i from the conditions B_j and W^a. As the next proof of Theorem 18.2 shows, this is indeed the case.

PROOF 3. Let $\mathbb{P} = \{\langle s, F\rangle : s \in [\omega]^{<\aleph_0}, F \in [A]^{<\aleph_0}\}$. We define a relation $\leq$ on $\mathbb{P}$ by:

$$\langle s, F\rangle \leq \langle s_1, F_1\rangle \text{ iff } s_1 \subseteq s, F_1 \subseteq F, \text{ and } (s \setminus s_1) \cap \bigcup F_1 = \emptyset.$$

Let us convince ourselves that $\leq$ is a partial order relation. Only transitivity requires an argument. Suppose $\langle s_0, F_0\rangle \leq \langle s_1, F_1\rangle$ and $\langle s_1, F_1\rangle \leq \langle s_2, F_2\rangle$. By transitivity of inclusion, $s_2 \subseteq s_0$ and $F_2 \subseteq F_0$. Moreover,

$$(s_0 \setminus s_2) \cap \bigcup F_2 = [(s_0 \setminus s_1) \cap \bigcup F_2] \cup [(s_1 \setminus s_2) \cap \bigcup F_2]$$

$$\subseteq [(s_0 \setminus s_1) \cap \bigcup F_1] \cup [(s_1 \setminus s_2) \cap \bigcup F_2] = \emptyset.$$

Thus $\langle s_0, F_0\rangle \leq \langle s_2, F_2\rangle$.

How do we want to use $\mathbb{P}$ in our proof? The idea is to specify a countable family $\mathcal{D}$ of dense subsets of $\mathbb{P}$ and pick a $\mathcal{D}$-generic filter $\mathbb{G} \subseteq \mathbb{P}$ whose existence is guaranteed by the Rasiowa-Sikorski Lemma. Then we shall put

$$d = \bigcup \{s : \exists F (\langle s, F\rangle \in \mathbb{G})\}$$

and show that d is as required.

For $j < \omega$ let

$$D_j = \{\langle s, F\rangle \in \mathbb{P} : |s| \geq j\}.$$

It is not hard to see that the D_j's correspond to the requirements (B_j). More precisely, we have:

CLAIM 18.4. *If $\mathbb{G} \cap D_j \neq \emptyset$, then $|d| \geq j$.*

PROOF. If $\langle s, F\rangle \in \mathbb{G} \cap D_j$, then $|s| \geq j$; and since $s \subseteq d$ we have $|d| \geq j$. □

CLAIM 18.5. *Each D_j is dense in $\mathbb{P}$.*

PROOF. Let $\langle s, F\rangle \in \mathbb{P}$. Since $|\omega \setminus \bigcup F| = \aleph_0$, there exists $t \subseteq \omega \setminus \bigcup F$ such that $|t| = j$. Then $\langle s \cup t, F\rangle \in D_j$, and $\langle s \cup t, F\rangle \leq \langle s, F\rangle$. □

For $a \in A$ let

$$D^a = \{\langle s, F\rangle \in \mathbb{P} : a \in F\}.$$

These sets are designed to take care of the requirements (W^a). More precisely:

CLAIM 18.6. *If $\mathbb{G} \cap D^a \neq \emptyset$, then $|d \cap a| < \aleph_0$.*

PROOF. Let $\langle s_0, F_0 \rangle \in \mathbb{G} \cap D^a$. Then for every $\langle s, F \rangle \in \mathbb{G}$ there exists an $\langle s_1, F_1 \rangle \in \mathbb{G}$ such that $\langle s_1, F_1 \rangle \leq \langle s, F \rangle$ and $\langle s_1, F_1 \rangle \leq \langle s_0, F_0 \rangle$. Since $(s_1 \setminus s_0) \cap \bigcup F_0 = \emptyset$, we also have $(s_1 \setminus s_0) \cap a = \emptyset$, which is equivalent to $s_1 \cap a \subseteq s_0$. Since $s \subseteq s_1$, we also have $s \cap a \subseteq s_0$. But s was the first coordinate of an arbitrary element of $\mathbb{G}$; therefore $d \cap a \subseteq s_0$, and thus $|d \cap a| < \omega$. □

CLAIM 18.7. *For each $a \in A$ the set D^a is dense in $\mathbb{P}$.*

PROOF. Let $\langle s, F \rangle \in \mathbb{P}$. Then $\langle s, F \cup \{a\} \rangle \in D^a$ and $\langle s, F \cup \{a\} \rangle \leq \langle s, F \rangle$. □

Now let

(18.1) $$\mathcal{D} = \{D_j : j < \omega\} \cup \{D^a : a \in A\}.$$

Since $\mathcal{D}$ is a countable set, the Rasiowa-Sikorski Lemma, in conjunction with Claims 18.5 and 18.7, tells us that there is a $\mathcal{D}$–generic filter $\mathbb{G} \subseteq \mathbb{P}$. If we construct $d \subseteq \omega$ from $\mathbb{G}$ as indicated above, then by Claims 18.4 and 18.6, d is as required. □

The last proof is much longer than each of the previous ones, but one could easily reuse it as a proof of Statement 18.3 (recycling is smart) *if only* there were a generalization of the Rasiowa-Sikorski Lemma to uncountable families of dense sets. Martin's Axiom, which we are now going to describe, is such a generalization.

Let κ be a cardinal. By MA(κ) we abbreviate the following statement:

(MA(κ))   Let $\langle \mathbb{P}, \leq \rangle$ be a p.o. that satisfies the c.c.c., and let $\mathcal{D}$ be a family of dense subsets of $\mathbb{P}$ with $|\mathcal{D}| \leq \kappa$. Then there exists a $\mathcal{D}$–generic filter in $\mathbb{P}$.

Martin's Axiom (abbreviated MA) is the following statement:

(MA)   MA(κ) holds for all $\kappa < 2^{\aleph_0}$.

Since MA($\aleph_0$) follows from the Rasiowa-Sikorski Lemma, the Continuum Hypothesis implies MA. Many authors use MA as an abbreviation for "MA + ¬CH." The negation of CH does not imply MA, but is consistent with it.

Now let us return to Statement 18.3. Note that the definition of $\mathbb{P}$ in Proof 3 of Theorem 18.2 does not depend on the size of A. Neither does the rest of the argument, except for the last two sentences. Since the family $\mathcal{D}$ defined in 18.1 will now be uncountable, the Rasiowa-Sikorski Lemma no longer applies. But $\langle \mathbb{P}, \leq \rangle$ satisfies the c.c.c. (see Fact 18.15 and Claim 18.16 below). Therefore, MA($\aleph_1$) implies Statement 18.3.

The latter observation can be generalized as follows: Let $\mathfrak{a}$ denote the minimum cardinality of an infinite mad family of subsets of ω. Then we have the following:

THEOREM 18.8. *Let κ be an infinite cardinal, and assume that MA(κ) holds. Then $\mathfrak{a} > \kappa$.*

Theorem 18.8 has an important consequence. By Exercise 9.20, there exists a mad family A. Since A is contained in $\mathcal{P}(\omega)$, its cardinality must be less than or equal to $2^{\aleph_0}$. Thus we get the following:

COROLLARY 18.9. MA($2^{\aleph_0}$) *fails.*

Now let us show that if we drop the countable chain condition in the formulation of MA($\aleph_1$), then we obtain a false statement.

DEFINITION 18.10. Let I, J be nonempty sets. Recall from Chapter 13 that $Fn(I, J)$ denotes the set of all functions p such that $|p| < \aleph_0$, $dom(p) \subseteq I$, $rng(p) \subseteq J$. When we speak of the *partial order* $F(I, J)$, the partial order relation is understood to be inverse inclusion $\supseteq$.

EXAMPLE 18.1. Consider $Fn(\omega, \omega_1)$. For $\alpha < \omega_1$, let $p_\alpha = \{(0, \alpha)\}$. Then $\{p_\alpha : \alpha < \omega_1\}$ is an antichain of cardinality $\aleph_1$ in $Fn(\omega, \omega_1)$, and it follows that $Fn(\omega, \omega_1)$ does not satisfy the c.c.c.

Now consider the sets $D_\alpha = \{p \in Fn(\omega, \omega_1) : \alpha \in rng(p)\}$, where $\alpha < \omega_1$. If $q \in Fn(\omega, \omega_1)$, then $q > q \cup \{\langle \min(\omega \setminus dom(q)), \alpha \rangle\} \in D_\alpha$. Thus, each D_α is a dense subset of $Fn(\omega, \omega_1)$. Let $\mathcal{D} = \{D_\alpha : \alpha < \omega_1\}$, and assume that $\mathbb{G}$ is a $\mathcal{D}$-generic filter.

EXERCISE 18.1 (G). Show that if $\mathbb{G}$ is as above, then $\bigcup \mathbb{G}$ is a function with $dom(\bigcup \mathbb{G}) \subseteq \omega$.

By $\mathcal{D}$-genericity of $\mathbb{G}$, if $\alpha < \omega_1$, then there exists some $p \in \mathbb{G} \cap D_\alpha$. This implies that $\alpha \in rng(\bigcup \mathbb{G})$. But then $rng(\bigcup \mathbb{G}) = \omega_1$, which is impossible.

Now let us discuss some techniques of showing that certain p.o.'s have the c.c.c. Let us start with a trivial case.

FACT 18.11. *Every countable p.o. has the c.c.c.*

DEFINITION 18.12. Let $\langle \mathbb{P}, \preceq \rangle$ be a p.o., and let $B \subseteq \mathbb{P}$. Then B is called *centered* if every finite $F \subseteq B$ has a lower bound[2] in $\mathbb{P}$. We call B *linked* if every two elements of B are compatible in $\mathbb{P}$.

Centered sets are always linked, but not vice versa.

EXERCISE 18.2 (PG). Consider $\langle \mathcal{P}(\omega) \setminus \emptyset, \supseteq \rangle$. Find a subset B of $\mathcal{P}(\omega) \setminus \emptyset$ that is linked but not centered in this p.o.

FACT 18.13. *If B is linked in a p.o. $\langle \mathbb{P}, \preceq \rangle$ and A is an antichain in $\langle \mathbb{P}, \preceq \rangle$, then $|A \cap B| \leq 1$.*

DEFINITION 18.14. A p.o. $\langle \mathbb{P}, \preceq \rangle$ is *σ-centered* (*σ-linked*) if $\mathbb{P}$ can be represented as $\mathbb{P} = \bigcup_{n \in N} \mathbb{P}_n$, where N is a countable set and each of the $\mathbb{P}_n$'s is centered (linked).

Note that we do not require that the $\mathbb{P}_n$'s are pairwise disjoint.

FACT 18.15. (a) *σ-centered p.o.'s satisfy the c.c.c.*
(b) *σ-linked p.o.'s satisfy the c.c.c.*

Many applications of MA use σ-centered p.o.'s. For example, consider the p.o. $\langle \mathbb{P}, \leq \rangle$ of Proof 3 of Theorem 18.2.

CLAIM 18.16. $\langle \mathbb{P}, \leq \rangle$ *is σ-centered.*

---

[2]I.e., there exists a $q \in \mathbb{P}$ such that $q \leq p$ for every $p \in F$.

PROOF. Let $s \in [\omega]^{<\aleph_0}$. If $F_0, \ldots, F_{k-1} \in [A]^{<\aleph_0}$, then $\langle s, F_0 \cup \ldots \cup F_{k-1}\rangle \leq \langle s, F_i\rangle$ for all $i < k$. Thus, the set $\mathbb{P}_s = \{\langle s, F\rangle : F \in [A]^{<\aleph_0}\}$ is centered. Since $\mathbb{P} = \bigcup_{s \in [\omega]^{<\aleph_0}} \mathbb{P}_s$ and $[\omega]^{<\aleph_0}$ is countable, the claim follows. □

Now we present an example of a σ-linked, not σ-centered p.o. Let μ_1 be one-dimensional Lebesgue measure, and let $\varepsilon > 0$. Define:

$$\mathbb{A}(\varepsilon) = \{U \subseteq \mathbb{R} : U \text{ is open in } \mathbb{R} \wedge \mu_1(U) < \varepsilon\}.$$

$\langle \mathbb{A}(\varepsilon), \supseteq\rangle$ is known among set theorists as *amoeba forcing*.

CLAIM 18.17. *For each $\varepsilon > 0$ the p.o. $\langle \mathbb{A}(\varepsilon), \supseteq\rangle$ is σ-linked.*

PROOF. Fix $\varepsilon > 0$ and let $\mathcal{B}$ be a countable open base for the topology on $\mathbb{R}$. For each $F \in [\mathcal{B}]^{<\aleph_0}$ let $\delta_F = \mu_1(\bigcup F)$, and define:

$$\mathbb{A}(\varepsilon)_F = \{U \in \mathbb{A}(\varepsilon) : \bigcup F \subseteq U \wedge \mu_1(U \setminus \bigcup F) < \frac{\varepsilon - \delta_F}{2}\}.$$

EXERCISE 18.3 (PG). (a) Show that $\mathbb{A}(\varepsilon) = \bigcup_{F \in [\mathcal{B}]^{<\aleph_0}} \mathbb{A}(\varepsilon)_F$.
(b) Show that each $\mathbb{A}(\varepsilon)_F$ is linked.
(c) Deduce Claim 18.17. □

EXERCISE 18.4 (PG). Fix $\varepsilon > 0$.
(a) Show that if B is a centered subset of $\langle \mathbb{A}(\varepsilon), \supseteq\rangle$, then $\bigcup B$ is open in $\mathbb{R}$ and $\mu_1(\bigcup B) \leq \varepsilon$.
(b) Show that if $\{B_n : n \in \omega\}$ is a countable family of centered subsets of $\mathbb{A}(\varepsilon)$, then there exists a $U \in \mathbb{A}(\varepsilon)$ such that $\mu_1(U \setminus \bigcup_{n \in \omega} B_n) > 0$ for every $n \in \omega$.
(c) Conclude that $\langle \mathbb{A}(\varepsilon), \supseteq\rangle$ is not σ-centered.

One can think of σ-linkedness as a particularly neat way of satisfying the c.c.c. Unfortunately, there are also much messier reasons for having the c.c.c. For example, let $I, J \neq \emptyset$, and consider the p.o. $Fn(I, J)$. For which I, J is $Fn(I, J)$ σ-centered or σ-linked? If J is uncountable, then $Fn(I, J)$ does not even satisfy the c.c.c. (see Example 18.1 for a special case). If $|J| = 1$, then $Fn(I, J)$ is simply centered, which is not very interesting. So the most interesting case occurs when $1 < |J| \leq \aleph_0$.

For $f : I \to J$, let $B_f = \{p \in Fn(I, J) : p \subseteq f\}$.

FACT 18.18. *Let $I, J \neq \emptyset$.*
(a) $p, q \in Fn(I, J)$ *are compatible iff* $p \cup q \in Fn(I, J)$.
(b) *Let $B \subseteq Fn(I, J)$. Then the following are equivalent:*
  (b1) B *is linked.*
  (b2) B *is centered.*
  (b3) $B \subseteq B_f$ *for some* $f \in {}^I J$.

EXERCISE 18.5 (G). Prove Fact 18.18. *Hint:* Use Exercise 13.7

Now consider J as a topological space with the discrete topology, note that ${}^I J = \prod_{i \in I} J$, and let τ be the product topology.

CLAIM 18.19. *Let $I, J \neq \emptyset$ and $F \subseteq {}^I J$. Then $Fn(I, J) = \bigcup\{B_f : f \in F\}$ if and only if F is a dense subset of $\langle {}^I J, \tau\rangle$.*

EXERCISE 18.6 (PG). Prove Claim 18.19.

COROLLARY 18.20. *Assume that $I \neq \emptyset$ and $1 < |J| \leq \aleph_0$. Then $Fn(I, J)$ is σ-centered if and only if $|I| \leq 2^{\aleph_0}$.*

PROOF. If J is as in the assumptions, then J is separable. Therefore, by the Hewitt-Marczewski-Pondiczery Theorem (Theorem 26.6) and by Theorem 26.7, $^I J$ is separable if and only if $|I| \leq 2^{\aleph_0}$. □

So what about $Fn(I, J)$ if $I > 2^{\aleph_0}$ and J is countable? Is this still a c.c.c. p.o.? Yes, but for a more subtle reason.

LEMMA 18.21. *Assume $I \neq \emptyset$, J is countable, and let A be an uncountable subset of $Fn(I, J)$. Then A contains an uncountable linked subset B.*

PROOF. Let I, J, A be as in the assumptions, and let $D = \{dom(p) : p \in A\}$. This is a family of finite subsets of I. Since for a given finite subset of I there are only countably many functions from I into J, the Pigeonhole Principle implies that D is uncountable. By the Δ-System Lemma, there is an uncountable Δ-system $D_1 \subseteq D$. Let r be the root of D_1, and for each function $p : r \to J$, let

$$A \lceil p = \{q \in A : dom(q) \in D_1 \land q \lceil r = p\}.$$

Since there are only countably many p's in $^r J$, applying the Pigeonhole Principle again, we conclude that some $A \lceil p$ must be uncountable. Now the following observation concludes the proof of Lemma 18.21:

FACT 18.22. *For each $p \in Fn(I, J)$, the set $A \lceil p$ is linked.*

EXERCISE 18.7 (G). Prove Fact 18.22. (Note that for some p, the set $A \lceil p$ may be empty, but the empty set is linked by default.) □

The property expressed by Lemma 18.21 deserves its own name.

DEFINITION 18.23. (a) A p.o. $\langle \mathbb{P}, \preceq \rangle$ has *Property K* (or *the Knaster property*) if every uncountable subset A of $\mathbb{P}$ contains an uncountable linked subset.

(b) A p.o. $\langle \mathbb{P}, \preceq \rangle$ has *precaliber $\aleph_1$* if every uncountable subset A of $\mathbb{P}$ contains an uncountable centered subset.

EXERCISE 18.8 (G). (a) Convince yourself that Precaliber $\aleph_1 \Rightarrow$ Property K $\Rightarrow$ c.c.c.

(b) Convince yourself that if J is countable, then $Fn(I, J)$ has precaliber $\aleph_1$.

Property K behaves nicely with respect to products.

LEMMA 18.24. *Let $\langle \mathbb{P}_0, \leq_0 \rangle$, $\langle \mathbb{P}_1, \leq_1 \rangle$ be p.o.'s.*

(a) *If both $\langle \mathbb{P}_0, \leq_0 \rangle$ and $\langle \mathbb{P}_1, \leq_1 \rangle$ have Property K, so does $\langle \mathbb{P}_0 \times \mathbb{P}_1, \leq_0 \otimes^s \leq_1 \rangle$.*

(b) *If $\langle \mathbb{P}_0, \leq_0 \rangle$ has Property K and $\langle \mathbb{P}_1, \leq_1 \rangle$ satisfies the c.c.c., then $\langle \mathbb{P}_0 \times \mathbb{P}_1, \leq_0 \otimes^s \leq_1 \rangle$ also satisfies the c.c.c.*

PROOF. We prove only point (b). Let $\langle \mathbb{P}_0, \leq_0 \rangle$ and $\langle \mathbb{P}_1, \leq_1 \rangle$ be as in the assumptions of (b), and let A be an uncountable subset of $\mathbb{P}_0 \times \mathbb{P}_1$. We show that A is not an antichain in $\langle \mathbb{P}_0 \times \mathbb{P}_1, \leq_0 \otimes^s \leq_1 \rangle$.

By Property K, we can choose an uncountable subset B of A such that the set of first coordinates of elements of B is linked in $\langle \mathbb{P}_0, \leq_0 \rangle$. Since $\langle \mathbb{P}_1, \leq_1 \rangle$ satisfies the c.c.c., the set of second coordinates of elements of B does not form an uncountable antichain in $\langle \mathbb{P}_1, \leq_1 \rangle$. Therefore, we can find $\langle p_0, p_1 \rangle, \langle p_0', p_1' \rangle \in B$ such that p_1 and p_1' are compatible in the sense of $\leq_1$. (In particular, this will be the case if $p_1 = p_1'$.) By the choice of B, p_0 and p_0' are compatible in the sense of $\leq_0$. Therefore $\langle p_0, p_1 \rangle$ and $\langle p_0', p_1' \rangle$ are compatible in the sense of $\leq_0 \otimes^s \leq_1$, and we are done. □

EXERCISE 18.9 (G). Prove Lemma 18.24(a).

Do all c.c.c. p.o.'s have Property K? Not necessarily. If $\langle T, \leq_T^* \rangle$ is a Suslin tree with the inverted order, then $\langle T, \leq_T^* \rangle$ has the c.c.c., but $\langle T \times T, \leq_T^* \otimes^s \leq_T^* \rangle$ does not (see Theorem 14.15). Thus, by Lemma 18.24, $\langle T, \leq_T^* \rangle$ does not have Property K. On the other hand, as we will show in the next chapter, MA + ¬CH implies that every c.c.c. p.o. has Property K.

Mathographical Remarks

Various modifications of Martin's Axiom have been proposed. For example, if P is a property of p.o.'s, then MA_P is the statement: "If $\langle \mathbb{P}, \leq \rangle$ is a c.c.c. p.o. that has Property P, and if $\mathcal{D}$ is a family of dense subsets of $\mathbb{P}$ with $|\mathcal{D}| < 2^{\aleph_0}$, then there exists a $\mathcal{D}$-generic filter in $\mathbb{P}$." We have:

$$MA \to MA_{\text{Property K}} \to MA_{\sigma-\text{linked}} \to MA_{\sigma-\text{centered}} \to MA_{\text{countable}}.$$

One can show that none of the above implications can be reversed, and that even $MA_{\text{countable}}$ is not a theorem of ZFC.

Although Example 18.1 shows that one cannot drop the c.c.c. altogether, it is possible to strengthen MA by extending it to *some* p.o.'s that do not satisfy the c.c.c. The most important strengthening in this direction is the *Proper Forcing Axiom*, which is commonly abbreviated by PFA. It is obtained by taking our definition of $MA(\aleph_1)$ and replacing the phrase "p.o. that satisfies the c.c.c." by "proper p.o." This is not a proper place to define the notion of a proper p.o.; let us just remark that c.c.c. p.o.'s are proper and, for example, the set of all countable partial functions from ω_1 into ω partially ordered by reverse inclusion is proper, but does not satisfy the c.c.c. PFA is equiconsistent with the existence of certain large cardinals. Somewhat surprisingly, PFA implies $2^{\aleph_0} = \aleph_2$. ($MA(\aleph_1)$ only implies that $2^{\aleph_0} \geq \aleph_2$.) J. E. Baumgartner's article *Applications of the Proper Forcing Axiom*, in the Handbook of Set-Theoretic Topology, K. Kunen and J. E. Vaughan, editors, North-Holland, 1984, is a good place to start learning more about PFA.

CHAPTER 19

Martin's Axiom

19.1. MA essentials

We begin with the historically first[1] application of Martin's Axiom.

THEOREM 19.1. $\mathrm{MA}(\aleph_1) \to \mathrm{SH}$.

PROOF. Suppose towards a contradiction that $\mathrm{MA}(\aleph_1)$ holds and that $\langle S, \leq_S \rangle$ is a Suslin tree. By Exercise 14.33, the dual p.o. $\langle S, \leq_S^* \rangle$ satisfies the c.c.c. By Lemma 14.24 we may assume without loss of generality that $\langle S, \leq_S \rangle$ is tall. This means that for each $\alpha < \omega_1$ the set $D_\alpha = S \setminus S_{(\alpha)}$ is dense in $\langle S, \leq_S^* \rangle$. Thus, by $\mathrm{MA}(\aleph_1)$, there exists a filter $\mathbb{G}$ in $\langle S, \leq_S^* \rangle$ that intersects each of the sets D_α. The following claim yields a contradiction.

CLAIM 19.2. $\mathbb{G}$ is an uncountable chain in $\langle S, \leq_S \rangle$.

PROOF. If $\mathbb{G}$ were countable, then $\mathbb{G} \subseteq S_{(\alpha)}$ for some $\alpha < \omega_1$. This is impossible, since $\mathbb{G} \cap D_\alpha \neq \emptyset$.

So it remains to show that $\mathbb{G}$ is a chain. Let $s, t \in \mathbb{G}$. Since $\mathbb{G}$ is a filter in $\langle S, \leq_S^* \rangle$, there is an $r \in S$ such that $r \leq_S^* s, t$. In other words, $s \leq_S r$ and $t \leq_S r$, i.e., $\{s, t\} \subseteq \hat{r}$. Since $\langle S, \leq_S \rangle$ is a tree, $\hat{r}$ is a chain, and hence s and t are $\leq_S$-comparable. □□

EXERCISE 19.1 (PG). Show that CH implies the following two statements:

(a) If $\mathcal{A}, \mathcal{B} \in [[\omega]^{\aleph_0}]^{<2^{\aleph_0}}$ are such that for each $a \in \mathcal{A}$ and each finite subset E of $\mathcal{B}$ we have $|a \setminus \bigcup E| = \aleph_0$, then there exists a set $d \subseteq \omega$ with

$$|a \cap d| = \aleph_0 \text{ for all } a \in \mathcal{A} \text{ and } |b \cap d| < \aleph_0 \text{ for all } b \in \mathcal{B}.$$

(b) If $\mathcal{A}, \mathcal{B} \in [[\omega]^{\aleph_0}]^{<2^{\aleph_0}}$ are such that $|a \cap b| < \aleph_0$ for each $a \in \mathcal{A}$ and $b \in \mathcal{B}$, then there exists a set $d \subseteq \omega$ with

$$|a \setminus d| < \aleph_0 \text{ for all } a \in \mathcal{A} \text{ and } |b \cap d| < \aleph_0 \text{ for all } b \in \mathcal{B}.$$

In the next chapter we will show that the statement in Exercise 19.1(b) is inconsistent with the negation of CH (Theorem 20.8). But in Exercise 19.1(a), CH can be replaced by MA.

LEMMA 19.3 (Solovay's Lemma). *Assume* $\mathrm{MA}_{\sigma\text{-centered}}$. *If* $\mathcal{A}, \mathcal{B} \in [[\omega]^{\aleph_0}]^{<2^{\aleph_0}}$ *are such that for each* $a \in \mathcal{A}$ *and each finite subset* E *of* $\mathcal{B}$ *we have* $|a \setminus \bigcup E| = \aleph_0$, *then there exists a set* $d \subseteq \omega$ *with*

$$|a \cap d| = \aleph_0 \text{ for all } a \in \mathcal{A} \text{ and } |b \cap d| < \aleph_0 \text{ for all } b \in \mathcal{B}.$$

---

[1] In a sense, Theorem 19.1 is even older than Martin's Axiom itself. In their paper *Iterated Cohen extensions and Souslin's problem*, Annals of Mathematics 94 (1971) 201–245, R. Solovay and S. Tennenbaum had given a consistency proof of SH. Upon seeing their proof, D. A. Martin realized that the argument showed something more general, namely the consistency of what would later become known as Martin's Axiom.

PROOF. This proof is similar to Proof 3 of Theorem 18.2. Let $\mathcal{A}, \mathcal{B}$ be as in the assumption, and assume $\text{MA}_{\sigma\text{-centered}}$. Define:

$$\mathbb{P} = \{\langle s, E\rangle : s \in [\omega]^{<\aleph_0}, E \in [\mathcal{B}]^{<\aleph_0}\},$$

$\langle s, E\rangle \leq \langle t, F\rangle$ if and only if $t \subseteq s, F \subseteq E$, and $(s \setminus t) \cap \bigcup F = \emptyset$.

We show that $\leq$ is a p.o. relation on $\mathbb{P}$. Only transitivity requires a proof.

Let $\langle s, E\rangle \leq \langle t, F\rangle$ and $\langle t, F\rangle \leq \langle u, G\rangle$. Then $u \subseteq s$ and $G \subseteq E$. In order to show that $(s \setminus u) \cap \bigcup G = \emptyset$, note that $(s \setminus u) \cap \bigcup G \subseteq ((s \setminus t) \cap \bigcup G) \cup ((t \setminus u) \cap \bigcup G)$. The second term $(t \setminus u) \cap \bigcup G$ is empty since $\langle t, F\rangle \leq \langle u, G\rangle$. Moreover, $F \supseteq G$; thus the first term $(s \setminus t) \cap \bigcup G$ is contained in $(s \setminus t) \cap \bigcup F$, and since $\langle s, E\rangle \leq \langle t, F\rangle$, this is again the empty set.

In order to see that $\langle \mathbb{P}, \leq\rangle$ is σ-centered, let $s \in [\omega]^{<\aleph_0}; E, F \in [\mathcal{B}]^{<\aleph_0}$. Then $\langle s, E \cup F\rangle \leq \langle s, E\rangle, \langle s, F\rangle$, and it follows that for each $s \in [\omega]^{<\aleph_0}$ the set $\mathbb{P}_s = \{\langle s, E\rangle : E \in [\mathcal{B}]^{<\aleph_0}\}$ is centered. Since the set $[\omega]^{<\aleph_0}$ is countable, we are done.

For $n \in \omega$, $a \in \mathcal{A}$, and $b \in \mathcal{B}$ we define:

$$D_{n,a} = \{\langle s, E\rangle \in \mathbb{P} : |a \cap s| \geq n\},$$
$$D_b = \{\langle s, E\rangle \in \mathbb{P} : b \in E\}.$$

EXERCISE 19.2 (G). Show that the assumptions about $\mathcal{A}$ and $\mathcal{B}$ imply that $D_{a,n}$ and D_b are dense in $\mathbb{P}$ for all $a \in \mathcal{A}$, $n \in \omega$, and $b \in \mathcal{B}$.

Now let $\mathcal{D} = \{D_{n,a} : n \in \omega, a \in \mathcal{A}\} \cup \{D_b : b \in \mathcal{B}\}$. This $\mathcal{D}$ is a family of dense subsets of $\mathbb{P}$ with $|\mathcal{D}| < 2^{\aleph_0}$. Let $\mathbb{G}$ be a $\mathcal{D}$-generic filter, and define:

$$d = \bigcup\{s : \exists E\,(\langle s, E\rangle \in \mathbb{G})\}.$$

Let us show that d is as required.

Fix $a \in \mathcal{A}$, $n \in \omega$. If $\langle s, E\rangle \in \mathbb{G} \cap D_{n,a}$, then $d \cap a \supseteq s \cap a$ and $|s \cap a| \geq n$; hence $|d \cap a| \geq n$. Thus $|d \cap a| \geq n$ for all $n \in \omega$, i.e., $|d \cap a|$ is infinite.

Now let $b \in \mathcal{B}$ and $\langle s, E\rangle \in \mathbb{G} \cap D_b$. Consider an arbitrary $\langle t, F\rangle \in \mathbb{G}$. There exists a common lower bound $\langle s', E'\rangle$ of $\langle s, E\rangle$ and $\langle t, F\rangle$. Then $(s' \setminus s) \cap \bigcup E = \emptyset$, and $b \in E$. Hence $(s' \setminus s) \cap b = \emptyset$, and thus $s \cap b = s' \cap b$. Since $t \cap b \subseteq s' \cap b$ and $\langle t, F\rangle$ was assumed an arbitrary element of $\mathbb{G}$, this in turn implies, in view of the definition of d, that $d \cap b = s \cap b$. Thus $|d \cap b| < \aleph_0$. □

Solovay's Lemma can be used to code subsets of κ for $\kappa < 2^{\aleph_0}$ by subsets of ω. This technique allows us to derive an important consequence of Martin's Axiom for cardinal exponentiation.

THEOREM 19.4 ($\text{MA}_{\sigma\text{-centered}}$). If $\kappa < 2^{\aleph_0}$, then $2^\kappa \leq 2^{\aleph_0}$.

PROOF. Assume $\text{MA}_{\sigma\text{-centered}}$ and let $\kappa < 2^{\aleph_0}$. Let C be an almost disjoint family of subsets of ω such that $|C| = \kappa$. The existence of such a family is an easy consequence of Theorem 17.18. For each $B \subseteq C$ pick $d_B \subseteq \omega$ such that $|d_B \cap a| = \aleph_0$ for $a \in C \setminus B$ and $|d_B \cap a| < \aleph_0$ for $a \in B$. Solovay's Lemma implies the existence of suitable d_B's.

EXERCISE 19.3 (G). Convince yourself that $B \neq B'$ implies that $d_B \neq d_{B'}$.

Since $|\mathcal{P}(C)| = 2^\kappa$, Exercise 19.3 implies that the set $\{d_B : B \subseteq C\}$ has cardinality 2^κ. But the latter set is contained in $\mathcal{P}(\omega)$, and thus has cardinality at most $2^{\aleph_0}$. The inequality $2^\kappa \leq 2^{\aleph_0}$ follows. □

COROLLARY 19.5 ($\text{MA}_{\sigma\text{-centered}}$). $2^{\aleph_0}$ is a regular cardinal.

PROOF. Suppose not and let $\lambda < 2^{\aleph_0}$ be the cofinality of $2^{\aleph_0}$. By Theorem 19.4, $2^\lambda = 2^{\aleph_0}$, so $cf(2^\lambda) = \lambda$, which contradicts König's Theorem (Corollary 11.24). □

Next we are going to discuss several equivalent formulations of Martin's Axiom. The first of these is topological. Consider the following statement.

(MA_{top})   No compact Hausdorff space with the c.c.c. is
a union of less than $2^{\aleph_0}$ nowhere dense subsets.

LEMMA 19.6. MA → MA_{top}.

PROOF. Let X be a compact Hausdorff space, and let $\mathbb{P}$ be the family of all closed subsets $F \subseteq X$ such that F has nonempty interior. If X is a c.c.c. topological space, then $\langle \mathbb{P}, \subseteq \rangle$ also satisfies the c.c.c. Let $\kappa < 2^{\aleph_0}$, and let $\{N_\alpha : \alpha < \kappa\}$ be a family of nowhere dense subsets of X. For $\alpha < \kappa$, let $D_\alpha = \{F \in \mathbb{P} : F \cap N_\alpha = \emptyset\}$. Then D_α is dense in $\langle \mathbb{P}, \subseteq \rangle$. To see this, consider $F \in \mathbb{P}$. Since N_α is nowhere dense in X, we can pick a nonempty open $U \subseteq F$ such that $U \cap N_\alpha = \emptyset$. Let $x \in U$. By regularity of X, there is a closed neighborhood V of x such that $V \subseteq U$. Clearly, $V \in D_\alpha$, and $V \subseteq F$.

Now assume MA holds, and let $\mathbb{G}$ be a filter in $\langle \mathbb{P}, \subseteq \rangle$ such that $\mathbb{G} \cap D_\alpha \neq \emptyset$ for each $\alpha < \kappa$. Then $\mathbb{G}$ is a family of closed subsets of X with the finite intersection property (it is even true that $\bigcap H$ has nonempty interior for every $H \in [\mathbb{G}]^{<\omega}$). By compactness of X, there exists some $x \in \bigcap \mathbb{G}$. Clearly, $x \notin \bigcup_{\alpha < \kappa} N_\alpha$. □

Recall that $cov(\mathcal{M})$ is the minimum number of meager sets that are needed to cover the real line. Note that $\mathbb{R}$ is homeomorphic to the open interval $(0, 1)$, and if $\mathcal{A}$ is a covering of $(0, 1)$ by meager sets, then $\mathcal{A} \cup \{\{0, 1\}\}$ is a a covering of the compact space $[0, 1]$ by meager sets. Thus, by Lemma 19.6, MA implies that $cov(\mathcal{M}) = 2^{\aleph_0}$. Even more is true.

EXERCISE 19.4 (G). Show that $\text{MA}_{\text{countable}} \to cov(\mathcal{M}) = 2^{\aleph_0}$. Hint: Let $\mathbb{P} = \{[a, b] : 0 \leq a < b \leq 1, a, b \in \mathbb{Q}\}$, and mimic the proof of Lemma 19.6.

Now let us consider two versions of Martin's Axiom for Boolean algebras.

(MA_{Ba})   If B is a Boolean algebra that satisfies the c.c.c.,
and if $\mathcal{D}$ is a family of dense subsets of B with $|\mathcal{D}| < 2^{\aleph_0}$,
then there exists a $\mathcal{D}$-generic filter in B.

(MA_{cBa})   If B is a complete Boolean algebra that satisfies the c.c.c.,
and if $\mathcal{D}$ is a family of dense subsets of B with $|\mathcal{D}| < 2^{\aleph_0}$,
then there exists a $\mathcal{D}$-generic filter in B.

Recall from Section 13.3 that if B is a Boolean algebra, then its Stone space $st\, B$ is the topological space $\langle Ult\, B, \tau \rangle$, where $Ult\, B$ is the set of all ultrafilters in B and τ is the topology generated by the base $\mathcal{U} = \{U_a : a \in B\}$, where $U_a = \{F \in Ult\, B : a \in F\}$. We have shown in Section 13.3 that Stone spaces of Boolean algebras are compact Hausdorff spaces.

EXERCISE 19.5 (PG). Let B be a Boolean algebra. Show that a set $D \subseteq B^+$ is dense in B if and only if the set $D^* = \{F \in Ult\, B : F \cap D \neq \emptyset\}$ is dense and open in $st\, B$.

Now suppose $\mathrm{MA_{top}}$ holds, let B be a Boolean algebra that satisfies the c.c.c., and let $\mathcal{D}$ be a family of fewer than $2^{\aleph_0}$ dense subsets of B. By Exercise 19.5, for each $D \in \mathcal{D}$ the set $Ult\,B \backslash D^*$ is a nowhere dense subset of $st\,B$. Since $st\,B$ is a compact Hausdorff space that satisfies the c.c.c., there exists $F \in Ult\,B$ such that $F \in D^*$ for all $D \in \mathcal{D}$. Then $F \cap D \neq \emptyset$ for all $D \in \mathcal{D}$, i.e., F is $\mathcal{D}$-generic. We have proved the following:

LEMMA 19.7. $\mathrm{MA_{top}} \to \mathrm{MA_{Ba}}$.

Clearly, $\mathrm{MA_{Ba}}$ implies $\mathrm{MA_{cBa}}$. The latter statement in turn implies the original version of Martin's Axiom.

LEMMA 19.8. $\mathrm{MA_{cBa}} \to \mathrm{MA}$.

PROOF. Assume $\mathrm{MA_{cBa}}$ and let $\kappa < 2^{\aleph_0}$. We want to show that $\mathrm{MA}(\kappa)$ holds. Consider the following statement:

($\mathrm{MA}^-(\kappa)$) Let $\langle \mathbb{P}, \leq \rangle$ be a c.c.c. p.o. of cardinality $\leq \kappa$, and let $\mathcal{D}$ be a family of dense subsets of $\mathbb{P}$ with $|\mathcal{D}| \leq \kappa$. Then there exists a $\mathcal{D}$-generic filter in $\mathbb{P}$.

$\mathrm{MA}^-(\kappa)$ appears to be weaker than $\mathrm{MA}(\kappa)$, but in Chapter 21 (Theorem 21.26) we will show that $\mathrm{MA}(\kappa)$ and $\mathrm{MA}^-(\kappa)$ are equivalent statements. Thus, in order to prove $\mathrm{MA}(\kappa)$, we can restrict our attention to p.o.'s of cardinality $\leq \kappa$.

So assume that $\langle \mathbb{P}, \leq \rangle$ is a c.c.c. p.o. such that $|\mathbb{P}| \leq \kappa$; let $\lambda \leq \kappa$, and consider a family $\mathcal{D} = \{D_\alpha : \alpha < \lambda\}$ of dense subsets of $\mathbb{P}$. By Theorem 13.34, there exist a complete Boolean algebra B and a function $i : \mathbb{P} \to B^+$ such that

(i) $rng(i)$ is dense in B^+;
(ii) $\forall p, q \in \mathbb{P}\,(p \leq q \to i(p) \leq_B i(q))$;
(iii) $\forall p, q \in \mathbb{P}\,(p \perp q \leftrightarrow i(p) \perp_B i(q))$.

Fix such i, B, and for $\alpha < \lambda$, let $i[D_\alpha]$ be denoted by E_α.

EXERCISE 19.6 (G). Convince yourself that B satisfies the c.c.c. and each E_α is dense in B.

By $\mathrm{MA_{cBa}}$, there exists an ultrafilter $\mathbb{F}$ in B such that $\mathbb{F} \cap E_\alpha \neq \emptyset$ for each $\alpha < \lambda$. Define $\mathbb{G} = i^{-1}[\mathbb{F}]$. It appears that we have found a $\mathcal{D}$-generic ultrafilter in $\mathbb{P}$: Clearly, $\mathbb{G} \cap D_\alpha \neq \emptyset$ for each $\alpha < \lambda$. If $p \in \mathbb{G}$ and $p \leq q$, then $i(p) \in \mathbb{F}$ and by (ii), $i(p) \leq_B i(q)$. Since $\mathbb{F}$ is a filter, $i(q) \in \mathbb{F}$, and hence $q \in \mathbb{G}$. Similarly, (iii) implies that the elements of $\mathbb{G}$ are pairwise compatible in $\mathbb{P}$. But this does not quite suffice: The definition of a filter requires that if $p, q \in \mathbb{G}$, then there exists a lower bound for p and q in $\mathbb{G}$, not just in $\mathbb{P}$.

EXERCISE 19.7 (G). Assume that i satisfies the following strengthening of (ii):

(ii)$^+$ $\forall p, q \in \mathbb{P}\,(p \leq q \leftrightarrow i(p) \leq_B i(q))$.

Show that in this case $\mathbb{G}$ is a $\mathcal{D}$-generic ultrafilter in $\mathbb{P}$.

Unfortunately, in general i satisfies just condition (ii), not (ii)$^+$. But by using a trick we can make sure that $\mathbb{G}$ is as required.

For $p \in \mathbb{P}$, let $D_p = \{r \in \mathbb{P} : r \leq p \vee r \perp p\}$.

EXERCISE 19.8 (G). (a) Show that for every $p \in \mathbb{P}$ the set D_p is dense and open in $\mathbb{P}$. (Recall that the latter means that $\forall q, r \in \mathbb{P}\, (q \in D_p \wedge r \leq q \to r \in D_p)$).
(b) Show that for all $p, q \in \mathbb{P}$ the intersection $D_p \cap D_q$ is dense in $\mathbb{P}$.

Here comes the trick: Since $|\mathbb{P}| \leq \kappa$, we may assume without loss of generality that for all $p, q \in \mathbb{P}$, the set $D_p \cap D_q$ is an element of $\mathcal{D}$! Now let $\mathbb{G}$ be as above, let $p, q \in \mathbb{G}$, and let $r \in \mathbb{G} \cap D_p \cap D_q$. Since $\mathbb{G}$ consists of pairwise compatible elements, we must have $r \leq p, q$. Thus, p and q have a lower bound in $\mathbb{G}$, and we are done. □

We have proved the following:

THEOREM 19.9. *Each of the statements* $\mathrm{MA}_{\mathrm{top}}$, $\mathrm{MA}_{\mathrm{Ba}}$, *and* $\mathrm{MA}_{\mathrm{cBa}}$ *is equivalent to Martin's Axiom.*

EXERCISE 19.9 (R). Show that $\mathrm{MA}_{\mathrm{countable}}$ is equivalent to $cov(\mathcal{M}) = 2^{\aleph_0}$. *Hint:* Use Corollary 25.4.

Now let us strengthen Theorem 19.1 by showing that $\mathrm{MA}(\aleph_1)$ implies that every Aronszajn tree is not only not Suslin, but even special.

THEOREM 19.10. *If* $\mathrm{MA}(\aleph_1)$ *holds, then every Aronszajn tree is special.*

PROOF. Assume $\mathrm{MA}(\aleph_1)$, and let $\langle T, \leq_T \rangle$ be an Aronszajn tree. We want to construct a function $f : T \to \omega$ such that $f(t) = f(s)$ implies that s and t are incomparable with respect to $\leq_T$. In order to use $\mathrm{MA}(\aleph_1)$, we need to work with a suitable c.c.c. partial order $\langle \mathbb{P}, \leq \rangle$. The elements of $\mathbb{P}$ will be approximations to our function f. More precisely, let $\mathbb{P}$ denote the set of all functions p such that

(i) $dom(p) \in [T]^{<\aleph_0}$;
(ii) $rng(p) \subseteq \omega$;
(iii) If $s, t \in dom(p)$ and $s <_T t$, then $p(s) \neq p(t)$.

We partially order $\mathbb{P}$ by reverse inclusion, i.e., we define:

$$p \leq q \text{ if and only if } q \subseteq p.$$

LEMMA 19.11. $\langle \mathbb{P}, \leq \rangle$ *satisfies the c.c.c.*

Before we prove the lemma, let us show how it implies Theorem 19.10. For $t \in T$ let

$$D_t = \{p \in \mathbb{P} : t \in dom(p)\}.$$

Each D_t is dense in $\mathbb{P}$. To see this, let $t \in T$ and consider $p \in \mathbb{P} \backslash D_t$. Pick $n \in \omega \backslash rng(p)$, and let $q = p \cup \{\langle t, n \rangle\}$. Then $q < p$ and $q \in D_t$.

Now consider the family

$$\mathcal{D} = \{D_t : t \in T\}.$$

This is a family of $\aleph_1$ dense subsets of $\mathbb{P}$, and by $\mathrm{MA}(\aleph_1)$, there exists a $\mathcal{D}$-generic filter $\mathbb{G}$. Let

$$f = \bigcup \mathbb{G}.$$

Then f is a function from T into ω and for each $n \in \omega$ the inverse image $f^{-1}\{n\}$ is an antichain in T: Otherwise, there would be s, t, n and $p, q \in \mathbb{P}$ with $s <_T t$, $\langle s, n \rangle \in p$, $\langle t, n \rangle \in q$. By the definition of a filter, there exists $r \in \mathbb{G}$ such that $r \leq s$ and $r \leq t$, but no such r can satisfy condition (iii) of the definition of $\mathbb{P}$.

It remains to prove Lemma 19.11.

PROOF OF LEMMA 19.11. Let D be an uncountable subset of $\mathbb{P}$. By the Δ-System Lemma, there exist $u \in [T]^{<\aleph_0}$ and an uncountable $C \subseteq D$ such that for all $\{p, q\} \in [C]^2$:
$$dom(p) \cap dom(q) = u.$$

By the Pigeonhole Principle, we can shrink C further to an uncountable subset $B \subseteq C$ such that $p\lceil u = q\lceil u$ for all $p, q \in B$.

If $p, q \in B$, can we assert that p and q are compatible? Not necessarily. The union $p \cup q$ will be a function that satisfies (i) and (ii), but if there are $s \in dom(p) \setminus u$ and $t \in dom(q) \setminus u$ such that $p(s) = q(t)$ and $s <_T t$, then condition (iii) is violated. This observation also works the other way round: If every $s \in dom(p) \setminus u$ is $\leq_T$-incomparable with every $t \in dom(q) \setminus u$, then $p \cup q \in \mathbb{P}$, and it follows that p and q are compatible. Thus, if we apply the next claim to the family $A = \{dom(p) \setminus u : p \in B\}$, Lemma 19.11 and hence Theorem 19.10 follow.

CLAIM 19.12. *Let $\langle T, \leq_T \rangle$ be an Aronszajn tree, A an uncountable family of pairwise disjoint finite subsets of T. There exists $\{a, b\} \in [A]^2$ such that if $s \in a$ and $s' \in b$, then s and s' are $\leq_T$-incomparable.*

PROOF. Assume the claim is false and let A be a counterexample. By the Pigeonhole Principle we may assume without loss of generality that all $a \in A$ have the same cardinality n. For each $a \in A$ choose an enumeration
$$a = \{a(0), a(1), \ldots, a(n-1)\}.$$

Our aim is to find an uncountable subset E of A and $s_e \in e$ for each $e \in E$ such that the set $\{s_e : e \in E\}$ consists of pairwise comparable nodes of T. This will yield a contradiction, since T was assumed to be an Aronszajn tree and thus not to contain uncountable chains. Note that if $n = 1$, we can simply take $E = A$. If $n > 1$, then we will use the following trick: We will find an uncountable subset $E \subseteq A$ and $s_c \in c$ for $c \in E$ such that if $d, e \in E$, then the set
$$S(d, e) = \{t \in T : t \text{ is comparable with both } s_d \text{ and } s_e\}$$
is uncountable. Note that $S(d, e) \subseteq \{s_e, s_d\} \cup \hat{s}_e \cup \hat{s}_d \cup \{t \in T : \{s_e, s_d\} \subseteq \hat{t}\}$. Since the set $\{s_e, s_d\} \cup \hat{s}_e \cup \hat{s}_d$ is countable, it follows that whenever $S(e, d)$ is uncountable, there exists $t \in T$ such that $\{s_e, s_d\} \subseteq \hat{t}$. Since $\hat{t}$ is wellordered by $\leq_T$, the latter implies that s_e and s_d are comparable.

So let us construct a set E and nodes s_e as above. Fix a uniform ultrafilter $\mathcal{F}$ on A. For $k < n$ and $s \in T$, let
$$B(s, k) = \{a \in A : s \text{ is comparable with } a(k)\}.$$

Since every two elements $a, b \in A$ contain comparable nodes, we have for each $a \in A$:
$$A = \{a\} \cup \bigcup_{s \in a} \{B(s, k) : k < n\}.$$

Thus for each $a \in A$ we find $s_a \in a$ and $k_a < n$ such that $B(s_a, k_a) \in \mathcal{F}$. Moreover, there exists $k^* < n$ such that the set
$$E = \{a \in A : k_a = k^*\}$$
is uncountable. For $d, e \in E$ we have
$$S(d, e) \supseteq \{a(k^*) : a \in B(s_d, k^*) \cap B(s_e, k^*)\}.$$

19.1. MA ESSENTIALS

Note that the map that assigns $a(k^*)$ to $a \in A$ is an injection. Since $\mathcal{F}$ is uniform and $B(s_d, k^*) \cap B(s_e, k^*) \in \mathcal{F}$, the set on the right hand side is uncountable, as required. □□□

THEOREM 19.13 (MA($\aleph_1$)). *Every c.c.c. partial order has precaliber $\aleph_1$.*

EXERCISE 19.10 (PG). Theorem 19.13 can be used to give yet another proof of Theorem 19.1. Can you see how?

PROOF OF THEOREM 19.13. Assume MA($\aleph_1$), let $\langle \mathbb{P}, \leq \rangle$ be a p.o. that satisfies the c.c.c., and let A be an uncountable subset of $\mathbb{P}$. We want to show that A contains an uncountable centered subset. Without loss of generality we may assume that $|A| = \aleph_1$. Note that every subset of a filter in $\mathbb{P}$ is centered. Thus, it suffices to find a filter $\mathbb{F}$ in $\mathbb{P}$ such that the intersection $\mathbb{F} \cap A$ is uncountable. In order to be able to apply MA($\aleph_1$), we need to find a family $\mathcal{D}$ of $\leq \aleph_1$ dense subsets of $\mathbb{P}$, or perhaps of a suitable c.c.c. suborder of $\mathbb{P}$, such that $\mathcal{D}$-genericity of $\mathbb{F}$ ensures that $\mathbb{F} \cap A$ is uncountable.

The most natural impulse is to try working with the p.o. $\langle \mathbb{P}, \leq \rangle$ itself. However, this may prove futile: For example, it could happen that $\mathbb{F}$ contains some p such that $p \perp q$ for every $q \in A$. In this case no amount of genericity of $\mathbb{F}$ will force $\mathbb{F}$ to intersect A, let alone uncountably often. So we need to consider a suitable subset of $\mathbb{P}$ instead. As a first step towards finding such a subset, define:

$$B = \{p \in \mathbb{P} : |\{q \in A : p \not\perp q\}| \leq \aleph_0 \vee \forall r \leq p (|\{q \in A : r \not\perp q\}| \geq \aleph_1)\}.$$

By Zorn's Lemma, there exists an antichain C in $\mathbb{P}$ that is maximal with respect to the property that $C \subseteq B$. Let us show that C is actually a maximal antichain in $\langle \mathbb{P}, \leq \rangle$: Suppose not, and let $p \in \mathbb{P}$ be such that $p \perp p'$ for all $p' \in C$. Then $p \notin B$; hence there exists $r < p$ such that $|\{q \in A : r \not\perp q\}| \leq \aleph_0$. But then $r \in B$ and $C \cup \{r\}$ is an antichain in $\mathbb{P}$ that extends C, which is ruled out by the choice of C.

Since C is maximal, every $q \in A$ must be compatible with some $p \in C$. By the c.c.c., C is countable. On the other hand, A is uncountable, and hence the Pigeonhole Principle implies that there exists $p_0 \in C$ such that p_0 is compatible with uncountably many $q \in A$. Since $p_0 \in B$, the latter implies that

$$\forall r \leq p_0 (|\{q \in A : r \not\perp q\}| \geq \aleph_1).$$

Now let p_0 be as above, and define $\mathbb{P}_0 = \{p \in \mathbb{P} : p \leq p_0\}$. Then $\langle \mathbb{P}_0, \leq \rangle$ is also a c.c.c. partial order. Enumerate A as $\{q_\alpha : \alpha < \omega_1\}$. For each $\alpha < \omega_1$ define:

$$D_\alpha = \{r \leq p_0 : \exists \beta > \alpha \, (r \leq q_\beta)\}.$$

EXERCISE 19.11 (G). Convince yourself that each of the sets D_α is dense in $\langle \mathbb{P}_0, \leq \rangle$.

Now apply MA($\aleph_1$) to find a filter $\mathbb{F}_0$ in $\langle \mathbb{P}_0, \leq \rangle$ that meets D_α for every $\alpha < \omega_1$. Let $\mathbb{F}$ be the filter in $\mathbb{P}$ generated by $\mathbb{F}_0$, i.e., let $\mathbb{F} = \{p \in \mathbb{P} : \exists r \in \mathbb{F}_0 \, (r \leq p)\}$. Then for every $\alpha < \omega_1$ there exists $\beta > \alpha$ such that $q_\beta \in \mathbb{F} \cap A$. Hence $\mathbb{F} \cap A$ is an uncountable centered subset of A, as required. □

Since precaliber $\aleph_1$ implies Property K, Lemma 18.24 and Theorem 19.13 yield the following:

COROLLARY 19.14 (MA($\aleph_1$)). *The simple product of any two c.c.c. p.o.'s satisfies the c.c.c. In particular, the product of any two topological spaces with the c.c.c. satisfies the c.c.c.*

Corollary 19.14 in turn gives together with Theorem 16.4:

COROLLARY 19.15 (MA($\aleph_1$)). *The Tychonoff product of any nonempty family of topological spaces with the c.c.c. satisfies the c.c.c.*

Corollary 19.14 can also be used to prove the following:

THEOREM 19.16 (MA($\aleph_1$)). *Every c.c.c. partial order of cardinality $\aleph_1$ is σ-centered.*

PROOF. Assume MA($\aleph_1$) and let $\langle \mathbb{P}, \leq \rangle$ be a c.c.c. partial order such that $|\mathbb{P}| = \aleph_1$. Let $\mathbb{Q} = {}^{<\omega}\mathbb{P}$, and define a p.o. relation $\preceq$ on $\mathbb{Q}$ as follows:

$q_1 \preceq q_0$ if and only if $dom(q_0) \subseteq dom(q_1) \wedge \forall i \in dom(q_0)\, (q_1(i) \leq q_0(i))$.

EXERCISE 19.12 (G). Show that that the p.o. $\langle \mathbb{Q}, \preceq \rangle$ satisfies the c.c.c. *Hint:* Note that for every uncountable $A \subseteq \mathbb{Q}$ there must be $n \in \omega$ such that $A \cap {}^n\mathbb{P}$ is uncountable. Then use Corollary 19.14 and induction over $n \in \omega$.

Now define for $p \in \mathbb{P}$:

$$D_p = \{q \in \mathbb{Q} : \exists i \in dom(q)\, (q(i) \leq p)\}.$$

It is not hard to see that each of the sets D_p is dense in $\langle \mathbb{Q}, \preceq \rangle$. Since $|\mathbb{P}| = \aleph_1$ and $\mathbb{Q}$ satisfies the c.c.c., MA($\aleph_1$) implies that there exists an ultrafilter $\mathbb{F}$ in $\mathbb{Q}$ such that $\mathbb{F}$ meets each of the sets D_p. Fix such $\mathbb{F}$ and define for $i \in \omega$:

$$\mathbb{P}_i = \{q(i) : q \in \mathbb{F}\}.$$

Then each of the sets $\mathbb{P}_i$ is centered and $\mathbb{P} = \bigcup_{i \in \omega} \mathbb{P}_i$. □

19.2. MA and cardinal invariants of the continuum

In this section we shall examine what Martin's Axiom implies about certain cardinal invariants of the continuum. One result in this direction was already proved in Chapter 18. Recall that

$\mathfrak{a} = \min\{|A| : A$ is an infinite maximal almost disjoint family of subsets of $\omega\}$.

In this new terminology, Theorem 18.2 asserts that $\mathfrak{a} > \aleph_0$, and Statement 18.3 translates into the inequality $\mathfrak{a} > \aleph_1$.

EXERCISE 19.13 (G). Convince yourself that Theorem 18.8 remains true if "MA(κ)" is replaced by "MA$_{\sigma\text{-centered}}(\kappa)$."

If $a, b \subseteq \omega$, then we write $a \subseteq^* b$ as shorthand for $|a \backslash b| < \aleph_0$. If $a \subseteq^* b$, then we say that a is *almost contained in* b. We say that a and b are *almost equal*, and write $a =^* b$, if $a \subseteq^* b$ and $b \subseteq^* a$. We write $a \subset^* b$ if $a \subseteq^* b$ and $a \neq^* b$. A subset b of ω is called a *pseudo-intersection* of a family $\mathcal{C} \subseteq \mathcal{P}(\omega)$ if b is infinite and $b \subseteq^* c$ for all $c \in \mathcal{C}$. We say that a family $\mathcal{C} \subseteq \mathcal{P}(\omega)$ has the *strong finite intersection property (sfip)* if $|\bigcap H| = \aleph_0$ for all $H \in [\mathcal{C}]^{<\aleph_0}$. Clearly, if $\mathcal{C}$ has a pseudo-intersection, then $\mathcal{C}$ has the sfip. However, the converse is not true.

EXERCISE 19.14 (G). Suppose $\mathcal{C}$ is a nonprincipal ultrafilter on ω. Show that $\mathcal{C}$ has the sfip but no pseudo-intersection.

Let us define our next cardinal invariant:

$\mathfrak{p} = \min\{|\mathcal{C}| : \mathcal{C} \subseteq [\omega]^{\aleph_0} \wedge \mathcal{C}$ has the sfip but no pseudo-intersection$\}$.

EXERCISE 19.15 (PG). Show that $\mathfrak{p} > \aleph_0$.

THEOREM 19.17. $\text{MA}_{\sigma\text{-centered}} \to \mathfrak{p} = 2^{\aleph_0}$.

PROOF. Exercise 19.14 implies that $\mathfrak{p}$ is well-defined and $\leq 2^{\aleph_0}$. So we only need to prove that every family $\mathcal{C} \subseteq [\omega]^{\aleph_0}$ with the sfip and of cardinality less than $2^{\aleph_0}$ has a pseudo-intersection. Assume $\text{MA}_{\sigma\text{-centered}}$, and fix $\mathcal{C}$ as above. Consider the families $\mathcal{A} = \{\omega\}$ and $\mathcal{B} = \{\omega \setminus c : c \in \mathcal{C}\}$. Note that the sfip of $\mathcal{C}$ implies that for every finite $E \subseteq \mathcal{B}$ we have $|\omega \setminus \bigcup E| = \aleph_0$. Hence the assumptions of Solovay's Lemma (Lemma 19.3) are satisfied and there exists $d \subseteq \omega$ such that d is infinite and $|d \cap b| < \aleph_0$ for all $b \in \mathcal{B}$. The latter means that $d \subseteq^* c$ for all $c \in \mathcal{C}$. Hence d is a pseudo-intersection of $\mathcal{C}$, as desired. □

A κ-*tower* is a sequence $\langle c_\xi : \xi < \kappa \rangle$ of elements of $[\omega]^{\aleph_0}$ such that $c_\xi \subset^* c_\eta$ for all $\eta < \xi < \kappa$ and the family $\{c_\xi : \xi < \kappa\}$ has no pseudo-intersection. We define:

$$\mathfrak{t} = \min\{\kappa : \text{there exists a } \kappa\text{-tower}\}.$$

EXERCISE 19.16 (G). (a) Convince yourself that for some $\kappa \leq 2^{\aleph_0}$ there exists a κ-tower.

(b) Convince yourself that $\mathfrak{t}$ is a regular cardinal.

Let us put down the following observation for later reference.

CLAIM 19.18. $\mathfrak{p} \leq \mathfrak{t}$.

One can use Theorem 19.17, Claim 19.18, and the next theorem to give an alternative proof of Theorem 19.4.

THEOREM 19.19 (Rothberger). *If $\kappa < \mathfrak{t}$, then $2^\kappa \leq 2^{\aleph_0}$.*

PROOF. Let $\kappa < \mathfrak{t}$. For $\alpha \leq \kappa$ and $f \in {}^\alpha 2$ we will construct sets $a_f \in [\omega]^{\aleph_0}$ such that:

(i) $\forall f, g \in {}^{\leq \kappa} 2 \, (f \subseteq g \to a_g \subseteq^* a_f)$;
(ii) $\forall f \in {}^{<\kappa} 2 \, (a_{f^\frown 0} \cap a_{f^\frown 1} = \emptyset)$.

EXERCISE 19.17 (G). Assume that the a_f's are constructed in such a way that (i) and (ii) hold. Convince yourself that if $\{f, g\} \in [{}^\kappa 2]^2$, then $|a_f \cap a_g| < \aleph_0$ and thus $a_f \neq a_g$. Infer that $2^\kappa \leq 2^{\aleph_0}$.

The a_f's can be constructed by recursion over α. If a_f has been constructed, we let $\{a_{f^\frown 0}, a_{f^\frown 1}\}$ be a partition of a_f into infinite disjoint subsets. Now suppose α is a limit ordinal $\leq \kappa$, $f \in {}^\alpha 2$, and the $a_{f \restriction \beta}$'s have been defined for all $\beta < \alpha$. Then $\langle a_{f \restriction \beta} : \beta < \alpha \rangle$ is a $\subseteq^*$-decreasing sequence of infinite subsets of ω of length $< \mathfrak{t}$. Thus, it is not a tower, and we can find a pseudo-intersection a_f of $\{a_{f \restriction \beta} : \beta < \alpha\}$. This construction yields a_f's that satisfy (i) and (ii). □

Now let us show that if $\text{MA}_{\sigma\text{-centered}}$ holds, then the cardinal invariants $\mathfrak{b}, \mathfrak{d}$, and $\mathfrak{s}$ that were introduced in Chapter 17 are all equal to $2^{\aleph_0}$.

THEOREM 19.20. $\text{MA}_{\sigma\text{-centered}} \to \mathfrak{s} = 2^{\aleph_0}$.

PROOF. Assume $\text{MA}_{\sigma\text{-centered}}$, and let $A \subseteq [\omega]^{\aleph_0}$ be a family of size $< 2^{\aleph_0}$ such that $\omega \setminus a$ is infinite for all $a \in A$. We want to show that A is not a splitting family, i.e., we want to construct $x \in [\omega]^{\aleph_0}$ such that for all $a \in A$, either $|x \cap a| < \aleph_0$ or $|x \setminus a| < \aleph_0$.

As in most applications of Martin's Axiom, the crucial step in the proof is finding a suitable p.o. to work with. Let $\mathcal{F}$ be a nonprincipal ultrafilter on ω, and define:
$$\mathbb{P} = \{\langle s, X\rangle : s \in [\omega]^{<\aleph_0} \wedge X \in [A]^{<\aleph_0}\};$$
$$\langle s', X'\rangle \preceq \langle s, X\rangle \text{ iff } s \subseteq s' \wedge X \subseteq X' \wedge s'\backslash s \subseteq \bigcap(X \cap \mathcal{F})\backslash \bigcup(X\backslash\mathcal{F}).$$
The x we are looking for will be defined as:
$$x_\mathbb{G} = \bigcup\{s : \exists X (\langle s, X\rangle \in \mathbb{G})\},$$
where $\mathbb{G}$ is a $\mathcal{D}$-generic filter in $\mathbb{P}$ for a suitable family $\mathcal{D}$ of fewer than $2^{\aleph_0}$ dense subsets of $\mathbb{P}$. The choice of $\mathcal{D}$ is dictated by the properties we want x to have:

(1) For $a \in A$, we want $x_\mathbb{G}$ to be almost contained in a or almost disjoint from a, i.e., we want $x_\mathbb{G} \subseteq^* a$ or $x_\mathbb{G} \cap a =^* \emptyset$. If $a \in X \cap \mathcal{F}$ for some $\langle s, X\rangle \in \mathbb{G}$, then x will be almost contained in a. Similarly, if $a \in X\backslash\mathcal{F}$ for some $\langle s, X\rangle \in \mathbb{G}$, then x will be almost disjoint from a. Thus, for each $a \in A$, we will want the set $D_a = \{\langle s, X\rangle \in \mathbb{P} : a \in X\}$ to be in $\mathcal{D}$.

(2) We want $x_\mathbb{G}$ to be infinite. As in our earlier applications of MA, this can be assured by putting the sets $E_n = \{\langle s, X\rangle \in \mathbb{P} : |s| \geq n\}$ into $\mathcal{D}$.

This describes the idea of the proof. Now we need to verify that it works. Once you get the knack of it, this part of a proof involving Martin's Axiom will become utterly boring, and you will be tempted to skip it. Don't. Practically all mistakes in proofs involving Martin's Axiom occur because the author skips one of the more obvious steps in the following exercise.

EXERCISE 19.18 (G). Let $\mathbb{P}$, $\preceq$, D_a, E_n, $\mathbb{G}$, and $x_\mathbb{G}$ be as above.
(a) Show that $\preceq$ is a p.o. relation on $\mathbb{P}$.
(b) Show that $\langle\mathbb{P}, \preceq\rangle$ is σ-centered.
(c) Show that for all $a \in A$ the set D_a is dense in $\mathbb{P}$.
(d) Show that for all $n \in \omega$ the set E_n is dense in $\mathbb{P}$.
(e) Show that if $\mathbb{G}$ is a filter in $\mathbb{P}$ and $a \in A$ is such that $\mathbb{G} \cap D_a \neq \emptyset$, then $x_\mathbb{G}$ is almost contained in or almost disjoint from a.
(f) Show that if $\mathbb{G}$ is a filter in $\mathbb{P}$ such that $\mathbb{G} \cap E_n \neq \emptyset$ for all $n \in \omega$, then $|x_\mathbb{G}| = \aleph_0$.

Now let $\mathcal{D} = \{D_a : a \in A\} \cup \{E_n : n \in \omega\}$. The assumption on A implies that $|\mathcal{D}| < 2^{\aleph_0}$. By points (a)–(d) of Exercise 19.18, MA$_{\sigma\text{-centered}}$ applies, and there exists a $\mathcal{D}$-generic filter $\mathbb{G}$ in $\mathbb{P}$. By points (e) and (f) of Exercise 19.18, the corresponding $x_\mathbb{G}$ witnesses that A is not splitting. $\square$

THEOREM 19.21. MA$_{\text{countable}} \to \mathfrak{d} = 2^{\aleph_0}$.

PROOF. Assume MA$_{\text{countable}}$, and let $\mathcal{F} \subseteq {}^\omega\omega$ be such that $|\mathcal{F}| < 2^{\aleph_0}$. Consider the countable set $Fn(\omega, \omega)$ partially ordered by reverse inclusion. The idea is to find a family $\mathcal{D}$ of fewer than $2^{\aleph_0}$ dense subsets of this p.o. such that for every $\mathcal{D}$-generic ultrafilter $\mathbb{G}$ the set $\bigcup \mathbb{G}$ is a function from ω into ω that is not eventually dominated by any function $f \in \mathcal{F}$.

EXERCISE 19.19 (PG). Finish the proof of Theorem 19.21. $\square$

It is not possible to find a family $\mathcal{D}$ of dense subsets of the p.o. $Fn(\omega, \omega)$ such that $f <^* \bigcup \mathbb{G}$ for every $f \in \mathcal{F}$ and every $\mathcal{D}$-generic filter $\mathbb{G} \subseteq Fn(\omega, \omega)$. Thus, if we want to derive the equality $\mathfrak{b} = 2^{\aleph_0}$ from Martin's Axiom, we must adorn the elements $s \in Fn(\omega, \omega)$ with so-called *side constraints*.

THEOREM 19.22. $MA_{\sigma\text{-centered}} \to \mathfrak{b} = 2^{\aleph_0}$.

EXERCISE 19.20 (PG). Prove Theorem 19.22. *Hint:* For $\mathcal{F}$ as in the proof of the previous theorem, consider the set
$$\mathbb{P} = \{\langle s, F \rangle : s \in {}^{<\omega}\omega \wedge F \in [\mathcal{F}]^{<\aleph_0}\}$$
and the binary relation $\preceq$ defined by:
$$\langle s, F \rangle \preceq \langle t, G \rangle \text{ iff } t \subseteq s \wedge G \subseteq F \wedge \forall n \in dom(s) \backslash dom(t) \forall f \in G\, (s(n) > f(n)).$$
Show that $\langle \mathbb{P}, \preceq \rangle$ is a σ-centered p.o., and find a family $\mathcal{D}$ of fewer than $2^{\aleph_0}$ dense subsets of $\mathbb{P}$ such that for every $\mathcal{D}$-generic filter $\mathbb{G}$ in $\langle \mathbb{P}, \preceq \rangle$ the set $g_\mathbb{G}$ defined by $g_\mathbb{G} = \bigcup \{s : \exists F\, (\langle s, F \rangle \in \mathbb{G})\}$ is a function from ω into ω with $f <^* g_\mathbb{G}$ for all $f \in \mathcal{F}$.

Theorem 19.22 can be used to derive further consequences of Martin's Axiom.

THEOREM 19.23. $MA_{\sigma\text{-centered}} \to add(\mathcal{M}) = 2^{\aleph_0}$.

PROOF. Assume $MA_{\sigma\text{-centered}}$, and let $\{M_\xi : \xi < \lambda < 2^{\aleph_0}\}$ be a family of fewer than $2^{\aleph_0}$ meager subsets of $\mathbb{R}$. Define $M = \bigcup_{\xi < \lambda} M_\xi$. We want to show that M is a meager subset of $\mathbb{R}$. Since $MA_{\sigma\text{-centered}}$ implies $MA_{\text{countable}}$, it follows from Exercise 19.4 that $M \neq \mathbb{R}$; in fact, it even follows that $\mathbb{R} \backslash M$ is dense in $\mathbb{R}$. So, let D be a countable dense subset of $\mathbb{R} \backslash M$. Enumerate D as $(d_n)_{n \in \omega}$. Since every meager set is contained in a meager F_σ-set, we may without loss of generality assume that each M_ξ is an F_σ-set. For every $\xi < \lambda$, choose a sequence $(F_{\xi,n})_{n \in \omega}$ of nowhere dense closed sets such that $\bigcup_{n \in \omega} F_{\xi,n} = M_\xi$, and define a function $f_\xi : \omega \to \omega$ by $f_\xi(n) = \min\{k \in \omega : (d_n - 2^{-k}, d_n + 2^{-k}) \cap F_{\xi,n} = \emptyset\}$. Since $D \cap F_{\xi,n} = \emptyset$ for all $\xi < \lambda$ and $n < \omega$, these functions are well defined. By Theorem 19.22, there exists a function $f : \omega \to \omega$ such that $f_\xi <^* f$ for all $\xi < \lambda$. Fix such f, and define for $m \in \omega$ a set $U_m = \bigcup_{n \in \omega} (d_n - 2^{-f(n)-m}, d_n + 2^{-f(n)-m})$. Then each U_m is a dense open subset of $\mathbb{R}$, and hence the set $K = \mathbb{R} \backslash \bigcap_{m \in \omega} U_m$ is meager.

EXERCISE 19.21 (G). Show that for each $\xi < \lambda$ and $n < \omega$ there exists $m \in \omega$ such that $U_m \cap F_{\xi,n} = \emptyset$. Conclude that $M \subseteq K$. □

The proofs of Theorems 19.20–19.22 presented above gave us opportunities to practice applying Martin's Axiom, but they do not tell us much about how the cardinals $\mathfrak{a}, \mathfrak{b}, \mathfrak{d}, \mathfrak{p}, \mathfrak{s}$, and $\mathfrak{t}$ are related to each other. The next theorem fills this gap and also indicates an alternative way of deriving some of the above results.

THEOREM 19.24. $\aleph_1 \leq \mathfrak{p} \leq \mathfrak{t} \leq \min\{\mathfrak{b}, \mathfrak{s}\} \leq \max\{\mathfrak{b}, \mathfrak{s}\} \leq \mathfrak{d} \leq 2^{\aleph_0}$ and $\mathfrak{b} \leq \mathfrak{a} \leq 2^{\aleph_0}$.

PROOF. The inequality $\aleph_1 \leq \mathfrak{p}$ was established in Exercise 19.15, the inequality $\mathfrak{p} \leq \mathfrak{t}$ is Claim 19.18, the inequality $\mathfrak{b} \leq \mathfrak{d}$ is Exercise 17.18(a), the inequalities $\min\{\mathfrak{b}, \mathfrak{s}\} \leq \max\{\mathfrak{b}, \mathfrak{s}\}$, $\mathfrak{a} \leq 2^{\aleph_0}$, and $\mathfrak{d} \leq 2^{\aleph_0}$ are obvious. Let us prove the remaining inequalities.

$\mathfrak{t} \leq \mathfrak{s}$: Consider $S \subseteq [\omega]^{\aleph_0}$ such that $|S| = \kappa < \mathfrak{t}$. We show that S is not a splitting family. Enumerate S as $(a_\alpha)_{\alpha < \kappa}$. By transfinite recursion we construct a sequence $(b_\alpha)_{\alpha < \kappa}$ of infinite subsets of ω such that for all $\alpha < \beta < \kappa$:

(1) $b_\beta \subseteq^* b_\alpha$; and
(2) $b_\alpha \subseteq^* a_\alpha$ or $|b_\alpha \cap a_\alpha| < \aleph_0$.

This can be done as follows: Let $b_0 = a_0$. If $0 < \alpha < \kappa$ and b_γ has already been constructed for all $\gamma < \alpha$, then let c_α be a pseudo-intersection of $\{b_\gamma : \gamma < \alpha\}$. The existence of c_α is assured by the assumption that κ, and hence α, is less than $\mathfrak{t}$. If $c_\alpha \cap a_\alpha$ is infinite, we define $b_\alpha = c_\alpha \cap a_\alpha$. If not, we define $b_\alpha = c_\alpha$. In both cases, condition (2) will be satisfied.

Since $\kappa < \mathfrak{t}$, condition (1) implies that there exists a pseudo-intersection b of $\{b_\alpha : \alpha < \kappa\}$.

EXERCISE 19.22 (G). Convince yourself that any pseudo-intersection b of $\{b_\alpha : \alpha < \kappa\}$ witnesses that S is not a splitting family.

$\mathfrak{t} \leq \mathfrak{b}$: With every $a \in [\omega]^{\aleph_0}$ we associate an increasing function $f_a \in {}^\omega\omega$ by letting $f_a(n)$ be the $(n+1)$st element of a.

EXERCISE 19.23 (PG). (a) Convince yourself that if $f : \omega \to \omega$ is increasing, then $f_{rng(f)} = f$.
(b) Show that if $a \in [\omega]^{\aleph_0}$ and $g \in {}^\omega\omega$, then there exits $b \in [a]^{\aleph_0}$ such that $g <^* f_b$.
(c) Show that if $a \subset^* b$ then $f_a <^* f_b$.

Now assume towards a contradiction that $\mathfrak{b} < \mathfrak{t}$, and let $\{g_\alpha : \alpha < \mathfrak{b}\}$ be unbounded in ${}^\omega\omega$. By recursion over $\alpha < \mathfrak{b}$ we construct a sequence $\langle a_\alpha : \alpha \leq \mathfrak{b}\rangle$ of infinite subsets of ω such that for all $\alpha < \beta \leq \mathfrak{b}$:

(1) $g_\alpha <^* f_{a_{\alpha+1}}$;
(2) $|a_\alpha \backslash a_{\alpha+1}| = \aleph_0$;
(3) $a_\beta \subseteq^* a_\alpha$.

Exercise 19.23(b) implies that given a_α, we can find a subset $a_{\alpha+1}$ of a_α such that (1) and (2) hold. At limit stages $\beta \leq \mathfrak{b}$, the assumption $\mathfrak{b} < \mathfrak{t}$ implies that we can find a pseudo-intersection a_β of $\{a_\alpha : \alpha < \beta\}$. This, together with taking subsets at successor stages, will take care of requirement (3).

Now consider $f_{a_\mathfrak{b}}$. It follows from (1)–(3) and Exercise 19.23(c) that $g_\alpha <^* f_{a_\mathfrak{b}}$ for all $\alpha < \mathfrak{b}$. But this contradicts the assumption that the family $\{g_\alpha : \alpha < \mathfrak{b}\}$ is unbounded.

$\mathfrak{b} \leq \mathfrak{a}$: Let $A \subseteq [\omega]^{\aleph_0}$ be an almost disjoint family of size κ, where $\omega \leq \kappa < \mathfrak{b}$. We show that A cannot be maximal. Enumerate A as $(a_\alpha)_{\alpha<\kappa}$. For each $\alpha \in [\omega, \kappa)$, define a function $f_\alpha : \omega \to \omega$ by $f_\alpha(n) = \max(a_\alpha \cap a_n)$, where the maximum of the empty set is arbitrarily defined as 0. Since $\kappa < \mathfrak{b}$, there exists a strictly increasing function $f \in {}^\omega\omega$ such that $f_\alpha <^* f$ for all $\alpha \in [\omega, \kappa)$. Define:
$a_\kappa = \{\min(a_n \backslash (f(n) + 1 \cup a_0 \cup a_1 \cup \cdots \cup a_{n-1})) : n \in \omega\}$.

EXERCISE 19.24 (G). Show that a_κ is an infinite subset of ω that is almost disjoint from a_α for all $\alpha < \kappa$.

$\mathfrak{s} \leq \mathfrak{d}$: This is a good exercise.

EXERCISE 19.25 (PG). Finish the proof of Theorem 19.24 by showing that $\mathfrak{s} \leq \mathfrak{d}$. *Hint:* The same idea as in the proof that in Lemma 17.14 CH can be replaced by $\mathfrak{d} = \aleph_1$ works here. □

By Theorem 19.24, all the consequences of $\mathrm{MA}_{\sigma\text{-centered}}$ for cardinal invariants that we have proved so far already follow from the equality $\mathfrak{p} = 2^{\aleph_0}$. Does this equality perhaps imply $\mathrm{MA}_{\sigma\text{-centered}}$ itself?

THEOREM 19.25 (Bell). $\mathrm{MA}_{\sigma\text{-centered}}$ *is equivalent to the equality* $\mathfrak{p} = 2^{\aleph_0}$.

PROOF. We have already shown that $\mathrm{MA}_{\sigma\text{-centered}} \to \mathfrak{p} = 2^{\aleph_0}$. To prove the converse, assume $\mathfrak{p} = 2^{\aleph_0}$, let $\langle \mathbb{P}, \leq \rangle$ be a σ-centered p.o., and let $\mathcal{D} = \{D_\alpha : \alpha < \kappa\}$ be a family of dense subsets of $\mathbb{P}$, where $\kappa < 2^{\aleph_0}$. We want to show that there exists a filter $\mathbb{F} \subseteq \mathbb{P}$ with $\mathbb{F} \cap D_\alpha \neq \emptyset$ for all $\alpha < \kappa$.

EXERCISE 19.26 (PG). (a) Convince yourself that we may assume without loss of generality that

(1) $|\mathbb{P}| \leq \kappa$;
(2) Every D_α is open, i.e., $\forall p, q \in \mathbb{P}\, (p \in D_\alpha \wedge q \leq p \to q \in D_\alpha)$;
(3) For all $p, q \in \mathbb{P}$, the set $D_p \cap D_q \in \mathcal{D}$, where $D_p = \{r \in \mathbb{P} : r \leq p \vee r \perp p\}$.

Hint: Recall the proof of Lemma 19.8.

(b) Conclude that it suffices to show that there is a linked subset $L \subseteq \mathbb{P}$ with $L \cap D_\alpha \neq \emptyset$ for all $\alpha < \kappa$.

For the remainder of this proof, let $\mathbb{P}$ be as in Exercise 19.26, and fix a partition $\mathbb{P} = \bigcup_{m \in \omega} P_m$ of $\mathbb{P}$ into centered subsets. For $p \in \mathbb{P}$ and $\alpha < \kappa$ we define:

$$A_{p,\alpha} = \{m : \exists q \in P_m \cap D_\alpha\, (q \leq p)\}.$$

EXERCISE 19.27 (G). (a) Show that if $\alpha < \kappa$ and $q \leq p$, then $A_{q,\alpha} \subseteq A_{p,\alpha}$.

(b) Show that for every $m \in \omega$ the family $\mathcal{A}_m = \{A_{p,\alpha} : p \in P_m \wedge \alpha < \kappa\}$ has the finite intersection property.

Now suppose we have a set $C \in [\omega]^{\aleph_0}$ and a set $L = \{p_\alpha : \alpha < \kappa\}$ such that:

(A) $\forall \alpha < \kappa\, (p_\alpha \in D_\alpha)$; and
(B) $\forall \alpha < \kappa\, \exists m_\alpha \in C\, (p_\alpha \in P_{m_\alpha} \wedge C \backslash (m_\alpha + 1) \subseteq A_{p_\alpha,\alpha})$.

Then L will be as required: It follows from (A) that L intersects every element of $\mathcal{D}$. If $p_\alpha, p_\beta \in L$, then by (B), for $n \in C$ such that $n > \max\{m_\alpha, m_\beta\}$, there are $q_\alpha, q_\beta \in P_n$ such that $q_\alpha \leq p_\alpha$ and $q_\beta \leq p_\beta$. Since P_n is centered, q_α and q_β are compatible, and so are p_α, p_β. It follows that L is linked.

Are there sets C and L that satisfy (A) and (B)? Not necessarily. It may happen that the families $\mathcal{A}_m$ defined in Exercise 19.27(b) only have the finite intersection property, but not the strong finite intersection property. If this is the case, then we cannot expect to find an infinite C as above. Fortunately, in such a situation Theorem 19.25 has a rather simple proof.

CLAIM 19.26. *Suppose that there exists* $m \in \omega$ *such that the family* $\mathcal{A}_m$ *does not have the sfip. Then there exists* $n \in \omega$ *such that* P_n *intersects all* D_α's.

PROOF. Let $m \in \omega$ be as in the assumptions. It suffices to show that $\bigcap \mathcal{A}_m \neq \emptyset$. Suppose this were not the case, and let $\{p_0, \ldots, p_{k-1}\} \subseteq P_m$ and $\{\alpha_0, \ldots, \alpha_{k-1}\} \subset \kappa$ be such that $\bigcap_{i<k} A_{p_i,\alpha_i} = \{n_0, \ldots, n_{\ell-1}\}$. If $\bigcap \mathcal{A}_m = \emptyset$, then there are $\{\alpha_k, \ldots, \alpha_{k+\ell-1}\}$ and $\{p_k, \ldots, p_{k+\ell-1}\} \subseteq P_m$ such that $n_j \notin A_{p_j,\alpha_j}$ for $j < \ell$. But then $\bigcap_{i<k+\ell} A_{p_i,\alpha_i} = \emptyset$, contradicting Exercise 19.27(b). □

So let us assume from now on that $\mathcal{A}_m$ has the sfip for each $m \in \omega$. By recursion over $s \in {}^{<\omega}\omega$ we will construct $B_s \in [\omega]^{\aleph_0}$ and $p_{s,\alpha} \in \mathbb{P}$ as follows:

Let $B_\emptyset$ be a pseudo-intersection of $\mathcal{A}_0$. Given B_s and $n \in \omega$, let $B_{s \frown n}$ be a pseudo-intersection of $\mathcal{A}_n$. Such a pseudo-intersection exists since $|\mathcal{A}_n| \leq \kappa < \mathfrak{p}$ for all $n \in \omega$.

For $\alpha < \kappa$, let $p_{\emptyset, \alpha}$ be an arbitrary element of P_0. Suppose $\{p_{s,\alpha} : \alpha < \kappa\}$ is given. If $n \in A_{p_{s,\alpha},\alpha}$, then we let $p_{s \frown n, \alpha} \in P_n \cap D_\alpha$ be such that $p_{s \frown n, \alpha} < p_{s,\alpha}$; otherwise we let $p_{s \frown n, \alpha}$ be an arbitrary element of P_n.

Note that for every $s \in {}^{<\omega}\omega$ and $\alpha < \kappa$ the set $\{n \in B_s : \neg \exists q \in P_n \, (q \leq p_{s,\alpha})\}$ is finite. Thus, for each $\alpha < \kappa$ we can define a function $f_\alpha : {}^{<\omega}\omega \to \omega$ as follows:

$$f_\alpha(s) = \min\{n \in B_s : \forall m \in B_s \setminus n \, (m \in A_{p_{s,\alpha},\alpha})\}.$$

Since $\kappa < \mathfrak{p} \leq \mathfrak{b}$, there exists a function $g : {}^{<\omega}\omega \to \omega$ such that for each $\alpha < \kappa$ the set $\{s : f_\alpha(s) \geq g(s)\}$ is finite. Moreover, we can choose g in such a way that $g(s) \in B_s$ and $g(s) > \max(rng(s))$ for all $s \in {}^{<\omega}\omega$. Fix such g and define recursively an increasing sequence $(\ell_k)_{k \in \omega}$ of natural numbers by letting $\ell_k = g(\langle \ell_0, \ldots, \ell_{k-1} \rangle)$ (in particular, $\ell_0 = g(\emptyset)$). Now define $C = \{\ell_k : k \in \omega\}$. By the choice of g, for each α there exists $k \in \omega$ such that $f_\alpha(\langle \ell_0, \ldots, \ell_j \rangle) < g(\langle \ell_0, \ldots, \ell_j \rangle)$ for all $j \geq k \in \omega$. Let m_α be the smallest such k, and define $p_\alpha = p_{\langle \ell_0, \ldots, \ell_{m_\alpha} \rangle, \alpha}$. Finally, let $L = \{p_\alpha : \alpha < \kappa\}$.

EXERCISE 19.28 (G). Convince yourself that C and L satisfy conditions (A) and (B). □

Not all implications of Martin's Axiom for cardinal invariants follow from $\mathrm{MA}_{\sigma\text{-centered}}$. For example, in the next theorem, $\mathrm{MA}_{\sigma\text{-linked}}$ cannot be weakened to $\mathrm{MA}_{\sigma\text{-centered}}$.

THEOREM 19.27. $\mathrm{MA}_{\sigma\text{-linked}} \to add(\mathcal{N}) = 2^{\aleph_0}$.

PROOF. Assume $\mathrm{MA}_{\sigma\text{-linked}}$ holds, let $\kappa < 2^{\aleph_0}$, and let $\{A_\alpha : \alpha < \kappa\} \subseteq \mathcal{N}$. We want to show that $\bigcup_{\alpha < \kappa} A_\alpha \in \mathcal{N}$. It suffices to show that for every $\varepsilon > 0$ there exists an open set U_ε such that $\mu(U_\varepsilon) \leq \varepsilon$ and $\bigcup_{\alpha < \kappa} A_\alpha \subseteq U_\varepsilon$.

Fix $\varepsilon > 0$. In Chapter 18 we introduced the set $\mathbb{A}(\varepsilon) = \{U \subseteq \mathbb{R} : U$ is open in $\mathbb{R} \wedge \mu(U) < \varepsilon\}$, and showed that the p.o. $\langle \mathbb{A}(\varepsilon), \supseteq \rangle$ is σ-linked, but not σ-centered. For $\alpha < \kappa$, we define:

$$D_\alpha = \{U \in \mathbb{A}(\varepsilon) : A_\alpha \subseteq U\},$$

and we let $\mathcal{D} = \{D_\alpha : \alpha < \kappa\}$.

EXERCISE 19.29 (G). Convince yourself that $\mathcal{D}$ is a family of dense subsets of $\mathbb{A}(\varepsilon)$.

Now let $\mathbb{G}$ be a $\mathcal{D}$-generic filter in $\mathbb{A}(\varepsilon)$, and consider $W = \bigcup \mathbb{G}$. Obviously, W is open and $A_\alpha \subseteq W$ for all $\alpha < \kappa$. It remains to show that $\mu(W) \leq \varepsilon$. For the latter we need a technical fact.

CLAIM 19.28. Let W be as above. Then there exists a countable $G \subseteq \mathbb{G}$ such that $W = \bigcup G$.

PROOF. Consider the set $I_\mathbb{G} = \{[a,b] : a, b \in \mathbb{Q} \wedge a < b \wedge [a,b] \subseteq W\}$. Since W is an open subset of $\mathbb{R}$, we must have $W = \bigcup I_\mathbb{G}$. Every interval $[a,b] \in I_\mathbb{G}$ is compact, and hence contained in $U_0 \cup U_1 \cup \cdots \cup U_k$, where $U_i \in \mathbb{G}$ for $i \leq k$. But since $\mathbb{G}$ is a filter in the p.o. $\langle \mathbb{A}(\varepsilon), \supseteq \rangle$, the union $U_0 \cup U_1 \cup \cdots \cup U_k$ is also an element of $\mathbb{G}$.

19.2. MA AND CARDINAL INVARIANTS OF THE CONTINUUM

Thus, for each $[a,b] \in I_{\mathbb{G}}$ we can pick $U_{[a,b]} \in \mathbb{G}$ such that $[a,b] \subseteq U_{[a,b]}$. Since $I_{\mathbb{G}}$ is countable, so is the family $G = \{U_{[a,b]} : [a,b] \in I_{\mathbb{G}}\}$. And, of course, $\bigcup G = W$. □

Now let G, W be as above, and assume towards a contradiction that $\mu(W) > \varepsilon$. Then there exists a finite $G_1 \subseteq G$ such that $\mu(\bigcup G_1) \geq \varepsilon$. But since $\mathbb{G}$ is a filter, we would have $\bigcup G_1 \in \mathbb{G}$, which is ruled out by the definition of $\mathbb{A}(\varepsilon)$. □

We conclude this section with an example that is not a theorem about cardinal invariants (although it has a similar flavor). Since the p.o. in this example satisfies the c.c.c. for rather unconventional reasons, we will have to use the full force of MA.

THEOREM 19.29 (MA). *Let κ and λ be infinite cardinals such that $\kappa, \lambda < 2^{\aleph_0}$, and let $\mathcal{A} = \{a_\alpha : \alpha < \kappa\}$ be a family of countable subsets of λ such that $|a_\alpha \cap a_\beta| < \aleph_0$ for all $\alpha < \beta < \kappa$. Then there exists $B \in [\lambda]^\lambda$ such that $|B \cap a_\alpha| < \aleph_0$ for all $\alpha < \kappa$.*

PROOF. Let us first deal with the easy cases.

EXERCISE 19.30 (G). Show that for $\kappa = \omega$ Theorem 19.29 can be proved in ZFC, and for the case $\lambda = \omega$ only $\text{MA}_{\sigma\text{-centered}}$ is needed.

From now on, let us assume that $\kappa, \lambda > \omega$. Let $\mathcal{A}$ be as in the assumptions. We define a p.o. $\langle \mathbb{P}, \leq \rangle$ as follows:

$$\mathbb{P} = \{\langle s, F \rangle : F \in [\mathcal{A}]^{<\aleph_0} \wedge s \in [\lambda]^{<\aleph_0}\}.$$

The partial order relation is defined by:

$$\langle t, G \rangle \leq \langle s, F \rangle \quad \text{if and only if} \quad F \subseteq G \wedge s \subseteq t \wedge (t \backslash s) \cap \bigcup F = \emptyset.$$

EXERCISE 19.31 (G). Convince yourself that $\leq$ is a p.o. relation on $\mathbb{P}$.

For $\alpha < \kappa$ and $\beta < \lambda$ let

$$D_\alpha = \{\langle s, F \rangle : a_\alpha \in F\},$$
$$E_\beta = \{\langle s, F \rangle : s \cap [\beta, \beta + \omega_1) \neq \emptyset\}.$$

Define

$$\mathcal{D} = \{D_\alpha : \alpha < \kappa\} \cup \{E_\beta : \beta < \lambda\}.$$

EXERCISE 19.32 (G). (a) Show that D_α and E_β are dense subsets of $\mathbb{P}$ for all $\alpha < \kappa$ and $\beta < \lambda$.
(b) Show that if $\mathbb{G}$ is a $\mathcal{D}$-generic filter in $\mathbb{P}$, then the set

$$B = \bigcup\{s : \exists F (\langle s, F \rangle \in \mathbb{G})\}$$

is as required.

It remains to show that $\langle \mathbb{P}, \leq \rangle$ has the c.c.c. Consider an uncountable set $\{\langle s_\alpha, F_\alpha \rangle : \alpha < \omega_1\} \subseteq \mathbb{P}$. We want to show that there are $\alpha < \beta < \omega_1$ such that $\langle s_\alpha, F_\alpha \rangle$ and $\langle s_\beta, F_\beta \rangle$ are compatible. Let us first reduce the general case to a more manageable situation. By the Δ-System Lemma, we may assume without loss of generality that there exist $s \in [\lambda]^{<\aleph_0}$ and $F \in [\mathcal{A}]^{<\aleph_0}$ such that $s_\alpha \cap s_\beta = s$ and $F_\alpha \cap F_\beta = F$ for all $\alpha < \beta < \omega_1$. Note that if $\langle s_\alpha, F_\alpha \rangle$ and $\langle s_\beta, F_\beta \rangle$ are incompatible, then so are $\langle s_\alpha \backslash s, F_\alpha \rangle$ and $\langle s_\beta \backslash s, F_\beta \rangle$. Thus, we may assume without loss of generality that $s = \emptyset$, i.e., that the s_α's are pairwise disjoint. Since $\bigcup F$ is countable, only countably many of the pairwise disjoint s_α's can intersect $\bigcup F$. By removing these s_α's if necessary, we may assume that if $\langle s_\alpha, F_\alpha \rangle$ and $\langle s_\beta, F_\beta \rangle$

are incompatible, then so are $\langle s_\alpha, F_\alpha \backslash F \rangle$ and $\langle s_\beta, F_\beta \backslash F \rangle$. In other words, we may further simplify the situation by assuming that $F = \emptyset$. Finally, by the Pigeonhole Principle, we may assume that $|s_\alpha| = n$ for a fixed $n \in \omega$ and all $\alpha < \omega_1$. Consider the set $\bigcup_{i \in \omega} \bigcup F_i$. This set is countable, and by thinning out the sequence if necessary, we may without loss of generality assume that $s_\alpha \cap \bigcup_{i \in \omega} \bigcup F_i = \emptyset$ for all α with $\omega \leq \alpha < \omega_1$.

Now if $i < \omega \leq \alpha < \omega_1$, then either $\langle s_\alpha, F_\alpha \rangle$ and $\langle s_i, F_i \rangle$ are compatible, or $s_i \cap \bigcup F_\alpha \neq \emptyset$. We show that there are $i < \omega \leq \alpha < \omega_1$ such that the former happens. Consider the sets $C_k = \bigcup F_{\omega+k}$ for $k \leq n$. Since the set $D = \bigcup \{C_k \cap C_\ell : k < \ell \leq n\}$ is finite, there exists $i_0 < \omega$ with $s_{i_0} \cap D = \emptyset$. The latter property of i_0 implies that the family $\{C_k \cap s_{i_0} : k \leq n\}$ consists of $n+1$ pairwise disjoint sets. Since $|s_{i_0}| = n$, we can find $k_0 \leq n$ such that $C_{k_0} \cap s_{i_0} = \emptyset$. This in turn implies that $\langle s_{i_0}, F_{i_0} \rangle$ and $\langle s_{\omega+k_0}, F_{\omega+k_0} \rangle$ are compatible. □

19.3. Ultrafilters on ω

In this section, we show how Martin's Axiom or some of its consequences can be used to construct nonprincipal ultrafilters with various interesting properties. Throughout this section, the word "ultrafilter" will be reserved for nonprincipal ultrafilters on ω.

DEFINITION 19.30. Let $\mathcal{F}$ be an ultrafilter. We say that $\mathcal{F}$ is *selective*, if for every partition $(a_i)_{i \in \omega}$ of ω into pairwise disjoint subsets such that none of the a_i's is in $\mathcal{F}$, there exists $b \in F$ with $|b \cap a_i| \leq 1$ for all $i \in \omega$. We say that $\mathcal{F}$ is a *P-point*, if for every partition $(a_i)_{i \in \omega}$ of ω into pairwise disjoint subsets such that none of the a_i's is in $\mathcal{F}$, there exists $b \in F$ with $|b \cap a_i| < \aleph_0$ for all $i \in \omega$. We say that $\mathcal{F}$ is a *Q-point*, if for every partition $(a_i)_{i \in \omega}$ of ω into pairwise disjoint finite subsets there exists $b \in F$ with $|b \cap a_i| \leq 1$ for all $i \in \omega$.

EXERCISE 19.33 (G). Convince yourself that an ultrafilter is selective if and only if it is simultaneously a P-point and a Q-point.

EXERCISE 19.34 (PG). (a) Show that an ultrafilter $\mathcal{F}$ is a P-point if and only if for every sequence $(a_i)_{i \in \omega}$ of elements of $\mathcal{F}$ there exists $b \in \mathcal{F}$ such that $b \subseteq^* a_i$ for all $i \in \omega$.

(b) A point x in a topological space X is called a P-point if x is in the interior of each G_δ-set that contains x. Consider the set ω^* of all nonprincipal ultrafilters on ω with the topology τ generated by the family $\{U_a : a \in [\omega]^{\aleph_0}\}$, where $U_a = \{\mathcal{F} \in \omega^* : a \in \mathcal{F}\}$. Show that an ultrafilter $\mathcal{F}$ is a P-point in the sense of Definition 19.30 if and only if it is a P-point in the topological space $\langle \omega^*, \tau \rangle$.

THEOREM 19.31. *If $\mathfrak{t} = 2^{\aleph_0}$, then there exists a selective ultrafilter.*

PROOF. Let $\{c_\alpha : \alpha < 2^{\aleph_0}\}$ be an enumeration of $\mathcal{P}(\omega)$, and let $\{(a_i^\delta)_{i \in \omega} : \delta \in 2^{\aleph_0} \cap \mathbf{LIM} \backslash \{0\}\}$ be an enumeration of all partitions of ω into pairwise disjoint finite subsets. We are going to construct recursively a transfinite sequence $\langle b_\alpha : \alpha < 2^{\aleph_0} \rangle$ of infinite subsets of ω such that $b_\beta \subseteq^* b_\alpha$ for all $\alpha < \beta < 2^{\aleph_0}$, and then we are going to define $\mathcal{F} = \{a \subseteq \omega : \exists \alpha < 2^{\aleph_0} (b_\alpha \subseteq^* a)\}$.

EXERCISE 19.35 (G). Show that if $\mathcal{F}$ is constructed as above, then $\mathcal{F}$ is a filter. Moreover, show that if $\mathcal{F}$ happens to be an ultrafilter, then it is a P-point. *Hint:* Use the result of Exercise 19.34(a).

Here is how we construct the b_α's:

We let $b_0 = \omega$.

Given b_α, we choose $b_{\alpha+1} \in [b_\alpha]^{\aleph_0}$ in such a way that $b_{\alpha+1} \subseteq c_\alpha$ or, if this is not possible, so that $b_{\alpha+1} \cap c_\alpha = \emptyset$, i.e., $b_{\alpha+1} \subseteq \omega\backslash c_\alpha$. This clause ensures that $\mathcal{F}$ will be an ultrafilter.

Suppose $\delta < 2^{\aleph_0}$ is a limit ordinal and the sequence $\langle b_\alpha : \alpha < \delta \rangle$ has already been constructed in such a way that $b_\beta \subseteq^* b_\alpha$ for all $\alpha < \beta < \delta$. Since $\delta < \mathfrak{t}$, we can find $B_\delta \in [\omega]^{\aleph_0}$ such that $B_\delta \subseteq^* b_\alpha$ for all $\alpha < \delta$. Now consider the partition $(a_i^\delta)_{i \in \omega}$ and define: $b_\delta = \{\min a_i^\delta : i \in d\}$.

EXERCISE 19.36 (PG). Verify that the family $\{b_\alpha : \alpha < 2^{\aleph_0}\}$ generates a selective ultrafilter $\mathcal{F}$. □

A technique similar to the proof of Theorem 19.31 can be used to construct a P-point that is not a Q-point (and hence not selective). We need a lemma.

LEMMA 19.32 (MA$_{\sigma\text{-centered}}$). *Let $(a_i)_{i \in \omega}$ be a partition of ω into pairwise disjoint sets with $|a_i| = i$ for all $i \in \omega$. Suppose $\mathcal{B} \subseteq \mathcal{P}(\omega)$ is a family such that $|\mathcal{B}| < 2^{\aleph_0}$, and for every $H \in [\mathcal{B}]^{<\aleph_0}$ the set $\{|\bigcap H \cap a_i| : i \in \omega\}$ is infinite. Then there exists a pseudo-intersection c of $\mathcal{B}$ such that $\{|c \cap a_i| : i \in \omega\}$ is infinite.*

PROOF. Consider the set
$$\mathbb{P} = \{\langle s, F\rangle : s \in [\omega]^{<\aleph_0} \wedge F \in [\mathcal{B}]^{<\aleph_0}\}.$$
Define a binary relation $\preceq$ on $\mathbb{P}$ by:
$$\langle s, F\rangle \preceq \langle t, G\rangle \quad \text{if and only if} \quad t \subseteq s \wedge G \subseteq F \wedge s\backslash t \subseteq \bigcap G.$$

EXERCISE 19.37 (G). (a) Verify that $\preceq$ is a p.o. relation on $\mathbb{P}$.
(b) Show that $\langle \mathbb{P}, \preceq \rangle$ is σ-centered.

We want to define $c = \bigcup\{s : \exists F(\langle s, F\rangle \in \mathbb{G})\}$, where $\mathbb{G}$ is a $\mathcal{D}$-generic filter for a suitable family of dense sets $\mathcal{D}$. "Suitable" means that $|\mathcal{D}| < 2^{\aleph_0}$, and that $\mathcal{D}$-genericity of $\mathbb{G}$ forces c to be as required.

EXERCISE 19.38 (G). Find a family $\mathcal{D}_0$ of cardinality $< 2^{\aleph_0}$ such that $\mathcal{D}_0$-genericity of $\mathbb{G}$ implies that $c \subseteq^* b$ for all $b \in \mathcal{B}$. Don't forget to verify that the elements of $\mathcal{D}_0$ are really dense in $\langle \mathbb{P}, \preceq \rangle$.

We have to ensure one more property of c. For each $n \in \omega$, let
$$E_n = \{\langle s, F\rangle \in \mathbb{P} : \exists i \in \omega \,(|a_i \cap s| \geq n)\}.$$

EXERCISE 19.39 (G). (a) Verify that each set E_n is dense in $\langle \mathbb{P}, \preceq \rangle$.
(b) Show that if $\mathcal{D}_0$ is as in Exercise 19.38, and $\mathbb{G}$ is $\mathcal{D}_0 \cup \{E_n : n \in \omega\}$-generic, then $c = \bigcup\{s : \exists F(\langle s, F\rangle \in \mathbb{G})\}$ is as postulated in Lemma 19.32. □

THEOREM 19.33. MA$_{\sigma\text{-centered}}$ *implies that there exists a P-point that is not a Q-point.*

EXERCISE 19.40 (PG). Prove Theorem 19.33. *Hint:* Fix a partition $(a_i)_{i \in \omega}$ of ω into pairwise disjoint sets such that $|a_i| = i$ for all $i \in \omega$. Then construct recursively a $2^{\aleph_0}$-tower $\langle b_\alpha : \alpha < 2^{\aleph_0}\rangle$ as in the proof of Theorem 19.31, but make sure that $\{|b_\alpha \cap a_i| : i \in \omega\}$ is infinite for all $\alpha < 2^{\aleph_0}$. At limit stages, use Lemma 19.32.

An ultrafilter $\mathcal{F}$ is called a *semi-Q-point* if for every partition $(a_i)_{i\in\omega}$ of ω into pairwise disjoint finite sets there exists $b \in \mathcal{F}$ such that $|b \cap a_i| \leq i$ for all $i \in \omega$.

EXERCISE 19.41 (R). Use $\mathrm{MA}_{\sigma\text{-centered}}$ to construct a P-point that is not a semi-Q-point.

THEOREM 19.34. $\mathrm{MA}_{\text{countable}}$ *implies the existence of a Q-point that is not a P-point.*

PROOF. Assume $\mathrm{MA}_{\text{countable}}$. For the duration of this proof, fix a partition $A = (a_i)_{i\in\omega}$ of ω into pairwise disjoint infinite sets. A set $b \subseteq \omega$ will be called *A-large* if $|\{i \in \omega : |b \cap a_i| = \aleph_0\}| = \aleph_0$. A family $\mathcal{B} \subseteq \mathcal{P}(\omega)$ will be called *A-large* if for every finite $H \subseteq \mathcal{B}$ the intersection $\bigcap H$ is A-large. In particular, every A-large family has the sfip. We are going to construct an A-large ultrafilter $\mathcal{F}$.

EXERCISE 19.42 (G). Convince yourself that an A-large ultrafilter cannot be a P-point.

LEMMA 19.35. *If $\mathcal{B} \subseteq \mathcal{P}(\omega)$ is an A-large family, then there exists an A-large ultrafilter $\mathcal{F}$ with $\mathcal{B} \subseteq \mathcal{F}$.*

EXERCISE 19.43 (PG). (a) Show that if $\mathcal{B}$ is A-large and $c \subseteq \omega$, then at least one of the families $\mathcal{B} \cup \{c\}$ and $\mathcal{B} \cup \{\omega \backslash c\}$ is A-large.
(b) Derive Lemma 19.35 from (a). □

By Lemma 19.35, the proof of Theorem 19.34 boils down to the following:

LEMMA 19.36. *There exists an A-large family $\mathcal{B} \subseteq \mathcal{P}(\omega)$ such that for every partition $(c_n)_{n\in\omega}$ of ω into pairwise disjoint finite sets there exists $b \in \mathcal{B}$ such that $|b \cap c_n| \leq 1$ for all $b \in \mathcal{B}$ and $n \in \omega$.*

PROOF. Let $\{(c_n^\alpha)_{n\in\omega} : \alpha < 2^{\aleph_0}\}$ be an enumeration of all partitions of ω into pairwise disjoint finite sets. We construct recursively a sequence $(b_\alpha)_{\alpha<2^{\aleph_0}}$ of subsets of ω such that $\{b_\alpha : \alpha \leq \beta\}$ is A-large and $|b_\beta \cap c_n^\beta| \leq 1$ for all $\beta < 2^{\aleph_0}$ and $n \in \omega$.

To see that this construction can be carried out, assume that $\beta < 2^{\aleph_0}$, and let $\{b_\alpha : \alpha < \beta\}$ be an A-large family. Consider the countable set $\mathbb{P} = \{s \in [\omega]^{<\aleph_0} : \forall n \in \omega\,(|s \cap c_n^\beta| \leq 1)\}$, partially ordered by reverse inclusion. For $k \in \omega$, $H \in [\{b_\alpha : \alpha < \beta\}]^{<\aleph_0}$, and i such that $|a_i \cap \bigcap H| = \aleph_0$, define:

$$D_{i,k}^H = \{s \in \mathbb{P} : |s \cap a_i| \geq k\}.$$

Let
$$\mathcal{D} = \{D_{i,k}^H : H \in [\{b_\alpha : \alpha < \beta\}]^{<\aleph_0} \wedge |a_i \cap \bigcap H| = \aleph_0 \wedge k \in \omega\}.$$

EXERCISE 19.44 (G). (a) Show that $\mathcal{D}$ consists of fewer than $2^{\aleph_0}$ dense subsets of $\langle \mathbb{P}, \supseteq \rangle$.
(b) Show that if $\mathbb{G}$ is a $\mathcal{D}$-generic filter and $b = \bigcup \mathbb{G}$, then b is as required.

$\mathrm{MA}_{\text{countable}}$ implies that $\mathbb{G}$ as in Exercise 19.44(b) exists, and we are done. □□

EXERCISE 19.45 (R). Show that $\mathrm{MA}_{\text{countable}}$ implies that there exists a semi-Q-point that is not a Q-point. *Hint:* Fix a partition $A = (a_i)_{i\in\omega}$ of ω such that $|a_i| = i$ for all $i \in \omega$, and call a family $\mathcal{B} \subseteq \mathcal{P}(\omega)$ A-large if for every $H \in [\mathcal{B}]^{<\aleph_0}$ the set $\{|\bigcap H \cap a_i| : i \in \omega\}$ is infinite. Then emulate the proof of Theorem 19.34. You will have to modify the partial order used in the proof of Lemma 19.36.

By Theorem 19.31, if $\mathfrak{t} = 2^{\aleph_0}$, then P-points exist. One can improve this result a little and show that the existence of P-points already follows from the equality $\mathfrak{d} = 2^{\aleph_0}$.

THEOREM 19.37 (Ketonen). *Assume $\mathfrak{d} = 2^{\aleph_0}$. If $\mathcal{C} \subseteq \mathcal{P}(\omega)$ is a family of cardinality $< 2^{\aleph_0}$ with the sfip, then there exists a P-point $\mathcal{F}$ such that $\mathcal{C} \subseteq \mathcal{F}$.*

PROOF. We need a lemma.

LEMMA 19.38. *Suppose $\mathcal{C} \subseteq \mathcal{P}(\omega)$ is a family of cardinality $< \mathfrak{d}$ with the sfip, and $(a_i)_{i \in \omega}$ is a partition of ω into pairwise disjoint infinite sets such that for each $H \in [\mathcal{C}]^{<\aleph_0}$ the set $\{i : \bigcap H \cap a_i \neq \emptyset\}$ is infinite. Then there exists $b \subseteq \omega$ such that for all $i \in \omega$ the intersection $b \cap a_i$ is finite and $\mathcal{C} \cup \{b\}$ has the sfip.*

PROOF. For every $H \in [\mathcal{C}]^{<\aleph_0}$ define a function $f_H : \omega \to \omega$ by: $f_H(n) = \min(\bigcap H \cap a_{i_n}) + 1$, where $i_n = \min\{i \geq n : \bigcap H \cap a_i \neq \emptyset\}$. Since $|[\mathcal{C}]^{<\aleph_0}| < \mathfrak{d}$, there exists an increasing function $f : \omega \to \omega$ such that for each $H \in [\mathcal{C}]^{<\aleph_0}$ the inequality $f(n) \geq f_H(n)$ holds infinitely often. Fix such f, and define $b = \bigcup \{a_i \cap f(i) : i \in \omega\}$.

EXERCISE 19.46 (PG). (a) Show that the family $\mathcal{C} \cup \{b\}$ has the sfip.
(b) Derive Theorem 19.37 from Lemma 19.38. □□

EXERCISE 19.47 (PG). Show that if $\mathfrak{d} < 2^{\aleph_0}$, then there exists a filter base $\mathcal{C}$ such that $|\mathcal{C}| < 2^{\aleph_0}$ and there is no P-point $\mathcal{F}$ that contains $\mathcal{C}$. *Hint:* Let $\{f_\alpha : \alpha < \mathfrak{d}\}$ be a dominating family, and let $(a_i)_{i \in \omega}$ be a partition of ω into pairwise disjoint infinite sets. For $\alpha < \mathfrak{d}$, let $c_\alpha = \bigcup_{i \in \omega}(a_i \setminus f_\alpha(i))$. Consider the family $\{\omega \setminus a_i : i \in \omega\} \cup \{c_\alpha : \alpha < \mathfrak{d}\}$.

We conclude this section with characterizations of P-points and selective ultrafilters in terms of partition properties. This material hasn't much to do with Martin's Axiom *per se*, but it is important and interesting in its own right.

THEOREM 19.39. *Let $\mathcal{F}$ be an ultrafilter. Then the following are equivalent:*
(i) *For each l.o. $\langle \omega, \preceq \rangle$ there exists $a \in \mathcal{F}$ such that either $ot(\langle a, \preceq \rangle) = \omega$ or $ot(\langle a, \preceq \rangle) = \omega^*$.*
(ii) *For every w.o. $\langle \omega, \preceq \rangle$ there exists $a \in \mathcal{F}$ such that $ot(\langle a, \preceq \rangle) = \omega$.*
(iii) *$\mathcal{F}$ is a P-point.*

PROOF. The implication (i) → (ii) is obvious. To prove the implication (ii) → (iii), assume that $\mathcal{F}$ is as in (ii). Let $(a_i)_{i \in \omega}$ be a partition of ω into pairwise disjoint infinite subsets such that $a_i \notin \mathcal{F}$ for all $i \in \omega$. Define a relation $\preceq$ on ω by:

$$n \preceq m \quad \text{iff} \quad \exists i < j < \omega \left((n \in a_i \wedge m \in a_j) \vee (n \in a_i \wedge m \in a_i \wedge n \leq m) \right).$$

EXERCISE 19.48 (G). (a) Convince yourself that $\langle \omega, \preceq \rangle$ is a w.o. of order type $\omega \cdot \omega$.
(b) Show that if $a \subseteq \omega$ is such that $\langle a, \preceq \rangle$ has order type ω, then $a \cap a_i$ is infinite for at most one $i \in \omega$.
(c) Conclude that $\mathcal{F}$ is a P-point.

For the proof of the implication (iii) → (i), assume that $\mathcal{F}$ is a P-point and $\langle \omega, \preceq \rangle$ is a l.o. Let $L = \{n \in \omega : \{m \in \omega : n \prec m\} \in \mathcal{F}\}$. We distinguish three cases.

Case 1: $L = \emptyset$.
Then there exists a partition $(a_i)_{i \in \omega}$ of ω into pairwise disjoint sets such that $a_i \notin \mathcal{F}$ for all $i \in \omega$, and $\forall i < j \forall n \in a_i \forall m \in a_j \, (m \prec n)$.

Case 2: $L \neq \emptyset$ and L has a largest element k_0.

Then there exists a partition $(a_i)_{i \in \omega}$ of ω into pairwise disjoint sets such that $a_i \notin \mathcal{F}$ for all $i \in \omega$, $a_0 = L$, and and $\forall 0 < i < j \, \forall n \in a_i \, \forall m \in a_j \, (k_0 \prec m \prec n)$.

Case 3: $L \neq \emptyset$ and L has no largest element.

Subcase 3.1: $\omega \setminus L \notin \mathcal{F}$.

Then there exists a partition $(a_i)_{i \in \omega}$ of ω into pairwise disjoint sets such that $a_i \notin \mathcal{F}$ for all $i \in \omega$, $a_0 = \omega \setminus L$, and $\forall 0 < i < j \, \forall k \in a_0 \, \forall n \in a_i \, \forall m \in a_j \, (n \prec m \prec k)$.

Subcase 3.2: $\omega \setminus L \in \mathcal{F}$.

Let $a_0 = L$. As in Case 1, there exists a partition $(a_i)_{i \in \omega \setminus \{0\}}$ of $\omega \setminus L$ into pairwise disjoint sets such that $a_i \notin F$ for all $i \in \omega \setminus \{0\}$, and $\forall 0 < i < j \, \forall n \in a_i \, \forall m \in a_j (m \prec n)$.

Since $\mathcal{F}$ is a P-point, in each case there exists $a \in \mathcal{F}$ such that $a \cap a_i$ is finite for each $i \in \omega$. Since ultrafilters are assumed nonprincipal in this section, we may choose a in such a way that $a \cap a_0 = \emptyset$.

EXERCISE 19.49. Let a be as above. Convince yourself that $\langle a, \preceq \rangle$ has order type ω^* in Case 1, Case 2, and Case 3.2, and has order type ω in Case 3.1. □

Now suppose $\mathcal{F}$ is a P-point, $\preceq$ is a l.o. relation on ω, and $a \in \mathcal{F}$ is such that $\langle a, \preceq \rangle$ has order type ω. Can we assume that the relation $\preceq$ agrees with the natural order $\leq$ on a? To make this question more precise, consider the following statement:

(i)$^+$ *for every l.o. relation $\preceq$ on ω, there exists $a \in \mathcal{F}$ such that*
- either $\forall n, m \in a \, (n \preceq m \leftrightarrow n \leq m)$;
- or $\quad \forall n, m \in a \, (n \preceq m \leftrightarrow m \leq n)$.

It turns out that Property (i)$^+$ implies more about $\mathcal{F}$ than just being a P-point.

LEMMA 19.40. *If $\mathcal{F}$ is an ultrafilter that satisfies* (i)$^+$, *then $\mathcal{F}$ is selective.*

PROOF. Let $\mathcal{F}$ be an ultrafilter that satisfies (i)$^+$, and let $(a_i)_{i \in \omega}$ be a partition of ω into pairwise disjoint subsets such that $a_i \notin \mathcal{F}$ for all $i \in \omega$. Define a relation $\preceq$ on ω as follows:

$n \preceq m$ if and only if $\exists i < j < \omega \, ((n \in a_i \wedge n \in a_j) \vee (n \in a_i \wedge m \in a_i \wedge m \leq n))$.

EXERCISE 19.50 (G). Convince yourself that if $a \in \mathcal{F}$ satisfies condition (i)$^+$, then $|a \cap a_i| \leq 1$ for all $i \in \omega$. Conclude that $\mathcal{F}$ is selective. □

Properties (i), (i)$^+$, and (ii) clearly are partition properties of sorts. Now let us define a collection of partition properties that resemble the ones considered in Chapter 15 even more closely. If $k, n \in \omega \setminus \{0\}$ and $\mathcal{F}$ is an ultrafilter, then the symbol $\omega \to (\mathcal{F})^n_m$ is an abbreviation of the following statement:

For every coloring $c : [\omega]^n \to m$ there exists a homogeneous set $a \in \mathcal{F}$.

EXERCISE 19.51 (G). (a) Show that if $\mathcal{F}$ is any ultrafilter and $m \in \omega \setminus \{0\}$, then $\omega \to (\mathcal{F})^1_m$.

(b) Convince yourself that $\omega \to (\mathcal{F})^2_2$ implies that $\mathcal{F}$ has Property (i)$^+$.

An ultrafilter $\mathcal{F}$ is called a *Ramsey ultrafilter* if $\omega \to (\mathcal{F})^n_m$ holds for all $n, m \in \omega \setminus \{0\}$.

It follows from Exercise 19.51(b) and Lemma 19.40 that every Ramsey ultrafilter is selective. The converse is also true.

LEMMA 19.41. *Every selective ultrafilter is a Ramsey ultrafilter.*

PROOF. Let $\mathcal{F}$ be a selective ultrafilter, and let $m \in \omega\setminus\{0\}$ be fixed. We prove by induction over n that $\omega \to (\mathcal{F})^n_m$.

For $n = 1$ the latter is true by Exercise 19.51(a).

Now let us assume $\omega \to (\mathcal{F})^n_m$ and prove that $\omega \to (\mathcal{F})^{n+1}_m$ holds as well. Toward this end, let $c : [\omega]^{n+1} \to m$ be an arbitrary coloring of all $(n+1)$-tuples of natural numbers with m colors. For $i \in \omega$, we define an induced coloring $c_i : [\omega]^n \to m$ by letting $c_i(x) = c(\{i\} \cup x)$ if $x \in [\omega\setminus(i+1)]^n$ and $c_i(x) = 0$ if $\min x \leq i$. By the inductive assumption, for each $i \in \omega$ we can find $A_i \in \mathcal{F}$ such that $c_i[[A_i]^n] = \{\ell_i\}$ for some $\ell_i < m$. Since $\mathcal{F}$ is a nonprincipal ultrafilter, we can choose the A_i's in such a way that $\min A_i > i$. Now suppose for a moment that we can find $b \in \mathcal{F}$ such that

(diag) $\qquad \forall i, j \in b \, (i < j \to j \in A_i).$

EXERCISE 19.52 (G). Suppose b satisfies (diag), and $x \in [b]^{n+1}$. Show that $c(x) = \ell_{\min x}$.

If $b \in \mathcal{F}$ satisfies (diag), then there exists $\ell < m$ such that $b_\ell \in \mathcal{F}$, where $b_\ell = \{j \in b : \ell_j = \ell\}$. By Exercise 19.52, $a = b_\ell$ is homogeneous for c.

It is not hard to find an infinite b such that (diag) holds. But can we find such a b in $\mathcal{F}$?

Let us call an ultrafilter *quasi-normal* if for every sequence $(A_i)_{i \in \omega}$ of elements of $\mathcal{F}$ there exists $b \in \mathcal{F}$ such that (diag) holds. The following sublemma provides the missing link in our proof:

SUBLEMMA 19.42. *Let $\mathcal{F}$ be a selective ultrafilter. Then $\mathcal{F}$ is quasi-normal.*

PROOF. Assume $\mathcal{F}$ is a selective ultrafilter, and let $(A_i)_{i \in \omega}$ be a sequence of elements of $\mathcal{F}$. We define a decreasing sequence $(b_n)_{n \in \omega}$ of elements of $\mathcal{F}$ by letting $b_0 = \omega$, and $b_n = \bigcap_{i<n} A_i \setminus n$ for $n \geq 1$. For each $n \in \omega$, we let $x_n = b_{n+1} \setminus b_n$. Then $(x_n)_{n \in \omega}$ is a partition of ω into pairwise disjoint sets such that none of the x_n's belongs to $\mathcal{F}$. Since $\mathcal{F}$ is selective, there exists $a \in \mathcal{F}$ such that $|a \cap x_n| \leq 1$ for all $n \in \omega$. We fix such an a and define recursively a function $f \in {}^\omega\omega$ by:

$$f(0) = 0; \qquad f(n+1) = \max\{f(n), \min(a \setminus b_{f(n)})\} + 1.$$

Note that the function f is strictly increasing. Thus, $\{[f(n), f(n+1)) : n \in \omega\}$ is a partition of ω into pairwise disjoint finite sets. Since $\mathcal{F}$ is selective, there exists $c \in \mathcal{F}$ such that for all $n \in \omega$, there is at most one $m \in c$ with $f(n) \leq m < f(n+1)$. Enumerate $a \cap c$ in increasing order as $(j_m)_{m \in \omega}$. Let

$$d_0 = \{j_{2m} : m \in \omega\}, \qquad d_1 = \{j_{2m+1} : m \in \omega\}.$$

There is $\ell < 2$ such that $d_\ell \in \mathcal{F}$. We fix this ℓ, and let $b = d_\ell$. It remains to show that b is as required. So suppose $i, j \in b$ are such that $i < j$. Then there exists $n \in \omega$ such that $i < f(n) < f(n+1) \leq j$. Note that the definition of f implies that if $j \in a$ is such that $j \geq f(n+1)$, then $j > \min(a \setminus b_{f(n)})$, and hence $j \in b_{f(n)}$. Since $b_{f(n)} \subseteq A_i$ for each $i < f(n)$, we conclude that $j \in A_i$, and hence b is an element of $\mathcal{F}$ that witnesses (diag). □

Let us summarize our findings.

THEOREM 19.43. *Let $\mathcal{F}$ be an ultrafilter. Then the following are equivalent:*
(a) $\mathcal{F}$ *is selective.*
(b) $\mathcal{F}$ *is Ramsey.*
(c) $\omega \to (\mathcal{F})_2^2$.
(d) $\mathcal{F}$ *is quasi-normal.*

Mathographical Remarks

The most comprehensive collection of applications of Martin's Axiom in print at the time of this writing is D. Fremlin's monograph *Consequences of Martin's Axiom*, Cambridge Tracts in Mathematics 84, Cambridge University Press, 1984. If you are interested in something less comprehensive, we recommend the following articles: W. Weiss, *Versions of Martin's Axiom*, In: Handbook of Set-Theoretic Topology, K. Kunen and J. E. Vaughan, eds., North-Holland, 1984, 827–886; K. Kunen and F. D. Tall, *Between Martin's Axiom and Souslin's Hypothesis*, Fundamenta Mathematicae 102 (1979), 173–181.

The converse of Theorem 19.16 is also true. In their paper *Martin's axiom and partitions*, Compositio Mathematica 63 (1987), 391–408, S. Todorčević and B. Veličković have shown that for every κ, $MA(\kappa)$ is equivalent to the statement: *Every c.c.c. partial order of size less than or equal to κ is σ-centered.*

Theorem 19.24 represents the current state of knowledge about inequalities between the cardinal invariants mentioned in it. It has been shown that each of the following inequalities is relatively consistent with ZFC: $\mathfrak{t} < \min\{\mathfrak{s}, \mathfrak{b}\}$, $\mathfrak{s} < \mathfrak{b}$, $\mathfrak{b} < \mathfrak{s}$, $\max\{\mathfrak{s}, \mathfrak{b}\} < \mathfrak{d}$, $\mathfrak{b} < \mathfrak{a}$, $\mathfrak{a} < \mathfrak{d}$, $\mathfrak{a} < \mathfrak{s}$. It is still open whether the inequalities $\mathfrak{d} < \mathfrak{a}$ and $\mathfrak{p} < \mathfrak{t}$ are consistent. Concerning the latter, Todorčević and Veličković have shown in the paper mentioned above that if $\mathfrak{p} = \aleph_1$, then $\mathfrak{p} = \mathfrak{t}$.

Neither the existence of P-points nor the existence of Q-points can be proved in ZFC alone. The former was shown by S. Shelah. For an exposition of this result, see E. L. Wimmers, *The Shelah P-point Independence Theorem*, Israel Journal of Mathematics 43(1) (1982), 28–48. The latter result was proved in A. Miller, *There are no Q-points in Laver's model for the Borel Conjecture*, Proceedings AMS 78(1) (1980), 103–106. It is still an open question whether there is a model of ZFC in which neither P-points nor Q-points exist.

CHAPTER 20

Hausdorff Gaps

After reading the previous chapter you may be under the impression that MA implies that every conceivable cardinal invariant of the continuum is equal to $2^{\aleph_0}$. Things are not quite that simple. In the present chapter, we are going to discuss a cardinal invariant of the continuum that is equal to $\aleph_1$—no matter how large $2^{\aleph_0}$ is and whether or not Martin's Axiom holds.

Let $\bar{a} = \langle a_\xi : \xi < \kappa \rangle$ and $\bar{b} = \langle b_\eta : \eta < \lambda \rangle$ be sequences of subsets of ω. We say that $\langle \bar{a}, \bar{b} \rangle$ is a $\langle \kappa, \lambda^* \rangle$-*pregap* if $a_{\xi_0} \subset^* a_{\xi_1} \subset^* b_{\eta_1} \subset^* b_{\eta_0}$ for all $\xi_0 < \xi_1 < \kappa$ and $\eta_0 < \eta_1 < \lambda$. A $\langle \kappa, \lambda^* \rangle$-pregap is called a $\langle \kappa, \lambda^* \rangle$-*gap* if there is no $c \in \mathcal{P}(\omega)$ such that $a_\xi \subset^* c \subset^* b_\eta$ for all $\xi < \kappa$ and $\eta < \lambda$.

We have already encountered gaps of one sort : A λ-tower is the same thing as a $\langle 1, \lambda^* \rangle$-gap with $a_0 = \emptyset$. Thus, one could redefine $\mathfrak{t}$ as the smallest λ such that there exists a $\langle 1, \lambda^* \rangle$-gap.

The next interesting case to look at is the situation where $\kappa = \omega$. Define:

$$\mathfrak{rb} = \min\{\lambda : \text{ there exists an } \langle \omega, \lambda^* \rangle\text{-gap}\}.$$

It turns out that the cardinal $\mathfrak{rb}$ is an old acquaintance of ours.

THEOREM 20.1 (Rothberger). $\mathfrak{rb} = \mathfrak{b}$.

PROOF. Let $\langle \langle a_n : n \in \omega \rangle, \langle b_\zeta : \zeta < \lambda \rangle \rangle$ be an $\langle \omega, \lambda^* \rangle$-pregap. For each $\zeta < \lambda$, define a function $f_\zeta : \omega \to \omega$ by letting $f_\zeta(n) = \max(a_n \backslash b_\zeta) + 1$, where the maximum of the empty set is arbitrarily defined as 0. Suppose that $\lambda < \mathfrak{b}$. Then there exists $g \in {}^\omega \omega$ such that $f_\zeta <^* g$ for all $\zeta < \lambda$. Define:

$$c = \bigcup (a_n \backslash g(n)).$$

Clearly, $a_n \subset^* c$ for all $n \in \omega$. Moreover, if $\zeta < \lambda$, then $a_n \backslash g(n) \subseteq b_\zeta$ for all but finitely many $n \in \omega$, and it follows that $c \subset^* b_\zeta$. Thus $\langle \langle a_n : n \in \omega \rangle, \langle b_\zeta : \zeta < \lambda \rangle \rangle$ is not a gap, and we have shown that $\mathfrak{rb} \geq \mathfrak{b}$.

To prove the inequality $\mathfrak{rb} \leq \mathfrak{b}$, let us construct an $\langle \omega, \mathfrak{b}^* \rangle$-gap. Fix a sequence $(d_k)_{k \in \omega}$ of pairwise disjoint infinite subsets of ω, and an unbounded sequence $(f_\zeta)_{\zeta < \mathfrak{b}}$ of elements of ${}^\omega \omega$ such that $f_{\zeta_0} <^* f_{\zeta_1}$ for all $\zeta_0 < \zeta_1 < \mathfrak{b}$. Such a sequence exists by Exercise 17.18(b). Now let $a_n = \bigcup_{k \leq n} d_k$ for $n \in \omega$, and $b_\zeta = \bigcup_{k \in \omega} d_k \backslash f_\zeta(k)$.

EXERCISE 20.1 (G). Convince yourself that $\langle \langle a_n : n \in \omega \rangle, \langle b_\zeta : \zeta < \mathfrak{b} \rangle \rangle$ is an $\langle \omega, \mathfrak{b}^* \rangle$-gap. $\square$

In particular, Theorem 20.1 implies that there are no $\langle \omega, \omega^* \rangle$-gaps. Are there $\langle \kappa, \kappa^* \rangle$ gaps for any κ? If so, what is the smallest κ for which such a symmetrical gap exists? One is tempted to define a new cardinal invariant of the continuum and use a fancy symbol to denote it. Surprisingly, this turns out to be unnecessary: This new cardinal invariant is always equal to $\aleph_1$.

THEOREM 20.2 (Hausdorff). *There exists an $\langle \omega_1, \omega_1^* \rangle$-gap.*

In honor of the author of Theorem 20.2, $\langle \omega_1, \omega_1^* \rangle$-gaps are often called *Hausdorff gaps*.[1] Before we prove Theorem 20.2, let us derive two important consequences of it.

CLAIM 20.3. *Let $\langle \langle a_\xi : \xi < \omega_1 \rangle, \langle b_\xi : \xi < \omega_1 \rangle \rangle$ be a Hausdorff gap. For $\xi < \omega_1$, let $J_\xi = \{c \in \mathcal{P}(\omega) : a_\xi \subseteq^* c \subseteq^* b_\xi\}$. Then $\langle J_\xi : \xi < \omega_1 \rangle$ is a decreasing sequence of nonempty F_σ-subsets of $\mathcal{P}(\omega)$ with $\bigcap_{\xi < \omega_1} J_\xi = \emptyset$.*

PROOF. Recall that the standard topology on $\mathcal{P}(\omega)$ is defined by identifying $\mathcal{P}(\omega)$ via characteristic functions with the Tychonoff product of ω copies of the discrete two-element space $\{0,1\}$. For every $\xi < \omega_1$ and $n \in \omega$, the set $K_{\xi,n} = \{c \subseteq \omega : a_\xi \setminus n \subseteq c \setminus n \subseteq b_\xi \setminus n\}$ is closed, and $J_\xi = \bigcup_{n \in \omega} K_{\xi,n}$. Thus, each J_ξ is an F_σ-set. Clearly, $J_\xi \neq \emptyset$ and $J_\xi \supset J_\eta$ for all $\xi < \eta < \omega_1$. If $c \in \bigcap_{\xi < \omega_1} J_\xi$, then $a_\xi \subset^* c \subset^* b_\xi$ for all $\xi < \omega_1$, which is ruled out by the assumption that $\langle \langle a_\xi : \xi < \omega_1 \rangle, \langle b_\xi : \xi < \omega_1 \rangle \rangle$ is a gap. □

Claim 20.3 is akin to Lemma 17.10, but the latter uses the extra assumption $\mathfrak{b} = \aleph_1$, while the former is a theorem of ZFC. Theorem 20.2, Claim 20.3, and Lemma 17.9 together show that uncountable, perfectly meager subsets of $\mathcal{P}(\omega)$ can be constructed using only the axioms of ZFC. More precisely, we have the following:

COROLLARY 20.4. *For any Hausdorff gap $\langle \langle a_\xi : \xi < \omega_1 \rangle, \langle b_\xi : \xi < \omega_1 \rangle \rangle$, the set $\{a_\xi : \xi < \omega_1\} \cup \{b_\xi : \xi < \omega_1\}$ is a perfectly meager subset of $\mathcal{P}(\omega)$.*

Corollary 20.4 has a measure-theoretic dual:

LEMMA 20.5. *For any Hausdorff gap $\langle \langle a_\xi : \xi < \omega_1 \rangle, \langle b_\xi : \xi < \omega_1 \rangle \rangle$, the set $\{a_\xi : \xi < \omega_1\} \cup \{b_\xi : \xi < \omega_1\}$ has universal measure zero.*

PROOF. Let $\langle \langle a_\xi : \xi < \omega_1 \rangle, \langle b_\xi : \xi < \omega_1 \rangle \rangle$ be a Hausdorff gap, and let μ be a Borel measure on $\mathcal{P}(\omega)$ that vanishes on the singletons. For $\xi < \omega_1$, let J_ξ be defined as in Claim 20.3. Since each J_ξ is μ-measurable and contains all but countably many elements of $\{a_\xi : \xi < \omega_1\} \cup \{b_\xi : \xi < \omega_1\}$, it suffices to show that $\mu(J_\xi) = 0$ for sufficiently large ξ.

Suppose towards a contradiction that this is not the case. Since the J_ξ's form a decreasing transfinite sequence, there must be $\xi_0 < \omega_1$ and $\varepsilon_0 > 0$ such that $\mu(J_\xi) = \varepsilon_0$ for all $\xi > \xi_0$. Define:

$$c = \{n \in \omega : \mu(\{d \in J_{\xi_0} : n \in d\}) \geq \varepsilon_0/2\},$$

and let $\xi_1 > \xi_0$ be such that $c \notin J_{\xi_1}$.

EXERCISE 20.2 (G). *Let c, ξ_1 be as above. Show that in the definition of c, the subscript ξ_0 can be replaced by ξ_1, i.e., show that $c = \{n \in \omega : \mu(\{d \in J_{\xi_1} : n \in d\}) \geq \varepsilon_0/2\}$.*

As in the proof of Claim 20.3, for $n \in \omega$, let $K_{\xi_1,n} = \{d \subseteq \omega : a_{\xi_1} \setminus n \subseteq d \setminus n \subseteq b_{\xi_1} \setminus n\}$. The $K_{\xi_1,n}$'s form an increasing sequence of μ-measurable sets with union J_{ξ_1}. Hence, there must exist $n_0 \in \omega$ such that $\mu(K_{\xi_1,n_0}) > \varepsilon_0/2$. Fix such n_0. Since $c \notin K_{\xi_1,n_0}$, there must exist $n_1 > n_0$ such that either $n_1 \in c \setminus b_{\xi_1}$ or $n_1 \in a_{\xi_1} \setminus c$. In the former case, by Exercise 20.2, the set $L = \{d \in J_{\xi_1} : n_1 \in d\}$ has measure

---

[1] Some authors reserve the term *Hausdorff gap* only for $\langle \omega_1, \omega_1^* \rangle$-gaps with special properties (like Property (iv) in the proof of Theorem 20.6).

at least $\varepsilon_0/2$ and is disjoint from K_{ξ_1,n_0}, which is impossible, since $\mu(K_{\xi_1} \cup L) \leq \mu(J_{\xi_1}) = \varepsilon_0$. In the latter case, the set $M = \{d \in J_{\xi_1} : n_1 \notin d\}$ has measure at least $\varepsilon_0/2$ and is disjoint from K_{ξ_1,n_0}, which again leads to a contradiction. □

We shall prove a version of Theorem 20.2 that differs from the one originally stated. Let $A, B \subseteq [\omega]^{\aleph_0}$. We say that A and B are *almost disjoint*, and write $A \perp B$, if $|a \cap b| < \aleph_0$ for all $a \in A$ and $b \in B$. We say that A and B can be *separated* if there exists $d \subseteq \omega$ such that $a \subseteq^* d$ and $|b \cap d| < \aleph_0$ for every $a \in A$ and $b \in B$. Of course, if A and B can be separated, then A and B are almost disjoint. However, the converse is not true.

EXERCISE 20.3 (G). Let $\langle\langle a_\xi : \xi < \kappa\rangle, \langle b_\eta : \eta < \lambda\rangle\rangle$ be a $\langle\kappa, \lambda^*\rangle$-pregap, let $c_\eta = \omega \backslash b_\eta$ for all $\eta < \lambda$, and let $A = \{a_\xi : \xi < \kappa\}$, $B = \{c_\eta : \eta < \lambda\}$.
(a) Convince yourself that $A \perp B$.
(b) Show that A and B can be separated if and only if $\langle\langle a_\xi : \xi < \kappa\rangle, \langle b_\eta : \eta < \lambda\rangle\rangle$ is *not* a $\langle\kappa, \lambda^*\rangle$-gap.

The above exercise implies that Theorem 20.2 is equivalent to the following statement:

THEOREM 20.6. *There exist transfinite sequences $\langle a_\xi : \xi < \omega_1\rangle$ and $\langle b_\xi : \xi < \omega_1\rangle$ of infinite subsets of ω such that $a_\xi \subset^* a_\zeta$ and $b_\xi \subset^* b_\zeta$ for all $\xi < \zeta < \omega_1$; and the families $\{a_\xi : \xi < \omega_1\}$ and $\{b_\xi : \xi < \omega_1\}$ are almost disjoint, but cannot be separated.*

PROOF. We shall construct sequences $\langle a_\xi : \xi < \omega_1\rangle$ and $\langle b_\xi : \xi < \omega_1\rangle$ simultaneously by recursion over ω_1 in such a way that for all $\xi < \zeta < \omega_1$:

(i) $|\omega \backslash (a_\xi \cup b_\xi)| = \aleph_0$;
(ii) $a_\xi \subset^* a_\zeta$ and $b_\xi \subset^* b_\zeta$;
(iii) $a_\xi \cap b_\xi = \emptyset$.

Condition (i) is a technical requirement that allows us to keep going. Conditions (ii) and (iii) together not only ensure that our sequences are increasing in the sense of $\subset^*$, but also that the families $\{a_\xi : \xi < \omega_1\}$ and $\{b_\xi : \xi < \omega_1\}$ will be almost disjoint. In order to ensure that these families cannot be separated, Hausdorff devised an ingeneous trick. Let us say that $b \subseteq \omega$ is *close to* $A \subseteq \mathcal{P}(\omega)$ if for every $n \in \omega$ the set $\{a \in A : a \cap b \subseteq n\}$ is finite.

EXERCISE 20.4 (G). (a) Suppose that b is close to a set A and $b \subseteq^* d \subseteq \omega$. Convince yourself that d is also close to A.
(b) Suppose that b is close to a set A and $B \subseteq A$. Convince yourself that b is also close to B.

We are going to require that for every $\zeta < \omega_1$:

(iv) b_ζ is close to $\{a_\xi : \xi < \zeta\}$.

If condition (iv) holds for all $\xi < \zeta < \omega_1$, then the sets $\{a_\xi : \xi < \omega_1\}$ and $\{b_\xi : \xi < \omega_1\}$ cannot be separated. This is a consequence of the following observation.

LEMMA 20.7. *Let $\{a_\xi : \xi < \omega_1\}$ and $\{b_\xi : \xi < \omega_1\}$ be separated families of infinite subsets of ω_1 such that $a_\xi \cap b_\xi = \emptyset$ for all $\xi < \omega_1$. Then there exists an uncountable set $X \subseteq \omega_1$ such that $(a_\xi \cap b_\zeta) \cup (a_\zeta \cap b_\xi) = \emptyset$ for all $\xi, \zeta \in X$.*

PROOF. Let $d \subseteq \omega$ be such that $a_\xi \subseteq^* d$ and $|b_\xi \cap d| < \aleph_0$ for all $\xi < \omega_1$. By the Pigeonhole Principle, there exists $n \in \omega$ such that the set $X_n = \{\xi < \omega_1 : (a_\xi \backslash d) \cup (b_\xi \cap d) \subseteq n\}$ is uncountable. Moreover, using the Pigeonhole Principle again we can see that there are $s, t \subseteq n$ such that the set $X = \{\xi \in X_n : a_\xi \backslash d = s \wedge b_\xi \cap d = t\}$ is uncountable. It is straightforward to verify that the above set X is as required. □

EXERCISE 20.5 (G). Use Lemma 20.7 to show that if condition (iv) holds for all $\xi < \zeta < \omega_1$, then the sets $\{a_\xi : \xi < \omega_1\}$ and $\{b_\xi : \xi < \omega_1\}$ cannot be separated. *Hint:* For X as in the lemma, consider the $(\omega + 1)$st element ζ of X and derive a contradiction.

Having all ideas in place, let us now describe the construction itself.

To get started, let $\{c_0, c_1, c_2\}$ be a partition of ω into three disjoint infinite subsets. Define:
$$a_0 = c_0, \qquad b_0 = c_1.$$
Now suppose $0 < \zeta < \omega_1$, and for all $\xi < \zeta$ the sets a_ξ, b_ξ have been defined in such a way that conditions (i)–(iv) hold.

If $\zeta = \eta + 1$ for some η, then we choose a partition $\{d_0, d_1, d_2\}$ of $\omega \backslash (a_\eta \cup b_\eta)$ into pairwise disjoint infinite sets and define:
$$a_\zeta = a_\eta \cup d_0, \qquad b_\zeta = b_\eta \cup d_1.$$
A straightforward argument shows that in this case, conditions (i)–(iv) continue to hold at stage ζ.

Now let us consider the case where ζ is a limit ordinal. Let us begin by choosing sets $x \subseteq \omega$ and $d \subseteq x$ such that $|\omega \backslash x| = \aleph_0$, $a_\xi \subseteq^* d$, and $b_\xi \subseteq^* x \backslash d$ for all $\xi < \omega_1$. This is possible by Theorem 20.1. Think of $\langle x \backslash d, d \rangle$ as a first approximation to $\langle a_\zeta, b_\zeta \rangle$. This approximation satisfies conditions (i)–(iii), but not necessarily (iv). To see to what extent (iv) fails, fix a strictly increasing sequence of ordinals $(\eta_n)_{n \in \omega}$ such that $\eta_0 = 0$ and $\zeta = \sup\{\eta_n : n \in \omega\}$, and define for each $n \in \omega$:
$$B_n = \{\eta : \eta_n \leq \eta < \eta_{n+1} \wedge d \cap a_\eta \subseteq n\}.$$
Since (iv) holds for every $\zeta = \eta_{n+1}$ and since $b_{\eta_{n+1}} \subseteq^* d$, each of the sets B_n is finite. Thus, if we define $B = \bigcup_{n \in \omega} B_n$, then B either is finite, or has order type ω. Let us consider these two cases separately.

Case 1: B is finite. Then for all but finitely many n the set $C_n = \{\xi < \zeta : d \cap a_\xi \subseteq n\}$ is contained in η_n. Since $b_{\eta_n} \subseteq^* d$ and, by the inductive assumption, b_{η_n} is close to $\{a_\xi : \xi < \eta_n\}$, so is d. Therefore, the set $\{\xi < \zeta : d \cap a_\xi \subseteq n\}$ is finite. Since this argument applies to all but finitely many $n \in \omega$, d is close to $\{a_\xi : \xi < \zeta\}$.

Case 2: B has order type ω. In this case, d may fail to be close to $\{a_\xi : \xi < \zeta\}$. However, d misses the mark of being close just barely.

EXERCISE 20.6 (G). Show that d is close to the set $\{a_\xi : \xi \in \zeta \backslash B\}$.

Let $(\xi_n)_{n \in \omega}$ be an enumeration of B in increasing order. For every $n \in \omega$, pick $p_n \in x \cap a_{\xi_n} \backslash \bigcup_{i < n} a_{\xi_i}$ such that $p_n \geq n$. Define:
$$b_\zeta = d \cup \{p_n : n \in \omega\}, \qquad a_\zeta = x \backslash b_\zeta.$$

EXERCISE 20.7 (PG). Convince yourself that the pair $\langle a_\zeta, b_\zeta \rangle$ satisfies conditions (i)–(iv). *Hint:* For the proof of (iv), use the results of Exercises 20.4 and 20.6. □

The technique of the proof of Theorem 20.6 can also be used to construct another interesting object. Let $A = \{a_\xi : \xi < \omega_1\}$ be an almost disjoint family of infinite subsets of ω. We say that A is a *Luzin gap* if there is no partition $\{B, C\}$ of A into two disjoint uncountable sets such that B can be separated from C.

THEOREM 20.8. *There exists a Luzin gap.*

PROOF. We are going to construct recursively a sequence $(a_\xi)_{\xi < \omega_1}$ of infinite subsets of ω such that for all $0 < \zeta < \omega$:

(1) $\forall \xi < \zeta \, |a_\xi \cap a_\zeta| < \aleph_0$;
(2) a_ζ is close to $\{a_\xi : \xi < \zeta\}$.

The construction is somewhat easier than the one in the proof of Theorem 20.6. We let $(a_n)_{n \in \omega}$ be a sequence of pairwise disjoint infinite subsets of ω. At stage $\zeta \geq \omega$, we arrange ζ into a sequence $(\xi_k)_{k \in \omega}$ and pick for every $k \in \omega$ a number $p_k \in a_{\xi_k} \setminus \bigcup_{i<k} a_{\xi_i}$. Then we define $a_\zeta = \{p_k : k \in \omega\}$.

EXERCISE 20.8 (G). Convince yourself that the sequence $(a_\xi)_{\xi < \omega_1}$ satisfies conditions (1) and (2) for every $0 < \zeta < \omega_1$.

It remains to show that the set $\{a_\xi : \xi < \omega_1\}$ constructed above is a Luzin gap. Clearly, this set consists of pairwise almost disjoint infinite subsets of ω.

Let B, C be two disjoint uncountable subsets of A (we do not even need to require that $B \cup C = A$). Suppose towards a contradiction that $d \subseteq \omega$ is such that $a_\xi \subseteq^* d$ for all $a_\xi \in B$ and $|a_\eta \cap d| < \aleph_0$ for all $a_\eta \in C$. By the Pigeonhole Principle, there exists $n \in \omega$ such that the set $B_n = \{a_\xi \in B : a_\xi \setminus d \subseteq n\}$ is uncountable. Since C is uncountable, we can find $a_\zeta \in C$ such that the set $B_n^\zeta = \{a_\xi \in B_n : \xi < \zeta\}$ is infinite. Fix such ζ and let $m \in \omega$ be such that $a_\zeta \cap d \subseteq m$. Then for all $a_\xi \in B_n^\zeta$ we have $a_\xi \cap a_\zeta \subseteq \max\{n, m\}$. It follows that a_ζ is not close to the set B_n^ζ. By Exercise 20.4(b), the latter implies that a_ζ is not close to $\{a_\xi : \xi < \zeta\}$, which contradicts (2). □

Let us briefly discuss a somewhat different technique for constructing $\langle \omega_1, \omega_1^* \rangle$-gaps. Let $\langle \langle a_\xi : \xi < \omega_1 \rangle, \langle b_\xi : \xi < \omega_1 \rangle \rangle$ be an $\langle \omega_1, \omega_1^* \rangle$-pregap. For $\xi < \zeta < \omega_1$, consider the following property:

(v) $(a_\xi \cap b_\zeta) \cup (b_\xi \cap a_\zeta) \neq \emptyset$.

EXERCISE 20.9 (PG). (a) Show that if $\langle \langle a_\xi : \xi < \omega_1 \rangle, \langle b_\xi : \xi < \omega_1 \rangle \rangle$ satisfies condition (iii) (from the proof of Theorem 20.6) for all $\xi < \omega_1$ and there is an uncountable $Y \subseteq \omega_1$ such that (v) holds for all $\{\xi, \zeta\} \in [Y]^2$, then it is an $\langle \omega_1, \omega_1^* \rangle$-gap. *Hint:* Use Lemma 20.7.

(b) Show that a Hausdorff gap that satisfies conditions (i)–(iii) (from the proof of Theorem 20.6) and (v) can be constructed by a slight modification of the proof of Theorem 20.6.

In a sense, the construction of a Hausdorff gap outlined in Exercise 20.9 is the *only* way to construct such objects in ZFC.

THEOREM 20.9. *Assume the Open Coloring Axiom, and assume that $G = \langle \langle a_\xi : \xi < \omega_1 \rangle, \langle b_\xi : \xi < \omega_1 \rangle \rangle$ satisfies conditions (i)–(iii) stated in the proof of Theorem 20.6 for all $\xi < \zeta < \omega_1$. Then the families $\{a_\xi : \xi < \omega_1\}$ and $\{b_\xi : \xi < \omega_1\}$ cannot be separated if and only if there exists an uncountable subset Y of ω_1 such that condition (v) holds for all $\{\xi, \zeta\} \in [Y]^2$.*

PROOF. Let G be as in the assumptions, and let X be the set $\{\langle a_\xi, b_\xi\rangle : \xi < \omega_1\}$ with the topology obtained by treating X as a subspace of $\mathcal{P}(\omega) \times \mathcal{P}(\omega)$. Consider the following partition $\{P_0, P_1\}$ of $[X]^2$:

$$P_0 = \{\{\langle a_\xi, b_\xi\rangle, \langle a_\zeta, b_\zeta\rangle\} \in [X]^2 : (a_\xi \cap b_\zeta) \cup (a_\zeta \cap b_\xi) \neq \emptyset\},$$
$$P_1 = [X]^2 \setminus P_0.$$

EXERCISE 20.10 (G). Recall the definition of an open partition from Chapter 15, and show that the partition $\{P_0, P_1\}$ defined above is open.

By the OCA, either there is an uncountable set Y such that $[Y]^2 \subseteq P_0$, or $X = \bigcup_{n\in\omega} X_n$, where $[X_n]^2 \subseteq P_1$ for all $n \in \omega$. In the former case, condition (v) holds for all $\{\xi, \zeta\} \in [Y]^2$, and by Exercise 20.9(a), G is a gap. In the latter case, at least one of the X_n's must be uncountable. Without loss of generality we may assume that X_0 has this property. Let $d = \bigcup_{\xi \in X_0} a_\xi$. This d separates $\{a_\xi : \xi < \omega_1\}$ and $\{b_\xi : \xi < \omega_1\}$.

EXERCISE 20.11 (G). Show that if d is as above, then $a_\xi \subseteq^* d$ and $|d \cap b_\xi| < \aleph_0$ for all $\xi < \omega_1$. □

Mathographical Remarks

If you would like to learn more about gaps, we recommend the survey article *Gaps in $^\omega\omega$*, by Marion Scheepers, in Israel Mathematical Conference Proceedings 6 (1993), AMS, 439–561.

CHAPTER 21

Closed Unbounded Sets and Stationary Sets

21.1. Closed unbounded and stationary sets of ordinals

Let γ be a limit ordinal. A set $C \subseteq \gamma$ is *bounded* in γ if there exists $\beta < \gamma$ such that $C \subseteq \beta$. Otherwise C is *unbounded* in γ.

C is *closed* in γ if $\alpha \in C$ for all limit ordinals α such that $0 < \alpha < \gamma$ and $C \cap \alpha$ is unbounded in α.

C is *club*[1] in γ if C is closed and unbounded in γ.

EXERCISE 21.1 (G). Show that C is closed in γ iff C is closed in the sense of the order topology on γ.

EXAMPLE 21.1. If $\beta < \gamma$ then the set $(\beta, \gamma) = \{\alpha : \beta < \alpha < \gamma\}$ is closed and unbounded in γ, but, if γ is a limit ordinal, then this set is not closed in any ordinal $\alpha > \gamma$. If $cf(\gamma) = \omega$, then every cofinal subset of order type ω is *club* in γ. The set $\gamma \cap \mathbf{LIM}$ of limit ordinals smaller than γ is closed in any ordinal γ, and, if $cf(\gamma) > \omega$, then this set is also unbounded in γ.

EXERCISE 21.2 (G). Find limit ordinals γ, δ of countable cofinality that are not cardinals so that $\mathbf{LIM} \cap \gamma$ is unbounded in γ and $\mathbf{LIM} \cap \delta$ is bounded in δ.

Let $\gamma \in \mathbf{ON}$. We define:

$$CLUB(\gamma) = \{A \subseteq \gamma : \exists C \subseteq A \ (C \text{ is } club \text{ in } \gamma)\}.$$

EXAMPLE 21.2. If $A = \omega \cup \{\beta < \omega_1 : \omega < \beta \text{ and } \beta \text{ is a limit ordinal}\}$, then $A \in CLUB(\omega_1)$, but A is not *club* since $\omega \subseteq A$ but $\sup(\omega) = \omega \notin A$.

EXAMPLE 21.3. Let κ be a cardinal of uncountable cofinality, and suppose A is an unbounded subset of κ. We let

$$A' = \{\alpha < \kappa : \alpha = \sup(A \cap \alpha)\}.$$

We call A' the *derived set of* A. Note that if we consider κ with the order topology, then this notion coincides with the concept of derived set used by topologists.

EXERCISE 21.3 (G). (a) Show that A' is *club* in κ whenever A is unbounded in κ.

(b) Show that if A is *club*, then $A' \subseteq A$.

EXERCISE 21.4 (PG). Let κ be an infinite cardinal such that $cf(\kappa) > \aleph_0$. Let λ denote $cf(\kappa)$, and let $\{c_\alpha : \alpha < \lambda\}$ and $\{d_\alpha : \alpha < \lambda\}$ be *club* subsets of κ with $c_\alpha < c_\beta$ and $d_\alpha < d_\beta$ for $\alpha < \beta < \lambda$. Show that the set $C = \{c_\alpha : \alpha < \lambda \wedge c_\alpha = d_\alpha\}$ is *club*. *Hint:* To show that C is unbounded, let $\beta < \kappa$ and define recursively a sequence $(\alpha_n)_{n \in \omega}$ by letting: $\alpha_0 = \min\{\alpha : c_\alpha \geq \beta\}$, $\alpha_{2n+1} = \min\{\alpha : \alpha \geq \alpha_{2n} \wedge d_\alpha \geq c_{\alpha_{2n}}\}$, $\alpha_{2n+2} = \min\{\alpha : \alpha \geq \alpha_{2n+1} \wedge c_\alpha \geq d_{\alpha_{2n+1}}\}$.

---

[1]Some authors write "cub." The reason why the word *club* is italic here will appear in Remark 21.29.

EXAMPLE 21.4. Let κ be a cardinal of uncountable cofinality, and let $f : \kappa \to \kappa$. We say that f is *normal* if it satisfies the following conditions:
  (i)   If $\alpha < \beta < \kappa$, then $f(\alpha) \leq f(\beta)$;
  (ii) If $\gamma < \kappa$ is a limit ordinal, then $f(\gamma) = \sup\{f(\alpha) : \alpha < \gamma\}$;
  (iii) $f(\alpha) \geq \alpha$ for all $\alpha < \kappa$.

Note that (i) says that f is nondecreasing, whereas (ii) means that f is continuous in the order topology. Note also that if f is strictly increasing, then (iii) is automatically satisfied.

EXERCISE 21.5 (G). Suppose $F \subseteq {}^\kappa\kappa$ is a set of normal functions such that for all $\alpha < \kappa$, $\sup\{f(\alpha) : f \in F\} < \kappa$ exists and is smaller than κ. Show that the function $g : \kappa \to \kappa$ defined by $g(\alpha) = \sup\{f(\alpha) : f \in F\}$ is also normal.

LEMMA 21.1. *If κ is a cardinal of uncountable cofinality and $f \in {}^\kappa\kappa$ is normal, then the set $Fix(f) = \{\alpha < \kappa : f(\alpha) = \alpha\}$ is club in κ.*

PROOF. In order to see that $Fix(f)$ is closed, either recall that the set of fixed points of any continuous function of a Hausdorff space into itself is closed (see Theorem 26.1), or note that if the set $\{\alpha : f(\alpha) = \alpha\}$ is cofinal in β, then $\sup\{f(\alpha) : \alpha < \beta\} = \sup\{\alpha : \alpha < \beta\} = \beta$.

We show that $Fix(f)$ is unbounded: Let $\delta < \kappa$. We want to show that there exists some $\alpha \in Fix(f)$ with $\delta < \alpha$. To this end, we construct recursively a sequence $(\beta_n)_{n \in \omega}$ as follows: $\beta_0 = \delta$. Given β_n, let $\beta_{n+1} > f(\beta_n)$. This construction yields an increasing sequence with
$$\delta = \beta_0 \leq f(\beta_0) < \beta_1 \leq f(\beta_1) < \cdots < \beta_n \leq f(\beta_n) < \beta_{n+1} < \cdots.$$
It follows that if $\alpha = \sup\{\beta_n : n \in \omega\}$, then $\alpha > \delta$ and $f(\alpha) = \sup\{f(\beta_n) : n \in \omega\} = \alpha$. □

EXERCISE 21.6 (PG). Let $D = [0,1) \cap \mathbb{Q}$, let $\preceq_a$ denote the antilexicographical order on $D \times \omega_1$ defined by:
$$\langle d, \alpha \rangle \leq_a \langle e, \beta \rangle \text{ iff } \alpha < \beta \vee (\alpha = \beta \wedge d \leq_\mathbb{Q} e)$$
(for the natural orders $\leq_\mathbb{Q}$ on $\mathbb{Q}$ and $\leq$ on ω_1 respectively), and let $A, B \subseteq \omega_1 \setminus \{0\}$. Prove that $\langle (D \times \omega_1) \setminus (\{0\} \times A), \preceq_a \rangle$ is order-isomorphic to $\langle (D \times \omega_1) \setminus (\{0\} \times B), \preceq_a \rangle$ if and only if there exists a *club* $C \subseteq \omega_1$ such that $A \cap C = B \cap C$.

EXAMPLE 21.5. Let κ be a regular uncountable cardinal, let $n \in \omega \setminus \{0\}$, and let g be any function from κ^n into κ. Then the set
$$\mathcal{M}_g = \{\alpha < \kappa : rng(g \restriction \alpha^n) \subseteq \alpha\}$$
is *club* in κ.

To see this, consider the function $f_g : \kappa \to \kappa$ defined by $f_g(\alpha) = \sup(\{g(\vec{\beta}) + 1 : \vec{\beta} \in \alpha^n\} \cup \{\alpha\})$. Note that f_g is a normal function. Moreover, f_g has been defined so that $Fix(f) = \mathcal{M}_g$.

We shall refer to $\mathcal{M}_g$ as the set of ordinals α that are *closed* under g. Note that this use of the word "closed" has nothing to do with the order topology.

In a certain sense, Examples 21.4 and 21.5 are the only examples of *club* sets.

THEOREM 21.2. *If γ is an ordinal and C is club in γ, then there exists a normal function $f : \gamma \to \gamma$ such that $C = Fix(f)$.*

PROOF. Let C be *club* in γ. Define a function $f : \gamma \to \gamma$ by $f(\alpha) = \min\{C \backslash \alpha\}$. A straightforward verification shows that f is as required. □

EXERCISE 21.7 (PG). (a) Show that for every *club* $C \subseteq \omega_1$ there exists a function $g : \omega_1 \to \omega_1$ such that $C = \mathcal{M}_g$.
(b) Let $C = [\omega_1, \omega_2)$. Then C is *club* in ω_2. Show that if $g : \omega_2^n \to \omega_2$ for some $n \in \omega \backslash \{0\}$, then $C \neq \mathcal{M}_g$.

THEOREM 21.3. *Let γ be an ordinal of uncountable cofinality λ. If $\kappa < \lambda$ and $\{C_\alpha : \alpha < \kappa\}$ is a family of club subsets of γ, then the intersection*

$$C = \bigcap_{\alpha < \kappa} C_\alpha$$

is club in γ.

PROOF. For each $\alpha < \kappa$, let $f_\alpha : \gamma \to \gamma$ be a normal function such that $C_\alpha = Fix(f_\alpha)$. For $\beta < \gamma$, let $f(\beta) = \sup\{f_\alpha(\beta) : \alpha < \kappa\}$. Since $\kappa < cf(\gamma)$, $f(\beta) < \gamma$ for each β. By Exercise 21.5, f is a normal function from γ into γ. Now the theorem follows from the next exercise.

EXERCISE 21.8 (G). Convince yourself that $Fix(f) = \bigcap_{\alpha < \kappa} C_\alpha$. □

COROLLARY 21.4. *Let $\gamma \in \mathbf{ON}$ and $\lambda = cf(\gamma) > \omega$. Then $CLUB(\gamma)$ is a λ-complete filter.*

PROOF. Let $\kappa < \lambda$, and suppose $\{A_\alpha : \alpha < \kappa\} \subseteq CLUB(\gamma)$. For each $\alpha < \kappa$, choose a *club* set $C_\alpha \subseteq A_\alpha$. Then

$$\bigcap_{\alpha < \kappa} C_\alpha \subseteq \bigcap_{\alpha < \kappa} A_\alpha.$$

Thus $\bigcap_{\alpha < \kappa} A_\alpha$ contains a *club*, i.e., $\bigcap_{\alpha < \kappa} A_\alpha \in CLUB(\gamma)$. □

COROLLARY 21.5. *If λ is a regular uncountable cardinal, $\nu < \lambda$, and $G = \{g_\alpha : \alpha < \nu\}$ is a family of fewer than λ functions, where g_α maps $\lambda^{n(\alpha)}$ into λ for some $n(\alpha) \in \omega \backslash \{0\}$, then $\mathcal{M}_G = \bigcap_{\alpha < \nu} \mathcal{M}_{g_\alpha}$ is club in λ.*

EXAMPLE 21.6. If $cf(\gamma) = \lambda > \omega$, then there exists a family $\{C_\alpha : \alpha < \lambda\}$ of *club* subsets of γ such that $\bigcap_{\alpha < \lambda} C_\alpha = \emptyset$. To see this, let $(\beta_\alpha)_{\alpha < \lambda}$ be a cofinal sequence of ordinals in γ, and define $C_\alpha = [\beta_\alpha, \gamma)$.

In particular, the family of closed unbounded subsets of a regular uncountable cardinal κ is not closed under intersections of size κ. However, as we will see in a moment, this family is closed under so-called *diagonal intersections* of size κ.

DEFINITION 21.6. Let γ be an ordinal, and let $(C_\alpha)_{\alpha < \gamma}$ be a sequence of subsets of γ. The *diagonal intersection of the sequence* $(C_\alpha)_{\alpha < \gamma}$ is the set

$$\Delta_{\alpha < \gamma} C_\alpha = \{\beta : \forall \alpha < \beta \, (\beta \in C_\alpha)\}.$$

EXAMPLE 21.7. Let γ be an ordinal. Then

$$\Delta_{\alpha < \gamma}(\alpha, \gamma) = \gamma \text{ and } \Delta_{\alpha < \gamma}(\alpha + 1, \gamma) = \gamma \cap \mathbf{LIM}.$$

EXERCISE 21.9 (G). Let $\gamma \in \mathbf{ON}$, and let $(X_\alpha)_{\alpha < \gamma}$ be a sequence of subsets of γ.
(a) For $\alpha < \gamma$ let $Y_\alpha = X_\alpha \backslash \alpha$. Show that $\Delta_{\alpha < \gamma} X_\alpha = \Delta_{\alpha < \gamma} Y_\alpha$.
(b) For $\alpha < \gamma$ let $Y_\alpha = \bigcap_{\beta \leq \alpha} X_\alpha$. Show that $\Delta_{\alpha < \gamma} X_\alpha = \Delta_{\alpha < \gamma} Y_\alpha$.

THEOREM 21.7. *Let $\kappa > \omega$ be a regular cardinal, and let $(C_\alpha)_{\alpha<\kappa}$ be a sequence of club subsets of κ. Then the diagonal intersection $\Delta_{\alpha<\kappa} C_\alpha$ is club in κ.*

PROOF. For each $\alpha < \kappa$, choose a normal function $f_\alpha : \kappa \to \kappa$ such that $C_\alpha = Fix(f_\alpha)$. Define $f : \kappa \to \kappa$ by:
$$f(\alpha) = \sup\{f_\beta(\alpha) : \beta < \alpha\}.$$
By regularity of κ, f is well defined.

EXERCISE 21.10 (G). (a) Show that f is normal.
(b) Convince yourself that $f(\alpha) = \alpha$ if and only if $\forall \beta < \alpha \, (f_\beta(\alpha) = \alpha)$, and conclude that $Fix(f) = \Delta_{\alpha<\kappa} C_\alpha$. □

Now we are going to discuss stationary sets.

DEFINITION 21.8. Let γ be an ordinal of uncountable cofinality. A set $S \subseteq \gamma$ is called *stationary in γ* (or just *stationary* if γ is implied by the context) if $\forall A \in CLUB(\gamma) \, (S \cap A \neq \emptyset)$.

Thus, S is stationary in γ if and only if $S \notin CLUB^*(\gamma)$, where $CLUB^*(\gamma) = \{T \subseteq \gamma : \gamma \setminus T \in CLUB(\gamma)\}$. Note that every stationary set must be cofinal in γ.

EXAMPLE 21.8. Let $S = \{\beta < \omega_2 : cf(\beta) = \omega\}$. We show that S is stationary in ω_2. Let A be *club* in ω_2, and let $(a_\alpha)_{\alpha<\omega_2}$ be an enumeration of A in strictly increasing order. Then $cf(a_\omega) = \omega$; thus $a_\omega \in S$ and $S \cap A \neq \emptyset$.

But S is not closed: Let $(s_\alpha)_{\alpha<\omega_2}$ be a strictly increasing enumeration of S, and let $\gamma = \sup\{s_\alpha : \alpha < \omega_1\}$. Then $cf(\gamma) = \omega_1$, and thus $\gamma \notin S$.

This example can be generalized as follows: Let γ be a limit ordinal with $\omega \leq \kappa = cf(\kappa) < \lambda = cf(\gamma)$. Then the set $S = \{\beta < \gamma : cf(\beta) = \kappa\}$ is stationary in γ.

COROLLARY 21.9. *If κ is weakly inaccessible, then there exists a family $\{S_\alpha : \alpha < \kappa\}$ of κ pairwise disjoint stationary subsets of κ.*

We shall see in Chapter 23 (Corollary 23.4) that the analogous result for successor cardinals also holds.

LEMMA 21.10. *Let $\gamma \in \mathbf{ON}$, $\kappa < \lambda = cf(\gamma)$, and let $\{A_\alpha : \alpha < \kappa\} \subseteq P(\gamma)$ be such that $S = \bigcup_{\alpha<\kappa} A_\alpha$ is stationary. Then there exists an $\alpha < \kappa$ such that A_α is stationary.*

EXERCISE 21.11 (G). Prove Lemma 21.10.

Now we are ready to prove Lemma 14.10.

LEMMA 21.11. *Let κ be a regular uncountable cardinal, and let NS_κ denote the family of nonstationary subsets of κ. Then NS_κ is an ideal on κ with the following properties:*

(i) $\forall A \in NS_\kappa \exists B \in [\kappa]^\kappa \, (A \cap B = \emptyset \wedge A \cup B \in NS_\kappa)$;
(ii) *If $\mathcal{A} \in [NS_\kappa]^{<\kappa}$, then $\bigcup \mathcal{A} \in NS_\kappa$;*
(iii) *If $\mathcal{A} \in [NS_\kappa]^\kappa$, then $\exists B \in NS_\kappa \forall A \in \mathcal{A} \, (|A \setminus B| < \kappa)$.*

PROOF. Let $A \in NS_\kappa$, and let $C \subseteq \kappa$ be a *club* such that $A \cap C = \emptyset$. If we set $B = C \setminus C'$, then B is as required in (i). Point (ii) is an immediate consequence of Lemma 21.10. For the proof of (iii), let $\{A_\alpha : \alpha < \kappa\} \subseteq NS_\kappa$, and consider the set $B = \{\beta \in \kappa : \exists \alpha < \beta\, (\beta \in A_\alpha)\}$. The set B is often called the *diagonal union* of the sequence $(A_\alpha)_{\alpha<\kappa}$ and is denoted by $\nabla_{\alpha<\kappa} A_\alpha$. It is immediate from the definition of B that $A_\beta \setminus B \subseteq \beta + 1$ for all $\beta < \kappa$. Moreover, Theorem 21.7 implies that $B \in NS_\kappa$. □

The next theorem, which is referred to in the literature as "Fodor's Lemma" or "the Pressing Down Lemma" is a tremendously useful tool.

Let f be a function such that $dom(f) \cup rng(f) \subseteq \mathbf{ON}$. We call f *regressive* if
$$\forall \alpha \in dom(f) \setminus \{0\}(f(\alpha) < \alpha).$$

THEOREM 21.12. *Let κ be a regular uncountable cardinal, S a stationary subset of κ, and $f : S \to \kappa$ a regressive function. Then there exists $\alpha < \kappa$ such that $f^{-1}\{\alpha\}$ is stationary in κ.*

PROOF. Assume towards a contradiction that for each $\alpha < \kappa$ there exists a *club* set C_α in κ such that $C_\alpha \cap f^{-1}\{\alpha\} = \emptyset$. Let $D = \Delta_{\alpha<\kappa} C_\alpha$ be the diagonal intersection of the family $\{C_\alpha : \alpha < \kappa\}$. If $\gamma \in D$, then $f(\gamma) \neq \alpha$ for all $\alpha < \gamma$, thus the inequality $f(\gamma) < \gamma$ does not hold; i.e., f is not regressive. □

Now we give some typical applications of the Pressing Down Lemma.

We begin with a Ramsey-type application. The Pressing Down Lemma can be used to improve the result of Exercise 15.27 as follows:

THEOREM 21.13. *Suppose λ is a regular uncountable cardinal, S is a stationary subset of λ, and $f : S \to \lambda$. Then there exists a stationary $T \subseteq S$ such that $f \restriction T$ is either constant or strictly increasing.*

PROOF. Let λ, f, S be as in the assumptions.

Let $C = \{\alpha \in \lambda : \forall \beta \in S \cap \alpha\, (f(\beta) < \alpha)\}$.

If $g : \lambda \to \lambda$ is such that $g(\beta) = f(\beta)$ for all $\beta \in S$ and $g(\beta) = 0$ for $\beta \notin S$, then $C = \mathcal{M}_g$, and thus C is *club* in λ (see Example 21.5). Thus $S \cap C$ is a stationary set.

Let $S_0 = \{\alpha \in S \cap C : f(\alpha) < \alpha\}$. If S_0 is stationary, then by the Pressing Down Lemma, there exists β such that the set $S_1 = \{\alpha \in S_0 : f(\alpha) = \beta\}$ is stationary. Setting $T = S_1$, we get what we want. If S_0 is not stationary, then we let $T = \{\alpha \in S \cap C : f(\alpha) \geq \alpha\}$. Again, T is as required. □

Let $\mathcal{A} = \{A_\beta : \beta < \alpha\}$ be a family of pairwise disjoint sets. A *transversal* for $\mathcal{A}$ is a set $X \subseteq \bigcup \mathcal{A}$ such that $|X \cap A_\beta| = 1$ for all $\beta < \alpha$. Recall from Chapter 17 that a family $\mathcal{B}$ of subsets of an infinite set B is *almost disjoint* if $|X| = |B|$ and $|X \cap Y| < |B|$ for all $\{X, Y\} \in [\mathcal{B}]^2$.

THEOREM 21.14. *Let κ be an infinite cardinal, $\mathcal{A} = \{A_\alpha : \alpha < \kappa^+\}$ a family of pairwise disjoint sets with $|A_\alpha| = \kappa$ for each $\alpha < \kappa^+$. Then there exists a maximal almost disjoint family $\mathcal{F}$ of transversals for $\mathcal{A}$ with $|\mathcal{F}| = \kappa^+$.*

PROOF. The A_α's are pairwise disjoint and have cardinality κ. So we can assume that
$$A_\alpha = \{\langle \alpha, \beta \rangle : \beta < \max\{\kappa, \alpha\}\}.$$
Thus each transversal for $\mathcal{A}$ can be treated as a function from κ^+ to κ^+ that is regressive on $\kappa^+ \setminus \kappa$. For $\beta < \kappa^+$ define:
$$F_\beta = \{\langle \alpha, 0 \rangle : \alpha \leq \beta\} \cup \{\langle \alpha, \beta \rangle : \beta < \alpha < \kappa^+\}.$$

Then the family $\mathcal{F}$ defined as:
$$\mathcal{F} = \{F_\beta : \beta < \kappa^+\}$$
is an almost disjoint family of transversals for $\mathcal{A}$ with $|\mathcal{F}| = \kappa^+$.

Let us show that $\mathcal{F}$ is maximal. Let G be a transversal for $\mathcal{A}$ such that $G \notin \mathcal{F}$. Then G is a regressive function on $\kappa^+ \backslash \kappa$, and Fodor's Lemma implies that there exists $\gamma < \kappa^+$ such that $G^{-1}\{\gamma\}$ is stationary. The latter implies that $|G \cap F_\gamma| = \kappa^+$, and hence $\{G\} \cup \mathcal{F}$ is not an almost disjoint family. $\square$

EXERCISE 21.12 (PG). Let κ be a regular uncountable cardinal, let $S \subseteq \kappa$ be stationary, and let $\{S_\alpha : \alpha < \kappa\}$ be a family of nonstationary, pairwise disjoint subsets of κ such that $S = \bigcup_{\alpha < \kappa} S_\alpha$. Use Fodor's Lemma to show that $\{\min S_\alpha : \alpha < \kappa\}$ is stationary in κ. Hint: Show that the set $S \backslash \{\min S_\alpha : \alpha < \kappa\}$ is not stationary.

EXERCISE 21.13 (G). Let S be a stationary subset of ω_1. Show that the set $S \cap S' = \{\alpha \in S : \sup(S \cap \alpha) = \alpha\}$ is stationary in ω_1.

If S is a stationary subset of ω_1, then S does not need to contain a *club*, i.e., a closed subset of order type ω_1, but S does contain closed subsets of order type α for every $\alpha < \omega_1$. Even more is true.

THEOREM 21.15. *Let S be a stationary subset of ω_1. Then the set*
$$B(S) = \{\alpha : \forall \gamma < \alpha \, \exists A \subseteq (S \cap \alpha) \backslash \gamma \, (A \text{ is closed in } \alpha \wedge ot(A) = \alpha)\}$$
is club in ω_1.

PROOF. Let S be a stationary subset of ω_1, and suppose towards a contradiction that $B(S)$ is nonstationary. Let $S_1 = S \backslash B(S)$, and define $S_2 = S_1 \cap S_1'$. By Lemma 21.10, S_1 is stationary, and Exercise 21.13 implies that S_2 is also stationary. Define a function $f : S_2 \to \omega_1$ by:
$$f(\alpha) = \sup\{\beta : \forall \gamma < \alpha \, \exists A \subseteq \alpha \backslash \gamma \, (ot(A) = \beta \wedge \sup A = \alpha)\}.$$
Clearly, $f(\alpha) \leq \alpha$ for all $\alpha \in S_2$. Thus $S_2 = S_3 \cup S_4$, where $S_3 = \{\alpha \in S_2 : f(\alpha) = \alpha\}$ and $S_4 = \{\alpha \in S_2 : f(\alpha) < \alpha\}$. The following claim implies that S_2 is not stationary in ω_1 and thus leads to a contradiction.

CLAIM 21.16. *Neither S_3 nor S_4 is stationary.*

PROOF. By contradiction. Suppose S_3 is stationary. By Exercise 21.13, there exists a strictly increasing sequence $(\alpha_n)_{n \in \omega}$ of elements of S_3 such that $\sup_{n \in \omega} \alpha_n \in S_3$. Pick such a sequence and let $\alpha_\omega = \sup_{n \in \omega} \alpha_n$. Let $\gamma < \alpha_\omega$, and let n_0 be such that $\gamma < \alpha_{n_0}$. For each $n \in \omega \backslash n_0$, choose a set $A_n \subseteq (\alpha_{n+1} \backslash \alpha_n) \cap S$ that is closed in α_n and of order type α_n. The the set $A = \bigcup_{n \in \omega \backslash n_0} A_n \cup \{\alpha_n : n \in \omega\}$ is closed in α_ω, and $ot(A) = \alpha_\omega$. Thus $\alpha_\omega \in B(S)$, and hence $\alpha_\omega \notin S_3$.

Now suppose S_4 is stationary. Since the restriction of f to S_4 is regressive, by the Pressing Down Lemma there exist a stationary $S_5 \subseteq S_4$ and a countable ordinal β such that $f(\alpha) = \beta$ for all $\alpha \in S_5$. Choose a strictly increasing sequence $(\alpha_n)_{n \in \omega}$ of elements of S_4 such that if $\alpha_\omega = \sup_{n \in \omega} \alpha_n$, then $\alpha_\omega \in S_4$. Let $\gamma < \alpha_\omega$, and proceed as above to produce a closed subset A of $\alpha_\omega \backslash \gamma$ of order type $(\beta+1) \cdot \omega > \beta$ such that A is cofinal in α_ω. $\square\square$

THEOREM 21.17. *Consider $X = \omega_1 \times (\omega_1 + 1)$ as a topological space with the product of the order topologies on ω_1 and $\omega_1 + 1$. Then X is not normal.*

PROOF. Let
$$A = \{\langle \alpha, \alpha \rangle : \alpha < \omega_1\}, \quad B = \{\langle \alpha, \omega_1 \rangle : \alpha < \omega_1\}.$$
It is not hard to see that A and B are disjoint closed subsets of X. Let U be an open neighborhood of A. Then for every β such that $0 < \beta < \omega_1$ there exists an ordinal $f(\beta) < \beta$ such that $(f(\beta), \beta+1) \times (f(\beta), \beta+1) \subseteq U$. By Fodor's Lemma there exist $\beta_0 < \omega_1$ and a stationary set $S \subseteq \omega_1$ such that $f(\gamma) = \beta_0$ for all $\gamma \in S$. Since S is unbounded in ω_1, it follows that
$$C = (\beta_0, \omega_1) \times (\beta_0, \omega_1) \subseteq U.$$
Now let V be an open neighborhood of B. There exists an α such that $\beta_0 < \alpha < \omega_1$ and $\langle \beta_0 + 1, \alpha \rangle \in V$. The latter implies that $U \cap V \neq \emptyset$, and hence X is not normal. □

EXERCISE 21.14 (PG). Show with the help of Fodor's Lemma that for every continuous $f : \omega_1 \to \mathbb{R}$ there exist a real number r and an ordinal $\alpha < \omega_1$ such that $f(\beta) = r$ for all $\beta \in (\alpha, \omega_1)$. *Hint:* As a first step, show that there exist an integer n and an $\alpha_0 < \omega_1$ such that $f(\beta) \in (n-1, n+1)$ for all $\beta \in (\alpha_0, \omega_1)$.

Our last application of Fodor's Lemma is the proof of a famous theorem of J. H. Silver. This theorem was already mentioned in the Mathographical Remarks at the end of Chapter 11. Silver's original proof used model-theoretic methods that are beyond the scope of this book.

THEOREM 21.18. *If β is a limit ordinal of uncountable cofinality such that $\beta < \aleph_\beta$, and if there exists a stationary subset $S \subseteq \beta$ such that $2^{\aleph_\alpha} = \aleph_{\alpha+1}$ for all $\alpha \in S$, then $2^{\aleph_\beta} = \aleph_{\beta+1}$.*

PROOF. We give the proof for the special case $\beta = \omega_1 = S$ and leave the general case as an exercise.

So let us assume that $\forall \alpha < \omega_1 \, (2^{\aleph_\alpha} = \aleph_{\alpha+1})$. First we code every subset of $\aleph_{\omega_1}$ by an ω_1-sequence of ordinals. To this end, choose for each $\alpha < \omega_1$ an injection $\pi_\alpha : \mathcal{P}(\aleph_\alpha) \to \aleph_{\alpha+1}$. A set $A \subseteq \aleph_{\omega_1}$ is coded by the sequence $\pi(A) = (\pi_\alpha(A \cap \aleph_\alpha))_{\alpha < \omega_1}$.

FACT 21.19. *If $A, B \subseteq \aleph_{\omega_1}$, then either $A = B$, or the set $\{\alpha < \omega_1 : \pi(A)(\alpha) = \pi(B)(\alpha)\}$ is bounded in ω_1.*

On $\mathcal{P}(\aleph_{\omega_1})$ we define a binary relation R as follows:
$$R(A, B) \text{ iff } \{\alpha < \omega_1 : \pi(A)(\alpha) \leq \pi(B)(\alpha)\} \text{ is stationary.}$$
Since it is impossible to split a set of the form (γ, ω_1) into two nonstationary subsets, Fact 21.19 implies that $\forall A, B \subseteq \aleph_{\omega_1} \, (R(A, B) \vee R(B, A))$.

For $B \subseteq \aleph_{\omega_1}$ let
$$D(B) = \{A \in \mathcal{P}(\aleph_{\omega_1}) : R(A, B)\}.$$

LEMMA 21.20. $|D(B)| \leq \aleph_{\omega_1}$ *for every* $B \subseteq \aleph_{\omega_1}$.

Let us first show how Theorem 21.18 follows from the lemma. For $X \subseteq \mathcal{P}(\aleph_{\omega_1})$ define $Y(X) = \bigcup \{D(A) : A \in X\}$.

COROLLARY 21.21. *If $X \in [\mathcal{P}(\aleph_{\omega_1})]^{\aleph_{\omega_1+1}}$, then $Y(X) = \mathcal{P}(\aleph_{\omega_1})$.*

PROOF. Note that if $Z \in \mathcal{P}(\aleph_{\omega_1}) \setminus Y(X)$ then $D(Z) \supseteq Y(X) \supseteq X$, and apply Lemma 21.20. □

Finally, note that it follows from Lemma 21.20 that $|Y(X)| = \aleph_{\omega_1+1}$ for every $X \in [\mathcal{P}(\aleph_{\omega_1})]^{\aleph_{\omega_1+1}}$.

It remains to prove the lemma.

PROOF OF LEMMA 21.20. Assume towards a contradiction that there exists $B^* \in \mathcal{P}(\aleph_{\omega_1})$ such that $|D(B^*)| \geq \aleph_{\omega_1+1}$. For $A \in D(B^*) \setminus \{B^*\}$ let $S_A = \{\alpha < \omega_1 : \pi(A)(\alpha) < \pi(B^*)(\alpha)\}$. Then S_A is a stationary subset of ω_1. Since $|D(B^*)| \geq \aleph_{\omega_1+1}$, the Pigeonhole Principle allows us to pick a stationary set $S \subseteq \omega_1$ such that $|\{A \in D(B^*) : S_A = S\}| \geq \aleph_{\omega_1+1}$. We define:

$$M = \{A \in D(B^*) : S_A = S\}.$$

Let $C = \omega_1 \cap \mathbf{LIM}$. Then $C \cap S$ is stationary. Consider three cases:[2]

Case 0: $\pi(B^*)(\gamma) < \aleph_\gamma$ for all $\gamma \in C \cap S$.

For $A \in M$ we define a function $r_A : C \cap S \to \omega_1$ by

$$r_A(\gamma) = \min\{\alpha < \omega_1 : \pi(A)(\gamma) < \aleph_\alpha\}.$$

By the assumption of Case 0, r_A is a regressive function. Hence there exists an α_A such that the set $r_A^{-1}\{\alpha_A\}$ is stationary. Let us assign such α_A to each $A \in M$. Since $\langle r_A^{-1}\{\alpha_A\}, \alpha_A \rangle \in \mathcal{P}(\aleph_1) \times \aleph_1$, the Pigeonhole Principle implies that there exists a pair $\langle T, \delta \rangle$ such that the set

$$N = \{A \in D(B^*) : \langle r_A^{-1}\{\alpha_A\}, \alpha_A \rangle = \langle T, \delta \rangle\}$$

has cardinality $\aleph_{\omega_1+1}$.

Note that if $A \in N$, then $\pi(A)\lceil T : T \to \aleph_\delta$. Since $\aleph_\delta^{\aleph_1} < \aleph_{\omega_1+1}$, there are different $A, B \in N$ such that $\pi(A)\lceil T = \pi(B)\lceil T$. This contradicts Fact 21.19.

Case 1: $\pi(B^*)(\gamma) < \aleph_{\gamma+1}$ for all $\gamma \in C \cap S$.

For each $\gamma \in C \cap S$ fix an injection $f_\gamma : \pi(B^*)(\gamma) \to \aleph_\gamma$. Let

$$r_A(\gamma) = \min\{\alpha < \omega_1 : f_\gamma(\pi(A)(\gamma)) < \aleph_\alpha\},$$

and proceed as in Case 0. To get the final contradiction, instead of $\pi(A)\lceil T$ consider functions $g_A : T \to \aleph_\delta$ defined by:

$$g_A(\gamma) = f_\gamma(\pi(A)(\gamma)).$$

Case 2: $\pi(B^*)(\gamma) \geq \aleph_{\gamma+1}$ for some $\gamma \in C \cap S$.

This is impossible, since $rng(\pi_\gamma) \subseteq \aleph_{\gamma+1}$. □□

EXERCISE 21.15 (G). Prove Theorem 21.18 in its full generality.

EXERCISE 21.16 (R). Prove that if $2^{\aleph_\alpha} \leq \aleph_{\alpha+2}$ for all $\alpha < \omega_1$, then $2^{\aleph_{\omega_1}} \leq \aleph_{\omega_1+2}$. *Hint:* GCH below $\aleph_{\omega_1}$ was only used to make Case 2 in the proof of Lemma 21.20 redundant. Under the weaker assumptions, let the π_α's be injections from $\mathcal{P}(\aleph_\alpha)$ into $\aleph_{\alpha+2}$, and for $A, B \subseteq \aleph_{\omega_1}$, consider the sets $D(A, B) = \{Y \subseteq \aleph_{\omega_1} : \exists$ stationary $S \subseteq \omega_1 \forall \alpha \in S(\pi(Y)(\alpha) < \pi(A)(\alpha) < \pi(B)(\alpha)\}$. Then prove the following modification of Lemma 21.20: $|D(B, A)| \leq \aleph_{\omega_1}$ for every $A, B \subseteq \aleph_{\omega_1}$.

---

[2] One could reduce the proof to Case 1. However, we want to highlight the place in the proof that requires modification for the solutions of Exercises 21.16 and 21.17.

EXERCISE 21.17 (X). (a) Prove that if $\gamma < \omega_1$ is such that $2^{\aleph_\alpha} \leq \aleph_{\alpha+\gamma}$ for all $\alpha < \omega_1$, then $2^{\aleph_{\omega_1}} \leq \aleph_{\omega_1+\gamma}$.

(b) Prove that if $2^{\aleph_\alpha} \leq \aleph_{\alpha+\alpha}$ for all $\alpha < \omega_1$, then $2^{\aleph_{\omega_1}} < \aleph_{\omega_1+\omega_1}$.

(c) Formulate and prove the most general version of Theorem 21.18 you can extract from its proof and points (a) and (b).

REMARK 21.22. It is possible and sometimes useful to generalize the notions of closed unbounded and stationary sets to classes. A class **C** of ordinals is *unbounded in* **ON** if it is proper. We say that **C** is *closed* if $\mathbf{C} \cap \gamma$ is closed in γ for every ordinal γ. A class $\mathbf{S} \subseteq \mathbf{ON}$ is *stationary in* **ON** if it has nonempty intersection with every class **C** that is closed and unbounded in **ON**. Many results of this section have analogues for proper classes, but it can be tricky to formalize these in L_S.

Let us consider just two examples. A proper functional class $\mathbf{F} \subseteq \mathbf{ON} \times \mathbf{ON}$ is *normal* if $dom(\mathbf{F}) = \mathbf{ON}$ and

(i) If $\alpha < \beta$, then $\mathbf{F}(\alpha) \leq \mathbf{F}(\beta)$;
(ii) If $\gamma \in \mathbf{LIM}$, then $\mathbf{F}(\gamma) = \sup\{\mathbf{F}(\alpha) : \alpha < \gamma\}$;
(iii) $\mathbf{F}(\alpha) \geq \alpha$ for all $\alpha \in \mathbf{ON}$.

We have the following analogue of Lemma 21.1:

CLAIM 21.23. *Suppose* **F** *is a proper, normal functional class. Then the class* $Fix(\mathbf{F}) = \{\alpha \in \mathbf{ON} : \mathbf{F}(\alpha) = \alpha\}$ *is closed and unbounded in* **ON**.

EXERCISE 21.18 (PG). Prove Claim 21.23. Don't forget to verify that $Fix(\mathbf{F})$ is a class in the first place.

EXERCISE 21.19 (PG). Show that if $\mathbf{C}_0, \ldots, \mathbf{C}_n$ are finitely many closed unbounded classes of ordinals, then $\mathbf{C}_0 \cap \cdots \cap \mathbf{C}_n$ is a closed unbounded class of ordinals.

EXERCISE 21.20 (R). Formulate and prove an analogue of Theorem 21.3. *Hint:* Of course, everybody's first impulse is to say: "Let $\{\mathbf{C}_i : i \in I\}$ be a set of classes of ordinals such that each $\mathbf{C}_i$ is closed and unbounded in **ON**." But this is *streng verboten*.

21.2. Closed unbounded and stationary subsets of $[X]^{<\kappa}$

Suppose that $\mathfrak{G} = \langle \omega_1, *, 0 \rangle$ is a group. Does there exist a countable subgroup $\mathfrak{H} = \langle H, *, 0 \rangle$ of $\mathfrak{G}$ such that $\omega \subseteq H$? Yes. To see this, consider functions $* : \omega_1 \times \omega_1 \to \omega_1$ and $i : \omega_1 \to \omega_1$, where i assigns to each $\alpha \in \omega_1$ its inverse in $\mathfrak{G}$. Let $\mathcal{M}_*$ and $\mathcal{M}_i$ be as in Example 21.5, and let $\mathcal{M}_{\{*,i\}}$ denote $\mathcal{M}_* \cap \mathcal{M}_i$. Then $\mathcal{M}_{\{*,i\}}$ is *club* in ω_1, and if $\alpha \in \mathcal{M}_{\{*,i\}} \backslash \omega$, then $\mathfrak{H} = \langle \alpha, *, 0 \rangle$ is as required.

This argument relies heavily on the fact that $\mathfrak{G}$ has cardinality $\aleph_1$. If the universe of $\mathfrak{G}$ were ω_2 instead, the above reasoning would yield a subgroup $\mathfrak{H}$ of $\mathfrak{G}$ whose universe is an ordinal $\alpha < \omega_2$, but not necessarily countable.

EXERCISE 21.21 (G). Let $\mathfrak{G} = \langle \omega_2, *, 0 \rangle$ be a group such that $2 * (\omega_1 + 7) = 0$, and define $\mathcal{M}_{\{*,i\}}$ as above. Show that the smallest $\alpha > \omega$ in $\mathcal{M}_{\{*,i\}}$ is uncountable.

But the group $\mathfrak{G}$ of Exercise 21.21 does have a countable subgroup whose universe contains ω. This is a consequence of the following general theorem.

THEOREM 21.24. *Let $X \neq \emptyset$, let $G = \{g_\xi : \xi < \eta\}$ be a family of functions $g : X^{n(\xi)} \to X$, where $n(\xi) \in \omega$ for all $\xi < \eta$, and let $A \subseteq X$. Then there exists a smallest $B \subseteq X$ such that $A \subseteq B$ and B is closed under all functions in G. Moreover, $|B| \leq \max\{\aleph_0, |A|, |\eta|\}$.*

The set B of Theorem 21.24 will be called the *closure of A under G*.

Theorem 21.24 has an important application to model theory. If
$$\mathfrak{X} = \langle X, (R_i)_{i \in I}, (F_j)_{j \in J}, (C_k)_{k \in K} \rangle$$
is a model of a first-order language L, and if $B \subseteq X$, then B is the universe of a submodel of $\mathfrak{X}$ if and only if $\{C_k : \in K\} \subseteq B$ and B is closed under all functions F_j. Since one can treat constants as functions of arity zero, Theorem 21.24 implies the following:

COROLLARY 21.25. *Let $\mathfrak{X} = \langle X, (R_i)_{i \in I}, (F_j)_{j \in J}, (C_k)_{k \in K} \rangle$ be a model of a first-order language L with nonlogical symbols $\{r_i : i \in I\} \cup \{f_j : j \in J\} \cup \{c_k : k \in K\}$, where r_i is a relational symbol of arity $\tau_0(i)$, f_j is a functional symbol with arity $\tau_1(j)$, and c_k is a constant symbol. Then for every $A \subseteq X$ there exists a smallest set $B \supseteq A$ such that $|B| \leq \max\{|A|, |J \cup K|, \aleph_0\}$ and*
$$\mathfrak{B} = \langle B, (R_i \cap B^{\tau_0(i)})_{i \in I}, (F_j \lceil B^{\tau_1(j)})_{j \in J}, (C_k)_{k \in K} \rangle$$
is a model of L.

In the sequel, we will rather informally write R_i instead of $R_i \cap B^{\tau_0(i)}$ and F_j instead of $F_j \lceil B^{\tau_1(j)}$. Also, we shall often say that B (rather than $\mathfrak{B}$) is the substructure (or submodel) of X (rather than of $\mathfrak{X}$) *generated by A*. For example, in the situation discussed at the beginning of this section, we would be looking for the *subgroup of ω_1 generated by ω*.

Before we prove Theorem 21.24, let us derive another important consequence of it. By $\mathrm{MA}^-(\kappa)$ we abbreviate the following statement:

(**MA$^-(\kappa)$**) Let $\langle \mathbb{P}, \leq \rangle$ be a c.c.c. p.o. of cardinality $\leq \kappa$, and let $\mathcal{D}$ be a family of dense subsets of $\mathbb{P}$ with $|\mathcal{D}| \leq \kappa$. Then there exists a $\mathcal{D}$-generic filter in $\mathbb{P}$.

THEOREM 21.26. *For each infinite cardinal κ, the statements $\mathrm{MA}(\kappa)$ and $\mathrm{MA}^-(\kappa)$ are equivalent.*

PROOF. Since $\mathrm{MA}^-(\kappa)$ is just $\mathrm{MA}(\kappa)$ restricted to a narrower class of p.o.'s, only the implication $\mathrm{MA}^-(\kappa) \to \mathrm{MA}(\kappa)$ needs a proof. This proof hinges on the following lemma.

LEMMA 21.27. *Let $\kappa \leq \lambda$ be infinite cardinals, let $\mathbb{P}$ be a set of cardinality λ, partially ordered by a relation $\leq$, and let $\mathcal{D}$ be a family of size κ of dense subsets of $\langle \mathbb{P}, \leq \rangle$. Then there exists $\mathbb{Q} \subseteq \mathbb{P}$ such that:*

(1) *If $p, q \in \mathbb{Q}$ and p is compatible with q in $\mathbb{P}$, then p is compatible with q in $\mathbb{Q}$, i.e., if there is some $r \in \mathbb{P}$ such that $r \leq p$ and $r \leq q$, then such an r also exists in $\mathbb{Q}$;*
(2) *$\mathbb{Q} \cap D$ is dense in $\langle \mathbb{Q}, \leq \rangle$ for all $D \in \mathcal{D}$;*
(3) *$|\mathbb{Q}| = \kappa$.*

EXERCISE 21.22 (G). Derive Theorem 21.26 from Lemma 21.27. *Hint:* For $\langle \mathbb{P}, \leq \rangle$ and $\mathcal{D}$ as in $\mathrm{MA}(\kappa)$, form $\mathbb{Q}$ as in Lemma 21.27. Note that by (1), $\langle \mathbb{Q}, \leq \rangle$ still satisfies the c.c.c. Then apply $\mathrm{MA}^-(\kappa)$ to $\langle \mathbb{Q}, \leq \rangle$.

PROOF OF LEMMA 21.27. Let $\mathcal{D} = \{D_\xi : \xi < \kappa\}$. For each $\xi < \kappa$, let $g_\xi : \mathbb{P} \to D_\xi$ be such that $g_\xi(p) \leq p$. Moreover, pick a function $g_\kappa : \mathbb{P}^2 \to \mathbb{P}$ in such a way that if p, q are compatible, then $g_\kappa(p, q) \leq p$ and $g_\kappa(p, q) \leq q$. Let $\mathbb{P}_0$ be an arbitrary subset of $\mathbb{P}$ of size κ, and let $\mathbb{Q}$ be the closure of $\mathbb{P}_0$ under $\{g_\xi : \xi < \kappa+1\}$. Then $\mathbb{Q}$ satisfies (1)-(3). □□

PROOF OF THEOREM 21.24. Let X, A, G be as in the assumptions. Consider the family:
$$\mathcal{M}_G = \{B \subseteq X : A \subseteq B \wedge B \text{ is closed under } G\}.$$
Then $\mathcal{M}_G \neq \emptyset$, since $X \in \mathcal{M}_G$, and hence, for every $B \in \mathcal{M}_G$, we have $A \subseteq \bigcap \mathcal{M}_G \subseteq B$. Therefore, if $\bigcap \mathcal{M}_G$ is closed under G, this must be the smallest element of $\mathcal{M}_G$.

EXERCISE 21.23 (G). Show that $\bigcap \mathcal{M}_G$ is closed under G.

It remains to prove the last sentence of Theorem 21.24. For this, it suffices to produce at least one element of $\mathcal{M}_G$ whose cardinality does not exceed $\max\{\aleph_0, |A|, |\eta|\}$. For $C \subseteq X$, let $G[C] = \bigcup_{\xi < \eta} g_\xi[C^{n(\xi)}]$. Now define recursively a sequence $(A_n)_{n \in \omega}$ of subsets of X by letting $A_0 = A$ and $A_{n+1} = A_n \cup G[A_n]$. Consider the set $B = \bigcup_{n \in \omega} A_n$. The following exercise concludes the proof of Theorem 21.24.

EXERCISE 21.24 (G). Verify the following properties of the A_n's and of B:
(a) $A_n \subseteq A_{n+1}$ for all $n \in \omega$.
(b) B is closed under all functions in G.
(c) For each $n \in \omega$, $|A_n| \leq \max\{\aleph_0, |A|, |\eta|\}$.
(d) $|B| \leq \max\{\aleph_0, |A|, |\eta|\}$. □

The next exercise is redundant for the proof of Theorem 21.24, but interesting on its own.

EXERCISE 21.25 (PG). Show that the set B constructed in the last part of the proof of Theorem 21.24 actually is equal to (rather than merely a superset of) the closure of A. *Hint:* This is similar to Exercise 4.7.

You will probably have noticed a certain similarity between the proof of Theorem 21.24 and the arguments used in describing Examples 21.4 and 21.5. In order to express the precise nature of this similarity, we need new concepts.

DEFINITION 21.28. Let κ be an uncountable regular cardinal, and let $X \neq \emptyset$. A subset $\mathcal{C} \subseteq [X]^{<\kappa}$ is *unbounded in* $[X]^{<\kappa}$ if for every $Y \in [X]^{<\kappa}$ there is $Z \in \mathcal{C}$ such that $Y \subseteq Z$. A set $\mathcal{C} \subseteq [X]^{<\kappa}$ is *closed* if for every increasing sequence $\langle Y_\alpha : \alpha < \nu \rangle$ of length $\nu < \kappa$ of elements of $\mathcal{C}$ the union $\bigcup_{\alpha < \nu} Y_\alpha$ of is also an element of $\mathcal{C}$. A set $\mathcal{C}$ that is both closed and unbounded in $[X]^{<\kappa}$ will be called club in $[X]^{<\kappa}$. If $\mathcal{D} \subseteq [X]^{<\kappa}$ contains a club $\mathcal{C}$, then we write $\mathcal{D} \in \text{CLUB}([X]^{<\kappa})$.

REMARK 21.29. Note that we use different fonts for clubs and CLUBs, depending on whether their members belong to λ or $[X]^{<\kappa}$. Also note that what we call *unbounded subsets of* $[X]^{<\kappa}$ should really be called *cofinal subsets of* $[X]^{<\kappa}$. Again, we are following a bad habit that is deeply entrenched in the literature.

If $|X| < \kappa$, then $\mathcal{C}$ is unbounded in $[X]^{<\kappa}$ if and only if $X \in \mathcal{C}$. Therefore, some authors would require in Definition 21.28 that $|X| \geq \kappa$. We do not want to make this restriction. If $|X| < \kappa$, the theorems of this section are not false, they are merely dull.

EXAMPLE 21.9. If $A \in [X]^{<\kappa}$, then the set $\check{A} = \{B \in [X]^{<\kappa} : A \subseteq B\}$ is club in $[X]^{<\kappa}$. The set $\check{A}$ is called the *cone of A*.

EXAMPLE 21.10. If κ is a regular uncountable cardinal, then κ itself is a club subset of $[\kappa]^{<\kappa}$, and so is every *club* subset of κ.

Now let X and $G = \{g_\xi : \xi < \eta\}$ be as in the assumptions of Theorem 21.24, and let κ be a regular uncountable cardinal such that $\kappa > \eta$. By Theorem 21.24, if $A \in [X]^{<\kappa}$, then the closure B of A under G is also in $[X]^{<\kappa}$. Since $B \supseteq A$, this shows that the set

$$\mathcal{M}_G = \{B \in [X]^{<\kappa} : A \subseteq B \land B \text{ is closed under } G\}$$

is unbounded in $[X]^{<\kappa}$. It turns out that $\mathcal{M}_G$ is also closed in $[X]^{<\kappa}$.

EXERCISE 21.26 (G). Show that if $\langle B_\alpha : \alpha < \nu \rangle$ is an increasing sequence of members of $\mathcal{M}_G$ of length $\nu < \kappa$, then $\bigcup_{\alpha < \nu} B_\alpha \in \mathcal{M}_G$.

We have just proved the following:

THEOREM 21.30. *Let* $X \neq \emptyset$, *let* $G = \{g_\xi : \xi < \eta\}$ *be a family of functions* $g : X^{n(\xi)} \to X$, *where* $n(\xi) \in \omega$ *for all* $\xi < \eta$, *and let* $A \subseteq X$. *Then for every regular uncountable cardinal* $\kappa > \eta$ *the set*

$$\mathcal{M}_G^\kappa = \{B \in [X]^{<\kappa} : A \subseteq B \land B \text{ is closed under } G\}$$

is club *in* $[X]^{<\kappa}$.

Note that the word "closed" is used in two utterly different meanings in Theorem 21.30. If κ is implied by the context, we will simply write $\mathcal{M}_G$ instead of $\mathcal{M}_G^\kappa$ (we did so already when we proved Theorem 21.30).

EXERCISE 21.27 (G). Where in the proof of Theorem 21.30 did we use the assumption that κ is regular? Why do we need to assume that κ is uncountable?

The observation that precedes Corollary 21.25 yields the following important result:

COROLLARY 21.31. *Let* $\mathfrak{X} = \langle X, (R_i)_{i \in I}, (F_j)_{j \in J}, (C_k)_{k \in K} \rangle$ *be a model of a first-order language L that has fewer than κ functional and constant symbols. Then the set* $\mathcal{B} = \{B \in [X]^{<\kappa} : B \text{ is the universe of a submodel of } \mathfrak{X}\}$ *is club in* $[X]^{<\kappa}$.

Club and CLUB subsets of $[X]^{<\kappa}$ behave in many ways like *club* and *CLUB* subsets of ordinals. Let us now state and prove some analogues of the theorems of Section 21.1. We begin with the analogue of Theorem 21.3.

THEOREM 21.32. *Let* $X \neq \emptyset$, *let* κ *be a regular uncountable cardinal, and let* $\{\mathcal{C}_\alpha : \alpha < \nu\}$ *be a family of* club *subsets of* $[X]^{<\kappa}$ *of size* $\nu < \kappa$. *Then* $\bigcap_{\alpha < \nu} \mathcal{C}_\alpha$ *is* club *in* $[X]^{<\kappa}$.

We shall not prove Theorem 21.32 at this point, but derive it later from other results. Of course, if you are eager for a little practice, you may prove it yourself before reading on.

COROLLARY 21.33. *If* X, κ *are as above, then* $\mathrm{CLUB}([X]^{<\kappa})$ *is a κ-complete filter in* $\mathcal{P}([X]^{<\kappa})$.

The diagonal intersection of a family $\{C_\alpha : \alpha \in \kappa\}$ of subsets of an ordinal κ can be defined as $\Delta_{\alpha \in \kappa} C_\alpha = \{\beta \in \kappa : \forall \alpha \in \beta \, (\beta \in C_\alpha)\}$. Thus it is natural to define the diagonal intersection of a family $\{\mathcal{C}_x : x \in X\}$ of subsets of $[X]^{<\kappa}$ as:

$$\Delta_{x \in X} \mathcal{C}_x = \{A \in [X]^{<\kappa} : \forall x \in A \, (A \in \mathcal{C}_x)\}.$$

For this type of diagonal intersection, the analogue of Theorem 21.7 holds.

THEOREM 21.34. *Let $X \neq \emptyset$, let κ be a regular uncountable cardinal, and let $\{\mathcal{C}_x : x \in X\}$ be a family of club subsets of $[X]^{<\kappa}$. Then $\Delta_{x \in X} \mathcal{C}_x$ is club in $[X]^{<\kappa}$.*

PROOF. First let us show that $\Delta_{x \in X} \mathcal{C}_x$ is closed. Let $\nu < \kappa$, and consider an increasing sequence $\langle A_\alpha : \alpha < \nu \rangle$ of elements of $\Delta_{x \in X} \mathcal{C}_x$. Fix $x \in \bigcup_{\alpha < \nu} A_\alpha$. There exists $\beta_x < \nu$ such that $x \in A_\alpha$ whenever $\beta_x \leq \alpha < \nu$. Since the A_α's were assumed to be in $\Delta_{x \in X} \mathcal{C}_x$, we have $A_\alpha \in \mathcal{C}_x$ whenever $\beta_x \leq \alpha < \nu$, and closedness of $\mathcal{C}_x$ implies that $\bigcup_{\alpha < \nu} A_\alpha = \bigcup_{\beta_x \leq \alpha < \nu} A_\alpha \in \mathcal{C}_x$. Since x was an arbitrary element of $\bigcup_{\alpha < \nu} A_\alpha$, we have shown that $\bigcup_{\alpha < \nu} A_\alpha \in \Delta_{x \in X} \mathcal{C}_x$.

To see that $\Delta_{x \in X} \mathcal{C}_x$ is unbounded in $[X]^{<\kappa}$, let $A \in [X]^{<\kappa}$, and construct recursively a sequence $(A_n)_{n \in \omega}$ as follows: Let $A_0 = A$. Given A_n, choose for every $x \in A_n$ an element B_n^x of $\mathcal{C}_x$ such that $B_n^x \supseteq A_n$. This is possible since the $\mathcal{C}_x$'s are unbounded in $[X]^{<\kappa}$. Then let $A_{n+1} = \bigcup_{x \in A_n} B_n^x$. Now consider $B = \bigcup_{n \in \omega} A_n$. Clearly, $A \subseteq B \in [X]^{<\kappa}$. Pick an arbitrary $x \in B$. There are $n_0 \in \omega$ and an increasing sequence $(B_n^x)_{n \in \omega \setminus n_0}$ such that $A_n \subseteq B_n^x \subseteq A_{n+1}$ and $B_n^x \in \mathcal{C}_x$ for all $n > n_0$. Thus $\bigcup_{n \in \omega \setminus n_0} B_n^x = B$, and closedness of $\mathcal{C}_x$ implies that $B \in \Delta_{x \in X} \mathcal{C}_x$. □

EXERCISE 21.28 (PG). Derive Theorem 21.32 from Theorem 21.34.

The notion of stationary sets has a straightforward generalization to the $[X]^{<\kappa}$ setting: We say that $\mathcal{S} \subseteq [X]^{<\kappa}$ is *stationary in* $[X]^{<\kappa}$ if $\mathcal{S} \cap \mathcal{C} \neq \emptyset$ for every club subset $\mathcal{C}$ of $[X]^{<\kappa}$.

EXERCISE 21.29 (G). Let κ be a regular uncountable cardinal.
(a) Show that every stationary subset of $[X]^{<\kappa}$ is unbounded in $[X]^{<\kappa}$.
(b) Show that $\mathcal{S}$ is a stationary subset of $[\kappa]^{<\kappa}$ if and only if $\mathcal{S} \cap \kappa$ is a stationary subset of the ordinal κ in the sense of Definition 21.8. *Hint:* Recall Example 21.10.
(c) Let $\nu < \kappa$ be an infinite regular cardinal, and let $\mathcal{S} \subseteq [X]^{<\kappa}$. Show that if $\mathcal{S}$ is closed under increasing unions of length ν and unbounded in $[X]^{<\kappa}$, then $\mathcal{S}$ is stationary in $[X]^{<\kappa}$. *Hint:* Compare Example 21.8.
(d) Show that the union of fewer than κ nonstationary subsets of $[X]^{<\kappa}$ is nonstationary in $[X]^{<\kappa}$.

The Pressing Down Lemma also has a generalization. Call a function whose domain is contained in $[X]^{<\kappa}$ *regressive* if $f(A) \in A$ for all $A \in dom(f)$.

THEOREM 21.35. *Let $X \neq \emptyset$, let κ be a regular uncountable cardinal, let $\mathcal{S} \subseteq [X]^{<\kappa}$ be stationary in $[X]^{<\kappa}$, and let $f : \mathcal{S} \to X$ be a regressive function. Then there exists an $x \in X$ such that $f^{-1}\{x\}$ is stationary in $[X]^{<\kappa}$.*

Theorem 21.35 is often called "Fodor's Lemma for $[X]^{<\kappa}$," although it was first formulated and proved by T. Jech.

EXERCISE 21.30 (G). Prove Theorem 21.35. *Hint:* Use Theorem 21.34.

It appears that the only major properties of *club* subsets of ordinals that we have not yet generalized to the $[X]^{<\kappa}$ setting are Lemma 21.1 and Theorem 21.2.

How about these? Let us call a function $F : [X]^{<\kappa} \to [X]^{<\kappa}$ *normal* if it satisfies the following conditions:
  (i)   If $A, B \in [X]^{<\kappa}$ are such that $A \subseteq B$, then $F(A) \subseteq F(B)$;
  (ii) If $\gamma < \kappa$ is a limit ordinal and $\langle A_\alpha : \alpha \in \gamma \rangle$ is an increasing sequence of members of $[X]^{<\kappa}$, then $F(\bigcup_{\alpha<\gamma} A_\alpha) = \bigcup_{\alpha<\gamma} F(A_\alpha)$;
  (iii) $A \subseteq F(A)$ for every $A \in [X]^{<\kappa}$.

For F as above, let $Fix(F) = \{A \in [X]^{<\kappa} : F(A) = A\}$. It is not hard to see that the $[X]^{<\kappa}$-analogue of Lemma 21.1 holds.

LEMMA 21.36. *Let $X \neq \emptyset$, let κ be a regular uncountable cardinal, and let $F : [X]^{<\kappa} \to [X]^{<\kappa}$ be a normal function. Then the set $Fix(F)$ is club in $[X]^{<\kappa}$.*

However, Theorem 21.2 cannot be literally translated into the $[X]^{<\kappa}$-context. To see this, note that if $F : [X]^{<\kappa} \to [X]^{<\kappa}$ is normal and $\langle A_\alpha : \alpha < \gamma \rangle$ is a *decreasing* sequence of fixed points of F, then $\bigcap_{\alpha<\gamma} A_\alpha$ is also a fixed point of F. On the other hand, club subsets of $[X]^{<\kappa}$ do not need to be closed under decreasing unions. For example $[X]^{<\kappa} \setminus \{\emptyset\}$ is club in $[X]^{<\kappa}$, but there is no normal function $F : [X]^{<\kappa} \to [X]^{<\kappa}$ such that $Fix(F) = [X]^{<\kappa} \setminus \{\emptyset\}$.

One can generalize Theorem 21.2 if one allows for a slight strengthening of the assumption, or a slight weakening of the conclusion.

THEOREM 21.37. *Let $X \neq \emptyset$, let κ be a regular uncountable cardinal, and let $\mathcal{C}$ be a club subset of $[X]^{<\kappa}$.*
  (a) *If for every $A \in [X]^{<\kappa}$ there exists a minimum $B \in \mathcal{C}$ above A (i.e., there exists $B \in \mathcal{C}$ such that $A \subseteq B$ and $\check{A} \cap \mathcal{C} = \check{B} \cap \mathcal{C}$), then there exists a normal function $F : [X]^{<\kappa} \to [X]^{<\kappa}$ such that $Fix(F) = \mathcal{C}$.*
  (b) *There exists a normal function $F : [X]^{<\kappa} \to [X]^{<\kappa}$ such that $Fix(F) \subseteq \mathcal{C}$.*

PROOF. To prove (a), simply let $F(A)$ be the minimum $B \in \mathcal{C}$ above A. A straightforward verification shows that F is normal and $Fix(F) = \mathcal{C}$.

For the proof of (b), construct recursively a sequence $(G_n)_{n\in\omega}$ of functions as follows: Let $G_0(\emptyset)$ be any element of $\mathcal{C}$, and let $G_1 : [X]^1 \to \mathcal{C}$ be such that $x \in G_1(\{x\})$ and $G_1(\{x\}) \supseteq G_0(\emptyset)$ for all $x \in X$. Given $G_n : [X]^n \to \mathcal{C}$, choose $G_{n+1} : [X]^{n+1} \to \mathcal{C}$ in such a way that $G_{n+1}(b) \supseteq G_n(a)$ for all $a \in [b]^n$. Then define $F : [X]^{<\kappa} \to [X]^{<\kappa}$ by: $F(A) = \bigcup_{n\in\omega} G_n[[A]^n]$.

EXERCISE 21.31 (PG). (a) Show that $F(A) = \bigcup_{a \in [A]^{<\aleph_0}} F(a)$ for each $A \in [X]^{<\kappa}$.
  (b) Conclude that F is a normal function with $Fix(F) \subseteq \mathcal{C}$. □

The extra requirement in part (a) of Theorem 21.37 is not as exotic as it may seem. For example, if $\mathfrak{X} = \langle X, (R_i)_{i\in I}, (F_j)_{j\in J}, (C_k)_{k\in K} \rangle$ is a model of a first-order language L that has fewer than κ functional and constant symbols, then the club set $\mathcal{B} = \{B \in [X]^{<\kappa} : B \text{ is the universe of a submodel of } \mathfrak{X}\}$ satisfies the assumption of Theorem 21.37(a). The following is a partial converse of Theorem 21.30:

COROLLARY 21.38. *Let $\kappa = \lambda^+$, where λ is an infinite cardinal, let $X \neq \emptyset$, and suppose $\mathcal{C}$ is club in $[X]^{<\kappa}$. Then there exists a family $\{g_\xi : \xi < \lambda\}$ such that $g_\xi : X^{n(\xi)} \to X$ for some $n(\xi) \in \omega$, and the set $\mathcal{M} = \{B \in [X]^{<\kappa} : B \text{ is closed under all functions } g_\xi\}$ is contained in $\mathcal{C}$. Moreover, if for every $A \in [X]^{<\kappa}$ there exists a minimum $B \in \mathcal{C}$ above A, then the g_ξ's can be chosen in such a way that $\mathcal{M} = \mathcal{C}$.*

EXERCISE 21.32 (PG). Prove Corollary 21.38. *Hint:* Partition λ into sets L_n, each of cardinality λ. Construct functions G_n as in the proof of Theorem 21.37(b) and define the g_ξ's in such a way that if $a \in [X]^n$, then $G_n(a) = \{g_\xi(a) : \xi \in L_n\}$.

CHAPTER 22

The ◊–principle

In Chapter 14 we promised you a consistent construction of a Suslin tree. Now is the time to make good on this promise. So let us consider the question: *What does it take to construct a Suslin tree?*

Let us try a recursive construction in the spirit of Chapter 17. Since a Suslin tree has $\aleph_1$ nodes, we may as well try to build one whose set of nodes is ω_1. We will construct a partial order relation $\leq_T$ on ω_1 level by level. Since the levels of our tree are supposed to be countable, we may as well decide at the outset that $T(\alpha) = [\omega \cdot \alpha, \omega \cdot (\alpha + 1)) = \{\xi : \omega \cdot \alpha \leq \xi < \omega \cdot \alpha + \omega\}$ for all $\alpha < \omega_1$.

The restriction of the p.o. relation $\leq_T$ to $T_{(\alpha)}$ will be denoted by $\leq_\alpha$. The recursive construction boils down to deciding, in a situation where $\leq_\alpha$ is given, which elements of $T(\alpha)$ to put on top of which elements of $T_{(\alpha)}$.

For technical reasons, we want to construct a tall tree such that every node of it has two immediate successors. The latter will ensure that the resulting tree is splitting, and then we will have to worry only about nonexistence of uncountable antichains. By Exercise 14.34, the absence of uncountable chains will follow automatically. To get a tall tree, we need to make sure that the following condition holds at every level $\alpha < \omega_1$:

$(tall(\alpha))$ $\qquad \forall \beta < \alpha \, \forall s \in T(\beta) \exists t \in T(\alpha) \, (s <_T t).$

These constraints already determine to a large extent what we need to do. At successor stages, if $\langle T_{(\alpha+1)}, \leq_{\alpha+1} \rangle$ has already been constructed, then for each $t \in T(\alpha)$ we choose two ordinals $t_\ell, t_r \in [\omega \cdot (\alpha+1), \omega \cdot (\alpha+2))$ and make them immediate successors of t. Moreover, we take care that all ordinals in $[\omega \cdot (\alpha + 1), \omega \cdot (\alpha + 2))$ will be used up as t_ℓ or t_r for some $t \in T(\alpha)$. This should take care of all reasonable requirements at successor stages.

EXERCISE 22.1 (G). Convince yourself that the implication $tall(\alpha) \to tall(\alpha + 1)$ holds if we proceed as above at successor stages.

If $\alpha > 0$ is a countable limit ordinal and $\langle T_{(\alpha)}, \leq_\alpha \rangle$ has already been constructed, we need to put the ordinals in $[\omega \cdot \alpha, \omega \cdot (\alpha+1))$ on top of cofinal branches of $T_{(\alpha)}$ so as to make $tall(\alpha)$ true.

EXERCISE 22.2 (G). Suppose $\alpha > 0$ is a countable limit ordinal and the tree $\langle T_{(\alpha)}, \leq_\alpha \rangle$ is such that $tall(\beta)$ holds for all $\beta < \alpha$. Show that for every $s \in T_{(\alpha)}$ there exists a cofinal branch B_s of $T_{(\alpha)}$ with $s \in B_s$.

So let $\{B_s : s \in T_{(\alpha)}\}$ be a family of cofinal branches of $T_{(\alpha)}$ such that $s \in B_s$ for each $s \in T_{(\alpha)}$. If we take care to put one $t_s \in T(\alpha)$ on top of each B_s, i.e., if we extend the p.o. relation so that $r <_T t_s$ for all $r \in B_s$, then $tall(\alpha)$ will hold. By careful bookkeeping we can make sure that every $t \in [\omega \cdot \alpha, \omega \cdot (\alpha + 1))$ is being used as t_s for some s.

This describes what we will call the *generic blueprint* for the construction of T. The generic blueprint yields a tall, splitting tree $\langle T, \leq_T \rangle$ of height ω_1 with all levels countable. Moreover, we have the following connection between countable ordinals and subtrees of T:

EXERCISE 22.3 (G). Convince yourself that the generic blueprint yields a tree T with the property that if $C = \{\alpha < \omega_1 : T_{(\alpha)} = \alpha\}$, then

$$C = \{\alpha < \omega_1 : \omega \cdot \alpha = \alpha\} = \{\omega^\gamma : \gamma \in \omega_1 \cap \mathbf{LIM}\},$$

and hence C is *club* in ω_1.

But how can we ensure that the tree T will not have an uncountable antichain? Let us do some brainstorming: First of all, Zorn's Lemma implies that every uncountable antichain in T is contained in a maximal uncountable antichain. So it suffices to make sure that T has no maximal uncountable antichains. Since the set of nodes of T is ω_1, every uncountable antichain of T would have to be an uncountable subset of ω_1. Let us consider an arbitrary $A \in [\omega_1]^{\aleph_1}$. How can we make sure that A will not become a maximal antichain in T? Suppose we are at stage $\alpha \in \mathbf{LIM}$ of the construction, i.e., suppose that $\langle T_{(\alpha)}, \leq_\alpha \rangle$ has been defined, and we want to decide which $t \in T(\alpha)$ to put on top of which $s \in T_{(\alpha)}$. Clearly, $A \cap T_{(\alpha)}$ is an antichain in $\langle T_{(\alpha)}, \leq_\alpha \rangle$. If $A \cap T_{(\alpha)}$ happens to be a *maximal* antichain in $\langle T_{(\alpha)}, \leq_\alpha \rangle$, we will be able to make sure that the following holds:

(K) $\qquad \forall s \in T(\alpha) \exists r \in T_{(\alpha)} \cap A \, (r <_T s).$

This suffices to destroy A as a candidate for becoming an uncountable antichain in T: Since A is uncountable, $A \backslash T_{(\alpha)} \neq \emptyset$. But for each $t \in A \backslash T_{(\alpha)}$ there exist $s \in T(\alpha)$ and $r \in A \cap T_{(\alpha)}$ with $r <_T s \leq_T t$, and thus A cannot become an antichain.

Now we would like to arrange all candidates $A \in [\omega_1]^{\aleph_1}$ for maximal uncountable antichains in a sequence $\langle A_\alpha : \alpha < \omega_1 \rangle$ and destroy A_α at stage α as outlined above. Unfortunately, there are $2^{\aleph_1}$ candidates for maximal uncountable antichains in T, and we cannot arrange all these candidates in a sequence of length ω_1. But let us take a closer look at our mechanism for destroying A_α at stage α: If $B \in [\omega_1]^{\aleph_1}$ is such that $B \cap T_{(\alpha)} = A_\alpha \cap T_{(\alpha)}$, then B cannot become a maximal antichain either. So it may be sufficient for our construction to arrange all potential initial segments of uncountable maximal antichains into a sequence of length ω_1. Exercise 22.3 suggests a particularly elegant way of doing this. Let C be as in Exercise 22.3. Perhaps it is true or at least consistent that the following property holds:

$(\diamondsuit^\circ(C))$ $\quad$ There exists a sequence $\langle A_\alpha : \alpha \in C \backslash \{0\} \rangle$ such that A_α is a subset of α for all $\alpha \in C \backslash \{0\}$, and for each $A \in [\omega_1]^{\aleph_1}$ there exists $\alpha \in C \backslash \{0\}$ such that $A \cap \alpha = A_\alpha$.

The statement $\diamondsuit^\circ(C)$ does not look obviously inconsistent: If CH holds, there are only $\aleph_1$ candidates for $A \cap \alpha$, where $A \in [\omega_1]^{\aleph_1}$ and $\alpha < \omega_1$, and it is conceivable that a clever arrangement of some of these sets would contain an initial segment $A \cap \alpha$ for every $A \in [\omega_1]^{\aleph_1}$.

But would Property $\diamondsuit^\circ(C)$ be sufficient for our purposes? At stage α of the construction, we would want to extend $\leq_\alpha$ in such a way that

(K_α) $\qquad \forall s \in T(\alpha) \exists r \in T_{(\alpha)} \cap A_\alpha \, (r <_T s).$

22. THE ◇-PRINCIPLE

We hinted above that this is possible if A_α happens to be a maximal antichain in $\langle T_{(\alpha)}, \leq_\alpha \rangle$. But will an initial segment A_α of a maximal antichain A in T be a maximal antichain in $T_{(\alpha)}$? The following claim sheds some light on this question.

CLAIM 22.1. *Suppose $\langle T \leq_T \rangle$ is constructed according to the generic blueprint, and $A \subseteq T$ is a maximal antichain in T. Then the set*

$$D(A) = \{\alpha \in C : A \cap \alpha \text{ is a maximal antichain in } \langle T_{(\alpha)}, \leq_\alpha \rangle\}$$

contains a closed unbounded subset of ω_1.

PROOF. It is easy to see that if A is an antichain in T, then $A \cap \alpha$ is an antichain in $T_{(\alpha)}$ for all $\alpha < \omega_1$. Now assume A is maximal. Then there is a function $g : \omega_1 \to A$ such that $g(\beta) \leq_T \beta$ or $\beta \leq_T g(\beta)$ for all $\beta < \omega_1$. Let $\mathcal{M}_g = \{\alpha \in \omega_1 : rng(g\lceil\alpha) \subseteq \alpha\}$. By Example 21.5, $\mathcal{M}_g$ is a *club* subset of ω_1. By Exercise 22.3, the set $C \cap \mathcal{M}_g$ is also *club*. Now it suffices to observe that $D(A) \supseteq C \cap \mathcal{M}_g$. □

O.k., if $\Diamond^\circ(C)$ holds and A is going to be a maximal antichain in T, then once in a while, $A \cap \alpha$ will be a maximal antichain in $T_{(\alpha)}$, and at least for one $\alpha \in C\setminus\{0\}$, we will have $A \cap \alpha = A_\alpha$. Unfortunately, so far we don't have any reason to believe that there is an $\alpha \in C\setminus\{0\}$ such that these two things happen simultaneously. To eliminate this problem, we would like to have a sequence of A_α's such that for every $A \in [\omega_1]^{\aleph_1}$ there are *stationarily many* α's such that $A \cap \alpha = A_\alpha$. This leads to the following statement, which is called the $\Diamond$-*principle*.

($\Diamond$)   *There exists a sequence $\langle A_\alpha : \alpha < \omega_1 \rangle$ such that A_α is a subset of α for all $\alpha < \omega_1$, and for each $A \subseteq \omega_1$ the set $\{\alpha < \omega_1 : A \cap \alpha = A_\alpha\}$ is stationary in ω_1.*

Clearly, $\Diamond \to \Diamond^\circ(C)$. The construction outlined above gives the following:

THEOREM 22.2. $\Diamond \to \neg SH$.

PROOF. We need a technical lemma.

LEMMA 22.3. *Suppose a tree T is being constructed according to the generic blueprint, $\alpha > 0$ is a countable limit ordinal such that $T_{(\alpha)} = \alpha$, and A_α is a maximal antichain in $\langle T_{(\alpha)}, \leq_\alpha \rangle$. Then $\langle T_{(\alpha+1)}, \leq_{\alpha+1} \rangle$ can be constructed in such a way that both $tall(\alpha)$ and K_α hold.*

PROOF. Choose an increasing sequence $(\beta_n)_{n \in \omega}$ of ordinals such that $\alpha = \sup\{\beta_n : n \in \omega\}$. For each $r \in T_{(\alpha)}$, choose a sequence $(s_n^r)_{n \in \omega}$ of elements of $T_{(\alpha)}$ such that:

(1) $s_0^r = r$;
(2) $s_n^r <_T s_{n+1}^r$ for each $n \in \omega$;
(3) $ht(s_n^r) \geq \beta_n$ for each $n > 0$;
(4) $\exists s \in A_\alpha \, (s <_T s_1^r)$.

Note that since $tall(\beta_n)$ holds for all $n \in \omega$, we always can find s_n^r's that satisfy conditions (1)-(3). Also, maximality of A_α is exactly the property that makes condition (4) possible: For each $r \in T_{(\alpha)}$, either $s \leq_T r$ for some $s \in A_\alpha$, in which case condition (4) will hold for any s_1^r as long as condition (2) is also satisfied; or $r <_T s$ for some $s \in A_\alpha$, in which case we can make condition (4) true by choosing s_1^r above this s.

Now enumerate $T(\alpha) = \{t_r : r \in T_{(\alpha)}\}$ and define: $s <_t t_r$ if and only if $\exists n \in \omega \, (s <_T s_n^r)$. It is not hard to see that $T_{(\alpha+1)}$ with this extended order is as required. □

EXERCISE 22.4 (PG). Derive Theorem 22.2. □

THEOREM 22.4. $\diamondsuit \to$ CH.

PROOF. Let $\langle B_\alpha : \alpha < \omega_1 \rangle$ be a sequence that witnesses $\diamondsuit$ (such a sequence is often called a $\diamondsuit$-*sequence*). Each subset B of ω is also a subset of ω_1. Thus, for each $B \in \mathcal{P}(\omega)$ there exists $\alpha(B) < \omega_1$ such that $\omega \le \alpha(B)$ and $B_{\alpha(B)} = B \cap \alpha(B) = B \cap \omega$. In fact, we have uncountably many choices for $\alpha(B)$, but we need only one. So let $\alpha(B)$ be the least such ordinal. Note that if $B \ne C$, then $\alpha(B) \ne \alpha(C)$. In other words, the function that assigns $\alpha(B)$ to B is a one-to-one function from $\mathcal{P}(\omega)$ into ω_1. The existence of such a function implies CH. □

It follows from Theorem 22.4 (and also from Theorems 22.2 and 19.1) that $\diamondsuit$ is not a theorem of ZFC. However, the statement $\diamondsuit$ is relatively consistent with ZFC; in fact, if **V** = **L**, then $\diamondsuit$ holds. It has also been proved that $\diamondsuit$ is not a consequence of CH. Therefore, whenever a theorem is shown to follow from $\diamondsuit$, the question whether this same theorem is also a consequence of CH deserves consideration. This question has been settled for the Suslin Hypothesis: R. B. Jensen has shown the relative consistency of CH + SH with ZFC.

EXERCISE 22.5 (PG). Prove that each of the following three statements is equivalent to $\diamondsuit$:

(a) There exists a sequence $\langle A_\alpha : \alpha < \omega_1 \rangle$ such that A_α is a subset of $\alpha \times \omega$ for each $\alpha < \omega_1$, and for all $A \subseteq \omega_1 \times \omega$, the set $\{\alpha < \omega_1 : A \cap \alpha \times \omega = A_\alpha\}$ is stationary in ω_1.

(b) There exists a sequence $\langle A_\alpha : \alpha < \omega_1 \rangle$ such that A_α is a subset of $\alpha \times \alpha$ for all $\alpha < \omega_1$, and for each $A \subseteq \omega_1 \times \omega_1$, the set $\{\alpha < \omega_1 : A \cap \alpha \times \alpha = A_\alpha\}$ is stationary in ω_1.

(c) There exists a sequence $\langle f_\alpha : \alpha < \omega_1 \rangle$ such that f_α is a function from α into α for each $\alpha < \omega_1$ and for all $f \in {}^{\omega_1}\omega_1$ the set $\{\alpha < \omega_1 : f \restriction \alpha = f_\alpha\}$ is stationary in ω_1.

There are many useful variants of the $\diamondsuit$-principle. In the remainder of this chapter we will introduce some of these variants. We begin with one that is equivalent to $\diamondsuit$.

($\diamondsuit^-$) *There exists a sequence* $\langle \mathcal{A}_\alpha : \alpha < \omega_1 \rangle$ *such that* $\mathcal{A}_\alpha \in [\mathcal{P}(\alpha)]^{\le \aleph_0}$ *for all* $\alpha < \omega_1$, *and for each* $A \subseteq \omega_1$ *the set* $\{\alpha < \omega_1 : A \cap \alpha \in \mathcal{A}_\alpha\}$ *is stationary in* ω_1.

THEOREM 22.5. $\diamondsuit \leftrightarrow \diamondsuit^-$.

PROOF. The implication $\diamondsuit \to \diamondsuit^-$ is obvious.

To prove the converse, fix a $\diamondsuit^-$-sequence, i.e., a sequence $\langle \mathcal{A}_\alpha : \alpha < \omega_1 \rangle$ that witnesses $\diamondsuit^-$. Without loss of generality, we may assume that each $\mathcal{A}_\alpha$ is infinite. For each α, fix an enumeration $(A_\alpha^n)_{n \in \omega}$ of $\mathcal{A}_\alpha$. Let C be the *club* set $\{\alpha < \omega_1 : \omega \cdot \alpha = \alpha\}$. For each $n \in \omega$ and $\alpha \in C$, let $B_\alpha^n = \{\beta < \alpha : \omega \cdot \beta + n \in A_\alpha^n\}$. For $n \in \omega$ and $\alpha \notin C$, we let $B_\alpha^n = \emptyset$.

CLAIM 22.6. *There exists* $n \in \omega$ *such that* $\langle B_\alpha^n : \alpha < \omega_1 \rangle$ *is a* $\diamondsuit$-*sequence*.

PROOF. Suppose not. Then for every $n \in \omega$ there exists B_n such that the set $S_n = \{\alpha : B_n \cap \alpha = B_\alpha^n\}$ is nonstationary. Consider the set $A = \{\omega \cdot \beta + n : n \in \omega \wedge \beta \in B_n\}$. The following exercise leads to a contradiction with the assumption that $\langle \mathcal{A}_\alpha : \alpha < \omega_1 \rangle$ is a $\diamondsuit^-$-sequence.

EXERCISE 22.6 (G). Convince yourself that for every α outside the nonstationary set $(\omega_1 \setminus C) \cup \bigcup_{n \in \omega} S_n$, the set $A \cap \alpha$ is not a member of $\mathcal{A}_\alpha$. □□

Now let us present two strengthenings of $\diamondsuit$.

($\diamondsuit^*$)   *There exists a sequence $\langle \mathcal{A}_\alpha : \alpha < \omega_1 \rangle$ such that $\mathcal{A}_\alpha \in [\mathcal{P}(\alpha)]^{\leq \aleph_0}$ for all $\alpha < \omega_1$, and for each $A \subseteq \omega_1$ the set $\{\alpha < \omega_1 : A \cap \alpha \in \mathcal{A}_\alpha\}$ is CLUB in ω_1.*

($\diamondsuit^+$)   *There exists a sequence $\langle \mathcal{A}_\alpha : \alpha < \omega_1 \rangle$ such that $\mathcal{A}_\alpha \in [\mathcal{P}(\alpha)]^{\leq \aleph_0}$ for all $\alpha < \omega_1$, and for each $A \subseteq \omega_1$ there exists a club subset C of ω_1 such that $\forall \alpha \in C \, (A \cap \alpha \in \mathcal{A}_\alpha \wedge C \cap \alpha \in \mathcal{A}_\alpha)$.*

It is clear that $\diamondsuit^+ \to \diamondsuit^*$ and $\diamondsuit^* \to \diamondsuit$. Jensen has shown that $\mathbf{V} = \mathbf{L} \to \diamondsuit^+$, and also that $\diamondsuit \not\to \diamondsuit^*$ and $\diamondsuit^* \not\to \diamondsuit^+$.

EXERCISE 22.7 (G). (a) Show that the following statement is inconsistent with ZFC: *There exists a sequence $\langle A_\alpha : \alpha < \omega_1 \rangle$ such that $A_\alpha \subseteq \alpha$ for all $\alpha < \omega_1$, and for each $A \subseteq \omega_1$, the set $\{\alpha < \omega_1 : A \cap \alpha = A_\alpha\}$ is CLUB in ω_1.*

(b) Show that $\diamondsuit$ implies the existence of a sequence $\langle S_\xi : \xi < 2^{\aleph_1} \rangle$ of stationary subsets of ω_1 such that $S_\xi \cap S_\eta$ is countable whenever $\xi < \eta < 2^{\aleph_1}$.

Recall from Chapter 14 that a *Kurepa tree* is an $\aleph_1$-tree with at least $\aleph_2$ cofinal branches, and that the *Kurepa Hypothesis* asserts the existence of a Kurepa tree. A *Kurepa family* is a family $\mathcal{F} \subseteq \mathcal{P}(\omega_1)$ such that

$$|\mathcal{F}| \geq \aleph_2 \text{ and } \forall \alpha < \omega_1 \, (|\{A \cap \alpha : A \in \mathcal{F}\}| \leq \aleph_0).$$

LEMMA 22.7. *The following statements are equivalent:*
(i)   *There exists a Kurepa tree.*
(ii)   *There exists a Kurepa family.*

PROOF. To prove the implication (i) $\to$ (ii), let $\langle T, \leq_T \rangle$ be a Kurepa tree whose set of nodes is ω_1, and let $\mathcal{F}$ be the set of all cofinal branches of T. Then $|\mathcal{F}| \geq \aleph_2$. Moreover, if $\alpha < \omega_1$ and $B \in \mathcal{F}$, then $B \cap T_{(\alpha)} = \hat{t}$ for some $t \in T(\alpha)$. Hence $|\{B \cap T_{(\alpha)} : B \in \mathcal{F}\}| \leq |T(\alpha)| \leq \aleph_0$. Since each $\beta < \omega_1$ is contained in $T_{(\alpha)}$ for some $\alpha < \omega_1$, it follows that $\forall \beta < \omega_1 \, (|\{B \cap \beta : B \in \mathcal{F}\}| \leq \aleph_0)$, and hence $\mathcal{F}$ is a Kurepa family.

To prove the implication (ii) $\to$ (i), let $\mathcal{F}$ be a Kurepa family. For each $A \in \mathcal{F}$, let $\chi_A : \omega_1 \to 2$ be the characteristic function of A. Let $T = \{\chi_A \lceil \alpha : A \in \mathcal{F} \wedge \alpha < \omega_1\}$.

EXERCISE 22.8 (G). Convince yourself that $\langle T, \subseteq \rangle$ is a Kurepa tree. □

THEOREM 22.8. $\diamondsuit^+ \to$ KH.

PROOF. We need a few technical preliminaries. Recall from Chapter 17 that a family $\mathcal{A} \subseteq [\omega_1]^{\aleph_1}$ is *pairwise almost disjoint* if $|A \cap B| \leq \aleph_0$ for all $A, B \in \mathcal{A}$

such that $A \neq B$. By Exercise 22.7(b), $\diamondsuit$ (and hence $\diamondsuit^+$) implies that there exists a pairwise almost disjoint family $\mathcal{A} \subseteq [\omega_1]^{\aleph_1}$ of stationary subsets of ω_1 such that $|\mathcal{A}| = 2^{\aleph_1}$. For our argument we will not need the sets in $\mathcal{A}$ to be stationary; it will suffice to assume that they are uncountable.

For $A, B \subseteq \omega_1$ we let

$$S(A,B) = \{\alpha \in A : \exists \beta \in B \, (\beta \leq \alpha \wedge \neg \exists \delta \in A \, (\beta \leq \delta < \alpha))\}.$$

We call $S(A,B)$ the *shadow of B in A*. Note that $S(A,B) \subseteq A$, and if A, B are uncountable, then so is $S(A,B)$. Moreover, for all $\delta < \omega_1$ we have $S(A,B) \cap \delta = S(A \cap \delta, B \cap \delta)$.

For the rest of this argument, let us fix a $\diamondsuit^+$-sequence $\langle \mathcal{A}_\alpha : \alpha < \omega_1 \rangle$ and a pairwise almost disjoint family $\{A_\alpha : \alpha < 2^{\aleph_1}\} \subseteq [\omega_1]^{\aleph_1}$ of size $2^{\aleph_1}$. A *club* set $C \subseteq \omega_1$ will be called an *approximation* of a set $A \subseteq \omega_1$ if $\forall \alpha \in C \, (A \cap \alpha \in \mathcal{A}_\alpha \wedge C \cap \alpha \in \mathcal{A}_\alpha)$. For each $\alpha < 2^{\aleph_1}$ we choose an approximation C_α of A_α. Let

$$\mathcal{F} = \{S(A_\alpha, C_\alpha) : \alpha < 2^{\aleph_1}\}.$$

We show that $\mathcal{F}$ is a Kurepa family. Let $\alpha < \beta < 2^{\aleph_1}$. Since $|A_\alpha \cap A_\beta| \leq \aleph_0$ and $S(A_\alpha, C_\alpha) \in [A_\alpha]^{\aleph_1}$, we have $S(A_\alpha, C_\alpha) \neq S(A_\beta, C_\beta)$, and thus $|\mathcal{F}| = 2^{\aleph_1} > \aleph_1$. Now fix $\delta < \omega_1$ and consider the family

$$T(\delta) = \{S(A_\alpha, C_\alpha) \cap \delta : \alpha < 2^{\aleph_1}\} = \{S(A_\alpha \cap \delta, C_\alpha \cap \delta) : \alpha < 2^{\aleph_1}\}.$$

If $\delta \in C_\alpha$, then both $A_\alpha \cap \delta$ and $C_\alpha \cap \delta$ are elements of $\mathcal{A}_\alpha$, and hence $S(A_\alpha \cap \delta, C_\alpha \cap \delta)$ is a member of the countable set $\{S(A,B) : A, B \in \mathcal{A}_\alpha\}$.

If $\delta \notin C_\alpha$, let $\delta^* = \sup(C_\alpha \cap \delta)$. Then $\delta^* < \delta$ and $\delta^* \in C_\alpha$. Again, $S(A_\alpha, C_\alpha) \cap \delta^* \in \{S(A,B) : A, B \in \mathcal{A}_{\delta^*}\}$. If $A_\alpha \cap [\delta^*, \delta) = \emptyset$, then $S(A_\alpha \cap \delta, C_\alpha \cap \delta) = S(A_\alpha \cap \delta^*, C_\alpha \cap \delta^*)$. If $A_\alpha \cap [\delta^*, \delta) \neq \emptyset$, then δ^* casts a shadow on the minimum element of $A_\alpha \backslash \delta^*$. In the latter case we have

$$S(A_\alpha \cap \delta, C_\alpha \cap \delta) = S(A_\alpha \cap \delta^*, C_\alpha \cap \delta^*) \cup \{\min(A_\alpha \cap [\delta^*, \delta))\}.$$

We conclude that $T(\delta) \subseteq \{S(A,B) : \exists \delta^* \leq \delta \, (A, B \in \mathcal{A}_{\delta^*})\} \cup \{S(A,B) \cup \{\gamma\} : \exists \delta^* < \delta \, (A, B \in \mathcal{A}_{\delta^*} \wedge \delta^* \leq \gamma < \delta)\}$. Since the right hand side is countable, so is $T(\delta)$. $\square$

Now let us discuss two weakenings of $\diamondsuit$. The first of these is called the ♣-*principle*.

(♣)   There exists a sequence $\langle s_\alpha : \alpha \in \omega_1 \cap \mathbf{LIM} \backslash \{0\} \rangle$ such that for all α:
(i) $s_\alpha \subseteq \alpha$;
(ii) $ot(s_\alpha) = \omega$;
(iii) $\sup s_\alpha = \alpha$; and
(iv) Each uncountable subset of ω_1 contains an s_α.

A sequence $\langle s_\alpha : \alpha \in \omega_1 \cap \mathbf{LIM} \backslash \{0\} \rangle$ that witnesses ♣ is called a ♣-*sequence*.

EXERCISE 22.9 (G). Convince yourself that $\diamondsuit \to$ ♣.

LEMMA 22.9. *Let $\langle s_\alpha : \alpha \in \omega_1 \cap \mathbf{LIM} \backslash \{0\} \rangle$ be a ♣-sequence. Then for every $A \in [\omega_1]^{\aleph_1}$ the set $\{\alpha \in \omega_1 : s_\alpha \subseteq A\}$ is stationary.*

PROOF. Let $\langle s_\alpha : \alpha \in \omega_1 \cap \mathbf{LIM} \backslash \{0\} \rangle$ be a ♣-sequence, let $A \in [\omega_1]^{\aleph_1}$, and let C be a *club* subset of ω_1. We want to show that there exists $\alpha \in C$ such that $s_\alpha \subseteq A$.

EXERCISE 22.10 (G). Show that there exists a sequence $(a_\beta)_{\beta<\omega_1}$ of elements of A such that $\forall \beta \in \omega_1 \exists \gamma \in C \, (a_\beta < \gamma < a_\beta + 1)$.

Fix a sequence $(a_\beta)_{\beta<\omega_1}$ as in Exercise 22.10, define $B = \{a_\beta : \beta \in \omega_1\}$, and let α be such that $s_\alpha \subseteq B$. Then α is the limit of a sequence of elements of C, and hence $\alpha \in C$. Moreover, since $B \subseteq A$ we have $s_\alpha \subseteq A$. □

S. Shelah has proved that ♣ does not imply CH. Hence ♣ is strictly weaker than ◇. However, in the presence of CH these two principles are equivalent.

THEOREM 22.10. CH + ♣ ↔ ◇.

PROOF. We only need to prove the implication from the left to the right. Let $\langle s_\alpha : \alpha \in \omega_1 \cap \mathbf{LIM} \setminus \{0\} \rangle$ be a ♣-sequence. Let $(A_\alpha)_{\alpha<\omega_1}$ be an enumeration of all countable subsets of ω_1 such that every $A \in [\omega_1]^{\leq \aleph_0}$ is listed cofinally often. Let C be the closed unbounded set $\omega_1 \cap \mathbf{LIM} \setminus \{0\}$. For $\alpha \in C$ we define $D_\alpha = \bigcup_{\gamma \in s_\alpha} A_\gamma$. For $\alpha \notin C$, let $D_\alpha = \emptyset$. The following claim implies Theorem 22.10.

CLAIM 22.11. $\langle D_\alpha : \alpha \in \omega_1 \rangle$ is a ◇-sequence.

PROOF. Let $A \subseteq \omega_1$, and let D be club in ω_1. We need to show that there exists a $\gamma \in C \cap D$ such that $D_\gamma = A \cap \gamma$.

EXERCISE 22.11 (G). Construct recursively an increasing transfinite sequence of ordinals $\langle \delta_\beta : \beta < \omega_1 \rangle$ such that for all $\beta < \omega_1$:

(i) $\exists \gamma \in C \cap D \, (\delta_\beta < \gamma < \delta_{\beta+1})$;
(ii) $A_{\delta_{\beta+1}} = A \cap \delta_\beta$.

Let $B = \{\delta_\beta : 0 < \beta < \omega_1\}$, and let α be such that $s_\alpha \subseteq B$. Suppose $(\delta_{\beta_n})_{n\in\omega}$ is an enumeration of s_α. Then for each $n \in \omega$ there exists $\gamma_n \in C \cap D$ such that $\delta_{\beta_n} < \gamma_n < \delta_{\beta_n+1}$. Since $\alpha = \sup\{\delta_{\beta_n} : n \in \omega\}$, we can infer that $\alpha \in C \cap D$. Moreover, $A_{\delta_{\beta_n}} \subseteq A \cap \alpha$ and $A \cap \delta_{\beta_n} \subseteq A_{\delta_{\beta_n+1}}$ for each $n \in \omega$. It follows that $D_\alpha = A \cap \alpha$, and we have proved Claim 22.11. □□

The ♣-principle can be further weakened to the so-called *stick-principle* (abbreviated ↿):

(↿)   There exists a sequence $\langle A_\alpha : \alpha < \omega_1 \rangle$ of infinite subsets of ω_1 such that for every $A \in [\omega_1]^{\aleph_1}$ there exists $\alpha < \omega_1$ with $A_\alpha \subseteq A$.

EXERCISE 22.12 (G). (a) Convince yourself that ♣ → ↿.
(b) Show that CH → ↿.

Since CH ↛ ◇, it follows from Theorem 22.10 and Exercise 22.12(b) that ↿ is strictly weaker than ♣. However, ↿ is not a theorem of ZFC.

THEOREM 22.12. $MA(\aleph_1) \to \neg$↿.

PROOF. Let $\langle A_\alpha : \alpha < \omega_1 \rangle$ be a sequence of infinite subsets of ω_1. Consider the c.c.c. partial order $\langle Fn(\omega_1, 2), \supseteq \rangle$. For each $\alpha < \omega_1$, let

$$D_\alpha = \{p \in Fn(\omega_1, 2) : \exists \xi \in A_\alpha \cap dom(p) \, (p(\xi) = 0)\}.$$

EXERCISE 22.13 (PG). (a) Show that each D_α is a dense subset of $Fn(\omega_1, 2)$.
(b) Derive Theorem 22.12. □

Let us briefly mention another way of strengthening $\diamondsuit$. Let E be a stationary subset of ω_1. Then $\diamondsuit(E)$ stands for the following statement:

($\diamondsuit(E)$) There exists a sequence $\langle A_\alpha : \alpha \in E \rangle$ such that
$A_\alpha \subseteq \alpha$ for all $\alpha \in E$, and for each $A \subseteq \omega_1$
the set $\{\alpha \in E : A \cap \alpha = A_\alpha\}$ is stationary.

Of course, $\diamondsuit$ is the same as $\diamondsuit(\omega_1)$. Moreover, $\diamondsuit$ trivially implies $\diamondsuit(E)$ for every *club* set E. By Theorem 22.13 below, $\mathbf{V} = \mathbf{L}$ implies $\diamondsuit(E)$ for every stationary E. On the other hand, Shelah proved that if ZFC is consistent, then there exists a model of ZFC that contains two stationary subsets E, F of ω_1 such that $\diamondsuit(E)$ holds and $\diamondsuit(F)$ fails. We will show in the next chapter (Corollary 23.7) that $\diamondsuit(E)$ implies the existence of two disjoint stationary sets $F, G \subset E$ such that both $\diamondsuit(F)$ and $\diamondsuit(G)$ hold.

One can also consider $\diamondsuit$-like principles for cardinals larger than $\aleph_1$. If κ is an uncountable cardinal and $E \subseteq \kappa$, then $\diamondsuit_\kappa(E)$ stands for the statement:

($\diamondsuit_\kappa(E)$) There exists a sequence $\langle \mathcal{A}_\alpha : \alpha \in E \rangle$ such that
$A_\alpha \subseteq \alpha$ for all $\alpha \in E$, and for each $A \subseteq \kappa$
the set $\{\alpha \in E : A \cap \alpha = A_\alpha\}$ is stationary in κ.

Of course, $\diamondsuit(E)$ is the same as $\diamondsuit_{\omega_1}(E)$. Instead of $\diamondsuit_\kappa(\kappa)$ one usually writes $\diamondsuit_\kappa$.

THEOREM 22.13. *Assume* $\mathbf{V} = \mathbf{L}$. *Let κ be an uncountable regular cardinal, E a stationary subset of κ. Then $\diamondsuit_\kappa(E)$ holds.*

Mathographical Remarks

The $\diamondsuit$-principle is due to R. B. Jensen. The $\clubsuit$-principle is also called the *Ostaszewski Principle*. In his paper *On countably compact, perfectly normal spaces*, Journal of the London Mathematical Society 14 (1967), 505–516, A. Ostaszewski used $\clubsuit$ + CH to construct a first countable, perfectly normal, hereditarily separable, locally countable, locally compact, countably compact Hausdorff space that is not Lindelöf (and hence not compact). K. Devlin later noticed that $\clubsuit$ + CH already implies $\diamondsuit$. For variations on the $\clubsuit$-principle, see S. Broverman, J. Ginsburg, K. Kunen, and F. D. Tall, *Topologies determined by σ-ideals on ω_1*, Canadian Journal of Mathematics 30(6) (1978), 1306–1312. The construction of a model for $\clubsuit$ + $\neg$CH appeared in S. Shelah, *Whitehead groups may not be free even assuming CH, II*, Israel J. Math. 35 (1980), 257–285. The proof of Theorem 22.13 can be found in K. Devlin's book *Constructibility*, Springer Verlag, 1984. The latter reference also contains information on several other useful combinatorial principles that are consequences of $\mathbf{V} = \mathbf{L}$.

CHAPTER 23

Measurable Cardinals

DEFINITION 23.1. Let κ be a cardinal. A *probability measure on κ* is a function $\mu : \mathcal{P}(\kappa) \to [0,1]$ such that
(1) $\mu(\emptyset) = 0$;
(2) $\mu(\kappa) = 1$;
(3) $\mu(\bigcup_{n \in \omega} A_n) = \sum_{n \in \omega} \mu(A_n)$ for every countable family $\{A_n : n \in \omega\}$ of pairwise disjoint subsets of κ.

Property (3) has several names in the literature: *σ-additivity, countable additivity,* or *$\aleph_1$-additivity of μ.* If μ is such that $\mu(\{\xi\}) = 0$ for every $\xi \in \kappa$, then we say that *μ vanishes on the singletons.* The family $\mathcal{I}_\mu = \{N \subseteq \kappa : \mu(N) = 0\}$ is called *the ideal of null sets of μ.*

EXERCISE 23.1 (G). Show that $\mathcal{I}_\mu$ is indeed an ideal on κ.

EXAMPLE 23.1. Let $\xi \in \kappa$, and define:
$$\mu(A) = \begin{cases} 1, & \text{if } \xi \in A; \\ 0, & \text{if } \xi \notin A. \end{cases}$$
Then μ is a probability measure on κ. Of course, this measure does not vanish on the singletons.

EXAMPLE 23.2. Let $\mathcal{U}$ be a countably complete ultrafilter on κ. Define:
$$\mu_\mathcal{U}(A) = \begin{cases} 1, & \text{if } A \in \mathcal{U}; \\ 0, & \text{if } A \notin \mathcal{U}. \end{cases}$$
Then $\mu_\mathcal{U}$ is a measure on κ. The measure of Example 23.1 is a special case of $\mu_\mathcal{U}$ for the principal ultrafilter $\mathcal{U}$ generated by $\{\xi\}$. Note that $\mu_\mathcal{U}$ vanishes on the singletons if and only if $\mathcal{U}$ is a nonprincipal ultrafilter.

EXERCISE 23.2 (G). Show that if μ is any measure on a cardinal κ, then the family $\mathcal{U}_\mu = \{A \subseteq \kappa : \mu(A) = 1\}$ is a countably complete filter on κ. Moreover, show that $\mathcal{U}_\mu$ is an ultrafilter on κ if and only if $rng(\mu) = \{0,1\}$. (The latter property will be referred to by saying that *μ is a two-valued measure.*)

How is Definition 23.1 related to more familiar probability measures, like Lebesgue measure on $[0,1]$? Of course, the latter is not defined on sets of ordinals. But this is only a superficial difference, since any one-to-one correspondence between the cardinal $2^{\aleph_0}$ and the unit interval $[0,1]$ allows us to treat a measure defined on $\mathcal{P}(2^{\aleph_0})$ as one defined on $\mathcal{P}([0,1])$ and vice versa. More relevant for the present purpose is the fact that Lebesgue measure is defined only on a certain σ-algebra of subsets of $[0,1]$, whereas the measures considered here are defined for *all* subsets of a cardinal κ. We shall say that *Lebesgue measure can be extended to $\mathcal{P}([0,1])$* if there exists a probability measure $\mu : \mathcal{P}([0,1]) \to [0,1]$ such that $\mu([a,b]) = b - a$

for all a, b with $0 \leq a \leq b \leq 1$. It is clear that each such measure vanishes on the singletons (consider the degenerate intervals $[a,a] = \{a\}$). The following question provided the initial motivation for the notions studied in this chapter.

QUESTION 23.2. Does there exist a probability measure on $\mathcal{P}([0,1])$ that extends Lebesgue measure?

Note that this is not the same as asking whether there exists a subset of $[0,1]$ that is not Lebesgue measurable. The latter question has been answered in the negative by Theorem 9.23 of Volume I. Incidentally, the Vitali set constructed in the proof of that theorem also demonstrates that Lebesgue measure cannot be extended to a translation-invariant measure defined on all subsets of $\mathbb{R}$.

THEOREM 23.3 (Ulam). Let κ be an infinite cardinal, and let $\lambda = \kappa^+$. Then there exists an indexed family $\{A_{\alpha,\xi} : \alpha < \kappa, \xi < \lambda\}$ of subsets of λ such that:
 (1) $A_{\alpha,\xi} \cap A_{\beta,\xi} = \emptyset$ for all $\alpha < \beta < \kappa$ and $\xi < \lambda$;
 (2) $A_{\alpha,\xi} \cap A_{\alpha,\eta} = \emptyset$ for all $\alpha < \kappa$ and $\xi < \eta < \lambda$;
 (3) $\bigcup_{\alpha<\kappa} A_{\alpha,\xi} = \lambda \setminus (\xi+1)$ for all $\xi < \lambda$.

An indexed family $\{A_{\alpha,\xi} : \alpha < \kappa, \xi < \lambda\}$ that satisfies (1)-(3) is called an *Ulam matrix*.[1]

PROOF OF THEOREM 23.3. For each $\eta < \lambda$, let f_η be a one-to-one function from η into κ. For $\alpha < \kappa$ and $\xi < \lambda$ define:

$$A_{\alpha,\xi} = \{\zeta > \xi : f_\zeta(\xi) = \alpha\}.$$

EXERCISE 23.3 (PG). Verify that the $A_{\alpha,\xi}$'s defined above satisfy conditions (1)-(3). □

Before we discuss the impact of Theorem 23.3 on the existence of measures on cardinals, let us present an application to stationary sets.

COROLLARY 23.4. Let κ be an infinite cardinal, and let S be a stationary subset of $\lambda = \kappa^+$. Then there exists a family $\mathcal{S}$ of λ pairwise disjoint stationary subsets of λ such that $\bigcup \mathcal{S} \subseteq S$. In particular, there exists a family of λ pairwise disjoint stationary subsets of λ.

PROOF. Let S be as in the assumptions, and let $\{A_{\alpha,\xi} : \alpha < \kappa, \xi < \lambda\}$ be an Ulam matrix. For each $\xi < \lambda$ there exists $\alpha(\xi) < \kappa$ such that $S \cap A_{\alpha(\xi),\xi}$ is stationary in λ: Otherwise, the stationary set S would be the union of a family of fewer than $cf(\lambda)$ nonstationary sets, namely $\{S \cap A_{\alpha,\xi} : \alpha < \kappa\} \cup \{\xi + 1\}$, which is impossible by Lemma 21.10.

Now the Pigeonhole Principle implies that there exists $\alpha < \kappa$ such that the set $X_\alpha = \{\xi < \lambda : \alpha(\xi) = \alpha\}$ has cardinality λ. Fix such α, and let $\mathcal{S} = \{S \cap A_{\alpha,\xi} : \xi \in X_\alpha\}$. This is a family of size λ of stationary subsets of λ, and clearly $\bigcup \mathcal{S} \subseteq S$. By point (2) of Theorem 23.3, $\mathcal{S}$ consists of pairwise disjoint sets. □

COROLLARY 23.5. There is no probability measure on ω_1 that vanishes on the singletons. In particular, if CH holds, then Lebesgue measure cannot be extended to a measure on $\mathcal{P}([0,1])$.

---

[1] The Polish name Ulam is pronounced *Oolawm*.

PROOF. Let $\{A_{\alpha,\xi} : \alpha < \omega, \xi < \omega_1\}$ be an Ulam matrix. Assume towards a contradiction that μ is a probability measure on ω_1 that vanishes on the singletons. Note that the latter property implies that $\mu(\xi+1) = 0$ for each countable ordinal ξ. Let $\xi < \omega_1$. Since $\omega_1 = \bigcup(\{A_{\alpha,\xi} : \alpha < \omega\} \cup \{\xi+1\})$, we have $\sum_{\alpha<\omega} \mu(A_{\alpha,\xi}) = \mu(\xi+1) + \sum_{\alpha<\omega} \mu(A_{\alpha,\xi}) = \mu(\omega_1) = 1$. Therefore, it cannot be the case that $\mu(A_{\alpha,\xi}) = 0$ for all $\alpha < \omega$. In other words, there must be some $\alpha(\xi) < \omega$ such that $\mu(A_{\alpha(\xi),\xi}) > 0$.

Now the Pigeonhole Principle implies that for some $\alpha < \omega$ the set

$$X_\alpha = \{\xi : \alpha(\xi) = \alpha\}$$

is uncountable. Fix such an α. For $\xi \in X_\alpha$, let

$$n(\xi) = \min\{n \in \omega : \mu(A_{\alpha,\xi}) > 2^{-n}\}.$$

Using the Pigeonhole Principle once more, we find an uncountable subset $Y \subseteq X_\alpha$ and $n \in \omega$ such that $n(\xi) = n$ for all $\xi \in Y$. This leads to a contradiction: Let $\{\xi_i : 0 \leq i \leq 2^n\}$ be the first 2^n+1 elements of Y. Since $A_{\alpha,\xi_i} \cap A_{\alpha,\xi_j} = \emptyset$ whenever $i \neq j$, we have

$$\mu(A_{\alpha,\xi_0} \cup \cdots \cup A_{\alpha,\xi_{2^n}}) = \sum_{i=0}^{2^n} \mu(A_{\alpha,\xi_i}) \geq (2^n+1) \cdot 2^{-n} > 1,$$

contradicting the assumption that μ is a probability measure. $\square$

COROLLARY 23.6. *There does not exist a nonprincipal, countably complete ultrafilter on ω_1.*

PROOF. This follows from Example 23.2 by an argument as in the proof of Corollary 23.5. $\square$

COROLLARY 23.7. *Suppose E is a stationary subset of ω_1 such that $\Diamond(E)$ holds. There exist disjoint $F, G \subseteq E$ such that both $\Diamond(F)$ and $\Diamond(G)$ hold.*

PROOF. Let E be as in the assumption. Note that if $E^- = \{\alpha \in E : \omega \cdot \alpha = \alpha\}$, then $\Diamond(E^-)$ also holds. So let us assume without loss of generality that $E = E^-$, i.e., that $\omega \cdot \alpha = \alpha$ for all $\alpha \in E$. Define:

$$\mathcal{J} = \{F \subseteq E : \neg \Diamond(F)\}.$$

CLAIM 23.8. *$\mathcal{J}$ is a countably complete ideal in $\mathcal{P}(E)$.*

PROOF. It is clear that $\mathcal{J}$ is closed under subsets.

Now suppose $F_n \in \mathcal{J}$ for $n \in \omega$, and let $F = \bigcup_{n \in \omega} F_n$. We want to show that $F \in \mathcal{J}$. Let $\langle A_\alpha : \alpha \in F \rangle$ be a sequence such that $A_\alpha \subseteq \alpha$ for all $\alpha \in F$. We want to show that there exists $B \subseteq \omega_1$ such that the set $\{\alpha \in F : B \cap \alpha = A_\alpha\}$ is nonstationary.

EXERCISE 23.4 (PG). Show that for every $n \in \omega$ there exists $B_n \subseteq \omega_1$ such that the set $S_n = \{\alpha \in F_n : \forall \beta < \alpha \, (\omega \cdot \beta + n \in A_\alpha \leftrightarrow \beta \in B_n)\}$ is nonstationary.

Let $(B_n)_{n \in \omega}$ and $(F_n)_{n \in \omega}$ be as above. Define $B = \{\omega \cdot \beta + n : n \in \omega \wedge \beta \in B_n\}$ and $S = \{\alpha \in F : B \cap \alpha = A_\alpha\}$. Then S is contained in $\bigcup_{n \in \omega} S_n$, and since the S_n's are nonstationary, so is S. Thus B is as required. $\square$

By Claim 23.8, the dual filter $\mathcal{F} = \mathcal{J}^* = \{G \subseteq E : E\backslash G \in \mathcal{J}\}$ is countably complete. This filter is nonprincipal, and, since $E \notin \mathcal{J}$, it is also proper. By Corollary 23.6, $\mathcal{F}$ is not an ultrafilter in $\mathcal{P}(E)$. In other words, $\mathcal{J}$ is not a maximal ideal in $\mathcal{P}(E)$. Thus there exists $F \subseteq E$ such that neither F nor $E\backslash F$ is in $\mathcal{J}$. This is exactly what we wanted to prove. □

Let κ, λ be cardinals. An ideal $\mathcal{I}$ in $\mathcal{P}(\kappa)$ is said to be λ-*saturated* if for every family $\{A_\alpha : \alpha < \lambda\} \subseteq \mathcal{P}(\kappa)\backslash\mathcal{I}$ there exist $\alpha < \beta < \lambda$ such that $A_\alpha \cap A_\beta \notin \mathcal{I}$.

EXERCISE 23.5 (PG). (a) Show that if $\mu : \mathcal{P}(\kappa) \to [0, 1]$ is a probability measure on κ, then the ideal of null sets $\mathcal{I}_\mu$ is $\aleph_1$-saturated.

(b) Show that there is no $\aleph_1$-saturated countably complete ideal in $\mathcal{P}(\omega_1)$ that contains all the singletons. *Hint:* Use an Ulam matrix.

(c) Show that if κ is a regular infinite cardinal, then the family $[\kappa]^{<\kappa}$ is a κ-complete ideal on κ that is not κ^+-saturated.

DEFINITION 23.9. Let κ, λ be uncountable cardinals, and let μ be a measure on λ. We say that μ is κ-*additive* if $\mu(\bigcup \mathcal{N}) = 0$ for every family $\mathcal{N}$ of subsets of λ such that $|\mathcal{N}| < \kappa$ and $\mu(N) = 0$ for each $N \in \mathcal{N}$.

A few changes in the notation of the proof of Corollary 23.5 yield the following result:

THEOREM 23.10. *If λ is an uncountable successor cardinal, then there is no λ-additive probability measure on λ that vanishes on the singletons.*

EXERCISE 23.6 (PG). Show that a measure μ on λ is κ-additive if and only if for every family $\mathcal{M}$ of pairwise disjoint subsets of λ of cardinality $|\mathcal{M}| < \kappa$ the following equation holds:

$$\mu(\bigcup \mathcal{M}) = \sum_{M \in \mathcal{M}} \mu(M)$$

The right hand side of this equation is a common abbreviation for the number $\sup\{\mu(\bigcup \mathcal{F}) : \mathcal{F} \in [\mathcal{M}]^{<\omega}\}$. *Hint:* Use the fact that only countably many sets in $\mathcal{M}$ have positive measure.

DEFINITION 23.11. An uncountable cardinal λ is called *real-valued measurable* if there exists a λ-additive probability measure μ on λ that vanishes on all singletons; λ is called a *measurable cardinal* if, in addition, μ is a two-valued measure.

EXERCISE 23.7 (G). (a) Show that λ is a measurable cardinal if and only if there exists a λ-complete nonprincipal ultrafilter on λ.

(b) Convince yourself that if $\mathcal{U}$ is a λ-complete filter on λ such that $\bigcap \mathcal{U} = \emptyset$, then $\mathcal{U}$ is a *uniform ultrafilter* on λ, i.e., $|A| = \lambda$ for every $A \in \mathcal{U}$.

Theorem 23.10 tells us that real-valued measurable cardinals are a rather uncommon species; the next theorem makes a similar point. In combination, Theorems 23.10 and 23.12 imply that real-valued measurable cardinals are large cardinals.

THEOREM 23.12. *Every real-valued measurable cardinal is regular.*

PROOF. Let λ be a singular cardinal, and let $\langle \xi_\alpha : \alpha < \kappa \rangle$ be an increasing sequence of ordinals cofinal in λ, where $\kappa = cf(\lambda) < \lambda$. Suppose towards a contradiction that μ is a λ-additive probability measure on λ that vanishes on all singletons. By Exercise 23.7(b), $\mu(\xi_\alpha) = 0$ for every $\alpha < \kappa$. It follows that $\mu(\lambda) = \mu(\bigcup\{\xi_\alpha : \alpha < \kappa\}) = 0$, which contradicts the assumption that μ is a probability measure. □

COROLLARY 23.13. *Every real-valued measurable cardinal is weakly inaccessible.*

For measurable cardinals, we can say even more.

THEOREM 23.14. *Every measurable cardinal is strongly inaccessible.*

PROOF. Since every measurable cardinal is obviously real-valued measurable, it suffices to show that if λ is a measurable cardinal, then λ is a strong limit cardinal, that is, $2^\kappa < \lambda$ for each $\kappa < \lambda$.

So assume towards a contradiction that λ is measurable, $\kappa < \lambda$, and $2^\kappa \geq \lambda$. Let $F : \lambda \to {}^\kappa 2$ be a one-to-one function, and let $\mathcal{U}$ be a λ-complete nonprincipal ultrafilter on λ. For each $\alpha < \kappa$ and $i \in \{0,1\}$, let $X_\alpha^i = \{\xi \in \lambda : F(\xi)(\alpha) = i\}$. For each λ, $X_\alpha^0 \cap X_\alpha^1 = \emptyset$ and $X_\alpha^0 \cup X_\alpha^1 = \lambda$. Therefore, exactly one of these sets is in $\mathcal{U}$. Define a function $g : \kappa \to 2$ by choosing $g(\alpha)$ such that $X_\alpha^{g(\alpha)} \in \mathcal{U}$.

By λ-completeness of $\mathcal{U}$, the intersection of any κ members of $\mathcal{U}$ is still a member of $\mathcal{U}$; in particular, $X = \bigcap_{\alpha < \kappa} X_\alpha^{g(\alpha)} \in \mathcal{U}$. But, by our construction, $X = F^{-1}\{g\}$. Since F was assumed to be one-to-one, X is a singleton, contradicting the assumption that $\mathcal{U}$ is a nonprincipal filter. □

The next theorem improves and actually supersedes Theorem 23.14. However, the proof of Theorem 23.14 given above is of independent interest.

THEOREM 23.15. *Every measurable cardinal is weakly compact.*

PROOF. Let κ be measurable, and let $c : [\kappa]^2 \to 2$ be a coloring. We want to show that there exists $X \in [\kappa]^\kappa$ so that X is homogeneous for c. Let $\mathcal{U}$ be a κ-complete nonprincipal ultrafilter on κ. By Exercise 23.7(b), each set in $\mathcal{U}$ has cardinality κ. Thus, for each $\alpha \in \kappa$, the set of ordinals between α and κ is in $\mathcal{U}$. If we define:

$$X_\alpha^i = \{\beta : \alpha < \beta < \kappa \wedge c(\{\alpha, \beta\}) = i\},$$

then exactly one of the disjoint sets X_α^0, X_α^1 is in $\mathcal{U}$. This allows us to define an induced coloring $d : \kappa \to 2$ by choosing $d(\alpha)$ so that $X_\alpha^{d(\alpha)} \in \mathcal{U}$. Now we define recursively a sequence $\langle \alpha(\xi) : \xi < \kappa \rangle$ by letting $\alpha(0) = 0$ and $\alpha(\eta) = \min(\bigcap_{\xi < \eta} X_{\alpha(\xi)}^{d(\alpha(\xi))} \setminus (\alpha(\xi)+1))$ for $\eta > 0$. Since at each stage of the construction we take an intersection of fewer than κ elements of $\mathcal{U}$, the resulting sequence $\langle \alpha(\xi) : \xi < \kappa \rangle$ is a well-defined, strictly increasing sequence of ordinals less than κ. Also note that if $\xi < \eta$, then $\alpha(\eta) \in X_{\alpha(\xi)}^{d(\alpha(\xi))}$. It follows that $c(\{\alpha(\xi), \alpha(\eta)\}) = d(\alpha(\xi))$ for all $\xi < \eta < \kappa$. By the Pigeonhole Principle, there exist $Y \in [\kappa]^\kappa$ and $i \in \{0,1\}$ such that $d(\alpha(\xi)) = i$ for all $\xi \in Y$. It follows that the set $Y = \{\alpha(\xi) : \xi \in X\}$ is homogeneous for c, as required. □

Could we get a homogeneous set that is not only of cardinality κ, but also in the ultrafilter $\mathcal{U}$ that was used in the proof of Theorem 23.15? Yes, provided that $\mathcal{U}$ has a special property. An ultrafilter $\mathcal{U}$ on a cardinal κ is called *normal* if for every sequence $\langle A_\alpha : \alpha < \kappa \rangle$ of elements of $\mathcal{U}$ the diagonal intersection $\triangle_{\alpha < \kappa} A_\alpha$ ($= \{\alpha \in \kappa : \forall \beta < \alpha \, (\alpha \in A_\beta)\}$) is also an element of $\mathcal{U}$.

EXERCISE 23.8 (PG). (a) Convince yourself that every uniform, normal ultrafilter on an uncountable cardinal κ is κ-complete. *Hint:* If $\langle A_\alpha : \alpha < \lambda \rangle$ is a sequence of elements of U with $\lambda < \kappa$, let $A_\alpha = \kappa$ for each α with $\lambda \leq \alpha < \kappa$ and consider the diagonal intersection $\Delta_{\alpha < \kappa} A_\alpha$.

(b) Show that if $\mathcal{U}$ is a uniform, normal ultrafilter on an uncountable cardinal κ, then $\mathcal{U} \supseteq CLUB(\kappa)$. *Hint:* Show that every *club* subset of κ can be represented as the diagonal intersection of a sequence of sets of the form $\kappa \setminus \beta$.

In the next chapter[2], we will prove the following:

THEOREM 23.16. *Suppose κ is a measurable cardinal. Then there exists a uniform, normal ultrafilter on κ.*

Theorem 23.15 can be strengthened as follows:

THEOREM 23.17. *Let κ be a measurable cardinal, $n \in \omega \setminus \{0\}, \sigma < \kappa$, and let $\mathcal{U}$ be a uniform, normal ultrafilter on κ. Then for every coloring $c : [\kappa]^n \to \sigma$ there exists $A \in \mathcal{U}$ such that A is homogeneous for c.*

PROOF. For $n = 1$ the existence of A follows immediately from Exercise 23.8(a) and the assumption that $\sigma < \kappa$. We will prove the theorem for $n = 2$ and leave the general case as an exercise. So suppose $\sigma < \kappa$, and let $c : [\kappa]^2 \to \sigma$ be a coloring. For $\alpha < \kappa$ and $i < \sigma$, let $A_{\alpha,i} = \{\beta : \alpha < \beta < \kappa \wedge c(\{\alpha,\beta\}) = i\}$. Since $\mathcal{U}$ is κ-complete and since $\bigcup_{i<\sigma} A_{\alpha,i} = (\alpha, \kappa) \in \mathcal{U}$, there exists exactly one $i(\alpha) < \sigma$ such that $A_{\alpha,i(\alpha)} \in \mathcal{U}$.

Let $A = \Delta_{\alpha<\kappa} A_{\alpha,i(\alpha)}$. By normality of $\mathcal{U}$, $A \in \mathcal{U}$. Moreover, if $\alpha < \beta$ and $\alpha, \beta \in A$, then $c(\{\alpha,\beta\}) = i(\alpha)$. Using κ-completeness of $\mathcal{U}$ again, we find $i \in \sigma$ and $B \in \mathcal{U}$ such that $B = \{\alpha \in A : i(\alpha) = i\}$. This B is a homogeneous set for c, as required. □

EXERCISE 23.9 (PG). (a) Prove Theorem 23.17 for $n > 2$. *Hint:* Use induction and induced colorings.

(b) Prove the following consequence of Theorem 23.17:

COROLLARY 23.18. *Every measurable cardinal is a Ramsey cardinal.*

Corollary 23.13 implies that if ZFC is a consistent theory, then the existence of real-valued measurable cardinals cannot be proved in this theory. Theorems 23.14–23.17 and Corollary 23.18 reinforce this point. Moreover, even if the existence of measurable cardinals (and thus the existence of real-valued measurable cardinals) does not contradict any of the axioms of ZFC, we shall never be able to prove this fact in ZFC.

Didn't we perhaps create too exotic a notion somewhere along the way? We started out by asking for an extension of Lebesgue measure to $\mathcal{P}([0,1])$, not necessarily a $2^{\aleph_0}$-additive one. Does the assumption that on some cardinal κ there exists a (not necessarily κ-additive) probability measure that vanishes on the singletons already imply the existence of large cardinals?

THEOREM 23.19. *The smallest cardinal carrying a probability measure that vanishes on the singletons is real-valued measurable. Similarly, the smallest cardinal carrying a two-valued probability measure that vanishes on the singletons is measurable.*

---

[2] Theorem 24.42

PROOF. Let λ be a cardinal, and suppose μ is a probability measure on λ that vanishes on the singletons. Suppose that μ is not λ-additive, i.e., that there exists a family $\mathcal{N} = \{N_\xi : \xi < \kappa < \lambda\}$ of subsets of λ such that $\mu(N) = 0$ for each $N \in \mathcal{N}$ and $\mu(\bigcup \mathcal{N}) = \varepsilon > 0$.

We show that there exists a probability measure ν on κ that vanishes on the singletons. Moreover, if μ is two-valued, then so is ν. In either case, this cannot happen if λ is the smallest cardinal which carries a (two-valued) probability measure that vanishes on the singletons; therefore the smallest such λ must be real-valued measurable, or, if μ is two-valued, measurable.

Note that we can assume without loss of generality that the sets in $\mathcal{N}$ are pairwise disjoint: If not, replace each N_ξ by $N_\xi \setminus \bigcup\{N_\eta : \eta < \xi\}$. This gives a family of null sets indexed by ordinals less than κ and with the same union of measure ε. Now define for $X \subseteq \kappa$:

$$\nu(X) = \frac{1}{\varepsilon}\mu(\bigcup\{N_\xi : \xi \in X\}).$$

This is a probability measure on κ that vanishes on the singletons:

(1) $\nu(\emptyset) = \varepsilon^{-1}\mu(\emptyset) = 0$;
(2) $\nu(\kappa) = \varepsilon^{-1}\mu(\bigcup \mathcal{N}) = \varepsilon^{-1}\varepsilon = 1$;
(3) $\nu(\bigcup_{n \in \omega} A_n) = \varepsilon^{-1}\mu(\bigcup\{N_\xi : \xi \in \bigcup_{n \in \omega} A_n\}) =$
$\varepsilon^{-1}\mu(\bigcup_{n \in \omega} \bigcup\{N_\xi : \xi \in A_n\}) \stackrel{(i)}{=} \varepsilon^{-1}\sum_{n \in \omega}\mu(\bigcup\{N_\xi : \xi \in A_n\}) =$
$\sum_{n \in \omega} \nu(A_n)$ for every countable family $\{A_n : n \in \omega\}$
of pairwise disjoint subsets of κ (equality (i) follows from
the assumption that the N_ξ's are pairwise disjoint);
(4) $\nu(\{\xi\}) = \varepsilon^{-1}\mu(N_\xi) = 0$ for all $\xi \in \kappa$.

Moreover, if μ takes on only the values 0 and 1, then $\varepsilon = 1$, and ν is also two-valued. □

It is very important that the λ in the proof of Theorem 23.19 is the *first* cardinal that carries a (two-valued) measure. In fact, if κ carries a (two-valued) measure μ and $\lambda > \kappa$, then λ also carries a (two-valued) measure ν: Just define $\nu(X) = \mu(X \cap \kappa)$ for $X \subseteq \lambda$. Note that λ may even be a successor cardinal or singular; as long as we do not require λ-additivity of the measure, we do not contradict any of the Theorems 23.10, 23.12, or 23.14.

A cardinal κ that carries a (not necessarily κ-additive) two-valued probability measure μ that vanishes on the singletons is sometimes called *Ulam measurable*.

Now let us present a beautiful theorem of Ulam that delineates the realm of measurable cardinals among all real-valued measurable cardinals. To state it, we need one more concept.

DEFINITION 23.20. Let μ be a measure on κ. A subset $X \subseteq \kappa$ is called an *atom of* μ if $\mu(X) > 0$, and for each $Y \subseteq X$, either $\mu(Y) = 0$ or $\mu(Y) = \mu(X)$. A measure μ is called *atomless* if μ has no atoms. We call μ *atomic* if for every $Z \subseteq \kappa$ with $\mu(Z) > 0$ there exists an atom X of μ such that $X \subseteq Z$.

Note that being "atomic" is a stricter requirement than "not being atomless."

THEOREM 23.21. *Suppose κ is a real-valued measurable cardinal. Then exactly one of the following holds:*

(1) κ is measurable and every κ-additive measure on κ is atomic, or
(2) $\kappa \leq 2^{\aleph_0}$ and every probability measure on κ that vanishes on the singletons is atomless.

PROOF. This boils down to the following:

LEMMA 23.22. *If there exists a κ-additive atomless probability measure on κ, then $\kappa \leq 2^{\aleph_0}$.*

Note that if μ is an atomless probability measure on any cardinal, then μ vanishes on the singletons, so we did not need to make this assumption explicit in Lemma 23.22.

Let us show how Theorem 23.21 follows from the lemma. Consider a κ-additive measure μ on κ that vanishes on the singletons.

Case 1: μ has an atom X.

Note that X has size κ. So we can enumerate $X = \{x_\alpha : \alpha < \kappa\}$, and define a function $\nu : \mathcal{P}(\kappa) \to \{0, 1\}$ by

$$\nu(A) = \frac{1}{\mu(X)} \mu(\{x_\alpha : \alpha \in A\}).$$

As in the proof of Theorem 23.19, one can show that ν is a two-valued κ-additive probability measure on κ. Thus κ is measurable; and by Theorem 23.14, $\kappa > 2^{\aleph_0}$.

Case 2: μ is not atomic.

There exists $X \subseteq \kappa$ such that $\mu(X) > 0$ and there is no atom $Y \subseteq X$. Again, since μ is κ-additive and vanishes on the singletons $\{\xi\} \subseteq X$, it follows that $|X| = \kappa$. Thus, we can enumerate $X = \{x_\alpha : \alpha < \kappa\}$, and define an atomless probability measure $\nu : \mathcal{P}(\kappa) \to [0, 1]$ on κ by

$$\nu(A) = \frac{1}{\mu(X)} \mu(\{x_\alpha : \alpha \in A\}).$$

Now it follows from Lemma 23.22 that $\kappa \leq 2^{\aleph_0}$.

This proves the dichotomy expressed by Theorem 23.21.

PROOF OF LEMMA 23.22. We need a technical fact about atomless measures.

CLAIM 23.23. *Let μ be an atomless probability measure on κ, and suppose $X \subseteq \kappa$ is such that $\mu(X) > 0$. Then there exists a partition of X into two disjoint sets X_0, X_1 such that $\frac{1}{4}\mu(X) < \mu(X_0) < \frac{3}{4}\mu(X)$.*

REMARK 23.24. Note that if X_0, X_1 are as in the above claim, then $\mu(X_1) = \mu(X) - \mu(X_0)$. Thus the estimate $\frac{1}{4}\mu(X) < \mu(X_1) < \frac{3}{4}\mu(X)$ holds as well. With a little extra effort, one can prove stronger results along the line of Claim 23.23. But since Claim 23.23 suffices for the present purpose, we leave the improvements as an exercise for the interested reader.

PROOF OF CLAIM 23.23. Let μ, X be as in the assumptions, and let

$$\delta = \inf\{\varepsilon : \exists X_0, X_1 \, (X_0 \cup X_1 = X \wedge X_0 \cap X_1 = \emptyset \wedge \varepsilon = \mu(X_0) \geq \mu(X_1))\}.$$

It suffices to show that $\delta < \frac{3}{4}\mu(X)$. Suppose towards a contradiction that $\delta \geq \frac{3}{4}\mu(X)$, and choose a sequence $\langle \langle X_0^n, X_1^n \rangle : n \in \omega \rangle$ of partitions of X such that $\mu(X_1^n) \leq \mu(X_0^n) < \delta + 2^{-n}$. Note that $\mu(X_0^n \cap X_0^{n+1}) \geq \delta$ for each $n \in \omega$: If not, the pair $\langle X_0^n \cap X_0^{n+1}, X_1^n \cup X_1^{n+1} \rangle$ would be a partition of X that contradicts the choice of δ. Therefore, we may without loss of generality assume that $X_0^{n+1} \subseteq X_0^n$ for all $n \in \omega$. Now let $X_0 = \bigcap_{n \in \omega} X_0^n$ and $X_1 = X \setminus X_0$. Then $\mu(X_0) = \delta$. Since μ is atomless, there exists $X_2 \subseteq X_0$ such that $0 < \mu(X_2) < \mu(X_0)$. If $\mu(X_2) \geq \frac{1}{2}\mu(X)$, then the pair $\langle X_2, X \setminus X_2 \rangle$ contradicts the choice of δ. If $\mu(X_2) < \frac{1}{2}\mu(X)$, then one of the pairs $\langle X_0 \setminus X_2, X_1 \cup X_2 \rangle$ or $\langle X_1 \cup X_2, X_0 \setminus X_2 \rangle$ contradicts the choice of δ. □

EXERCISE 23.10 (PG). Show that if μ and X are as in the assumptions of Claim 23.23, then there exists $X_0 \subseteq X$ such that $\mu(X_0) = \frac{1}{2}\mu(X)$.

Now we are ready to prove Lemma 23.22. Let κ, μ be as in the assumptions. Claim 23.23 allows us to construct by recursion over $\subseteq$ (which is a well-founded relation on $^{<\omega}2$) a family $\{X_s : s \in {}^{<\omega}2\}$ with the following properties:

(i) $X_\emptyset = \kappa$;
(ii) If $dom(s) = n$, then we have $X_{s\frown 0} \cap X_{s\frown 1} = \emptyset$, $X_{s\frown 0} \cup X_{s\frown 1} = X_s$, and $\frac{1}{4}\mu(X_s) < \mu(X_{s\frown 0}) < \frac{3}{4}\mu(X_s)$.

For each $f \in {}^\omega 2$, let $X_f = \bigcap_{\{s \in {}^{<\omega}2:\, s \subset f\}} X_s$. Then $\kappa = \bigcup_{f \in {}^\omega 2} X_f$, and $\mu(X_f) = 0$ for each $f \in {}^\omega 2$. Thus, we have found $2^{\aleph_0}$ sets of measure 0 that add up to a set of measure 1, and κ-additivity of μ implies that $\kappa \leq 2^{\aleph_0}$. This concludes the proof of Lemma 23.22 and simultaneously of Theorem 23.21. □□

Let us return to Question 23.2. By Theorem 23.19, the existence of a real-valued measurable cardinal $\kappa \leq 2^{\aleph_0}$ is a necessary condition for Lebesgue measure to be extendable to $\mathcal{P}([0,1])$. It turns out that this condition is also sufficient.

THEOREM 23.25. *If there exists a real-valued measurable cardinal $\kappa \leq 2^{\aleph_0}$, then Lebesgue measure can be extended to a probability measure on $\mathcal{P}([0,1])$.*

PROOF. Let μ be a κ-additive measure on $\kappa \leq 2^{\aleph_0}$, and let $\{X_s : s \in {}^{<\omega}2\}$ be a family of subsets of κ like the one constructed in the last part of the proof of Lemma 23.22. By recursion over $\subseteq$, we construct families of reals $\{a_s : s \in {}^{<\omega}2\}$ and $\{b_s : s \in {}^{<\omega}2\}$ as follows:

(1) $a_\emptyset = 0$, $b_\emptyset = 1$;
(2) $a_{s\frown 0} = a_s$, $b_{s\frown 1} = b_s$ for all $s \in {}^{<\omega}2$;
(3) $a_{s\frown 1} = a_s + \mu(X_{s\frown 0}) = b_{s\frown 0}$ for all $s \in {}^{<\omega}2$.

Define a function $H : {}^\omega 2 \to [0,1]$ such that $H(f) = x$ if $\bigcap_{\{s \in {}^{<\omega}2:\, s \subset f\}} [a_s, b_s] = \{x\}$. By compactness of $[0,1]$, H is well-defined. Let us define $\nu : \mathcal{P}([0,1]) \to [0,1]$ by: $\nu(Y) = \mu(H^{-1}Y)$. This is a probability measure defined on all subsets of $[0,1]$. It follows from our definitions that $\nu([a_s, b_s]) = \mu(X_s) = b_s - a_s$ for every $s \in {}^{<\omega}2$.

In order to see that the same is true for arbitrary subintervals $[a, b]$ of $[0, 1]$, note that for $s \in {}^n 2$, each of the intervals $[a_s, b_s]$ has length $\mu(X_s) \leq (0.75)^n$. Moreover, for $s \neq t$, if both $s, t \in {}^n 2$, then $[a_s, b_s] \cap [a_t, b_t]$ contains at most one point and thus has Lebesgue measure zero. Also, for any fixed n, there are at most two $s \in {}^{<\omega}2$ such that $[a_s, b_s]$ has nonempty intersection with $[a, b]$ but is not contained in $[a, b]$. It follows that

$$\left| b - a - \sum_{\{s \in {}^n 2:\, [a_s, b_s] \subseteq [a,b]\}} (b_s - a_s) \right| \leq 2(0.75)^n.$$

In other words, we can approximate the interval $[a,b]$ by intervals of the form $[a_s, b_s]$ with any desired accuracy. Therefore, $\nu([a,b]) = b - a$, and it follows that ν extends Lebesgue measure. $\square$

Not surprisingly, the assumption that $2^{\aleph_0}$ is real-valued measurable has numerous implications for the structure of $\mathbb{R}$, $\mathcal{P}(\omega)$, and $^\omega\omega$. Let us give just one example that illustrates some of the techniques involved in the proofs of such consequences.

THEOREM 23.26. *If $2^{\aleph_0}$ is a real-valued measurable cardinal, then $\mathfrak{b} < 2^{\aleph_0}$.*

PROOF. Suppose towards a contradiction that μ is a $2^{\aleph_0}$-additive probability measure on $\mathcal{P}(2^{\aleph_0})$ that vanishes on the singletons, and $\mathfrak{b} = 2^{\aleph_0}$. By Exercise 17.18(b), there exists a sequence $\langle f_\xi : \xi < 2^{\aleph_0}\rangle$ of elements of $^\omega\omega$ such that $f_\xi <^* f_\eta$ for all $\xi < \eta < 2^{\aleph_0}$, and the set $\{f_\xi : \xi < 2^{\aleph_0}\}$ is unbounded in $^\omega\omega$. For each $n, k \in \omega$ let $B_{n,k} = \{\xi < 2^{\aleph_0} : f_\xi(n) \leq k\}$. Then $\mu(\bigcup_{k\in\omega} B_{n,k}) = 1$ for each $n \in \omega$, and since $(B_{n,k})_{k\in\omega}$ is an increasing sequence of sets, we can find for each $n \in \omega$ a number k_n such that $\mu(B_{n,k_n}) > 1 - 2^{-(n+2)}$. Then

$$\mu(\bigcap_{n\in\omega} B_{n,k_n}) = 1 - \mu(\bigcup_{n\in\omega}(2^{\aleph_0}\setminus B_{n,k_n})) \geq 1 - \sum_{n\in\omega} \mu(2^{\aleph_0}\setminus B_{n,k_n})$$

$$= 1 - \sum_{n\in\omega} 2^{-(n+2)} = 1 - \frac{1}{2} = \frac{1}{2}.$$

Thus $\bigcap_{n\in\omega} B_{n,k_n}$ is a set of positive measure, and since μ is $2^{\aleph_0}$-additive and vanishes on the singletons, $|\bigcap_{n\in\omega} B_{n,k_n}| = 2^{\aleph_0}$. It follows that the sequence $\langle f_\xi : \xi \in \bigcap_{n\in\omega} B_{n,k_n}\rangle$ is cofinal in $\langle f_\xi : \xi < 2^{\aleph_0}\rangle$, and hence unbounded in $^\omega\omega$. But consider the function $g \in {}^\omega\omega$ defined by $g(n) = k_n + 1$. For all $\xi \in \bigcap_{n\in\omega} B_{n,k_n}$ we have $f_\xi < g$, which leads to a contradiction. $\square$

We conclude this section with some remarks on consistency. As we already mentioned, one cannot show in ZFC that real-valued measurable cardinals exist. Moreover, if the theory "ZFC + there exists a real-valued measurable cardinal" is consistent, one cannot prove this fact in the theory "ZFC + CON(ZFC)." The reasons for this have been spelled out in Chapter 12 when we were discussing inaccessible cardinals. Since this makes real-valued measurable cardinal somewhat suspicious, the question arises: Should one study these objects at all? Most contemporary set theorists would answer: "Yes, but cautiously."

First, we cannot prove the consistency of the existence of a (real-valued) measurable cardinal, but neither can we prove the consistency of ZFC. However, the existence of (real-valued) measurable cardinals *might* be consistent, and as long as we have no proof to the contrary, we should study the consequences of this possibility.

Second, there is at least strong empirical evidence that the existence of these cardinals is consistent with ZFC. For one thing, many seemingly unrelated questions in mathematics turn out to be equivalent to[3] or equiconsistent with[4] the existence of measurable cardinals. This adds weight to the belief that there is more substance to the concept than just being inconsistency in disguise. Moreover, although the consequences of the existence of (real-valued) measurable cardinals have been extensively studied, no inconsistency has been found so far. This resembles the status

---

[3] Question 13.15 is a prime example here.
[4] The notion of equiconsistency was discussed in Chapter 12.

of several famous open conjectures in number theory, like the Riemann Hypothesis. There, we know that if a counterexample exists, it must involve big numbers. Here, we know at least that no *easy* proof of the inconsistency of a (real-valued) measurable cardinal exists.

But some caution never hurts. If a mathematical theorem has been proved under the assumption that there exists a (real-valued) measurable cardinal, one should investigate whether and how much this assumption can be weakened. The aim is to either prove the same result in ZFC, or to prove an equiconsistency result. We conclude this chapter with the perhaps most important equiconsistency results involving measurable cardinals.

THEOREM 23.27. *If any one of the following four theories is consistent, then so are the other three.*

(a) ZFC + *"there exists a measurable cardinal;"*
(b) ZFC + *"there exists a real-valued measurable cardinal $\kappa < 2^{\aleph_0}$;"*
(c) ZFC + *"$2^{\aleph_0}$ is a real-valued measurable cardinal;"*
(d) ZFC + *"some uncountable regular cardinal κ carries a κ-complete, κ^+-saturated proper ideal that contains all singletons."*

Mathographical Remarks

The equivalences between (a), (b), and (c) of Theorem 23.27 were announced by Solovay in 1966 and published in his article *Real-valued measurable cardinals*, in *Axiomatic Set Theory*, D. Scott, ed., Proceedings of Symposia in Pure Mathematics, vol. 13, part 1 (1971), AMS, 396–428. The implication (d)→(a) of Theorem 23.27 was proved by K. Kunen in *Some applications of iterated ultrapowers in set theory*, Annals of Mathematical Logic 1 (1970), 179–227.

For more information about measurable and other large cardinals we recommend the encyclopaedic treatise *The Higher Infinite*, by Akihiro Kanamori, Springer Verlag, 1994. D. Fremlin's article *Real-valued measurable cardinals*, in *Set Theory of the Reals*, H. Judah, ed., Israel Mathematical Conference Proceedings, vol. 6 (1993), AMS, 151–304, is a comprehensive survey of contemporary knowledge about real-valued measurable cardinals $\leq 2^{\aleph_0}$.

CHAPTER 24

Elementary Submodels

24.1. Elementary facts about elementary submodels

Throughout this section, L denotes a first-order language with nonlogical symbols $\{r_i : i \in I\} \cup \{f_j : j \in J\} \cup \{c_k : k \in K\}$, where r_i is a relational symbol of arity $\tau_0(i)$, f_j is a functional symbol of arity $\tau_1(j)$, and the c_k's are constant symbols. Recall from Chapter 6 of Volume I that if $\mathfrak{A} = \langle A, (R_i^A)_{i \in I}, (F_j^A)_{j \in J}, (C_k^A)_{k \in K} \rangle$ and $\mathfrak{B} = \langle B, (R_i^B)_{i \in I}, (F_j^B)_{j \in J}, (C_k^B)_{k \in K} \rangle$ are models of the language L, then a function $h : A \to B$ is a *homomorphism* from $\mathfrak{A}$ into $\mathfrak{B}$ if

(i) For each $i \in I$ and each $\vec{a} \in A^{\tau_0(i)}$, if $\vec{a} \in R_i^A$, then $h(\vec{a}) \in R_i^B$;
(ii) For each $j \in J$ and each $\vec{a} \in A^{\tau_1(j)}$, $h(F_i^A(\vec{a})) = F_i^B(h(\vec{a}))$;
(iii) For each $k \in K$, $h(C_k^A) = C_k^B$.

Also recall that a homomorphism h is an *embedding*, if h is injective and satisfies the following condition (iv), which is stronger than (i):

(iv) For each $i \in I$ and each $\vec{a} \in A^{\tau_0(i)}$, $\vec{a} \in R_i^A$ iff $h(\vec{a}) \in R_i^B$.

In this chapter, we will consider a special class of embeddings.

DEFINITION 24.1. Let
$$\mathfrak{A} = \langle A, (R_i^A)_{i \in I}, (F_j^A)_{j \in J}, (C_k^A)_{k \in K} \rangle$$
and
$$\mathfrak{B} = \langle B, (R_i^B)_{i \in I}, (F_j^B)_{j \in J}, (C_k^B)_{k \in K} \rangle$$
be models of L, and let $h : A \to B$ be an embedding of $\mathfrak{A}$ into $\mathfrak{B}$. We say that h is an *elementary embedding* if

(v) For every formula φ of L with free variables among $v_0, \ldots, v_n$ and arbitrary $a_0, \ldots, a_n \in A$,
$\mathfrak{A} \models \varphi[a_0, \ldots, a_n]$ if and only if $\mathfrak{B} \models \varphi[h(a_0), \ldots, h(a_n)]$.

We say that $\mathfrak{A}$ is an *elementary submodel* of $\mathfrak{B}$, and write $\mathfrak{A} \prec \mathfrak{B}$, if $A \subseteq B$ and the identity function $id : A \to B$ is an elementary embedding.

REMARK 24.2. Clearly, $\mathfrak{A} \prec \mathfrak{A}$. It would make more sense to use the symbol $\preceq$ instead of $\prec$, but we decided to adopt the illogical choice of symbols that is used throughout the literature on mathematical logic.

EXERCISE 24.1 (R). Recall that a formula is *atomic* if it is either of the form "$t_0 = t_1$" or of the form "$r_i(t_0, \ldots, t_{\tau_0(i)-1})$" where $i \in I$ and $t_0, \ldots, t_{\tau_0(i)-1}$ are terms of L. Show that if $\mathfrak{A}, \mathfrak{B}$ are as in Definition 24.1, and $h : A \to B$ is an injection, then h is an embedding if and only if the following holds:

(v)$^-$ For every atomic formula φ of L with free variables among $v_0, \ldots, v_n$ and arbitrary $a_0, \ldots, a_n \in A$,
$\mathfrak{A} \models \varphi[a_0, \ldots, a_n]$ if and only if $\mathfrak{B} \models \varphi[h(a_0), \ldots, h(a_n)]$.

Hint: Note that if $s : \omega \to A$ is a valuation for $\mathfrak{A}$, then $h \circ s : \omega \to B$ is a valuation for $\mathfrak{B}$. Use induction over the length of t to show that for every term t of L, every valuation $s : \omega \to A$, and each $a \in A$ we have: $t^s = a$ if and only if $t^{h \circ s} = h(a)$.

Recall that we write $\mathfrak{A} \equiv \mathfrak{B}$, and say that the models $\mathfrak{A}$ and $\mathfrak{B}$ are *elementarily equivalent*, if for every sentence φ of L,

$$\mathfrak{A} \models \varphi \text{ if and only if } \mathfrak{B} \models \varphi.$$

The following claim is an immediate consequence of condition (v).

CLAIM 24.3. *Let $\mathfrak{A}$ and $\mathfrak{B}$ be models of L such that there exists an elementary embedding from $\mathfrak{A}$ into $\mathfrak{B}$. Then $\mathfrak{A} \equiv \mathfrak{B}$.*

EXAMPLE 24.1. Let L_G be the language of group theory.[1] Let $\mathbb{R}^+$, $\mathbb{Z}^+$ be the sets of nonnegative reals and nonnegative integers respectively, and let $\mathbb{E}$ denote the set of even integers. Consider the following models of L_G:

$$\mathfrak{A} = \langle \mathbb{R}^+, \cdot, 1 \rangle; \quad \mathfrak{B} = \langle \mathbb{Z}, +, 0 \rangle; \quad \mathfrak{C} = \langle \mathbb{Z}^+, +, 0 \rangle; \quad \mathfrak{D} = \langle \mathbb{E}, +, 0 \rangle.$$

Note that $\mathfrak{C}$ is a submodel of $\mathfrak{B}$, despite the fact that $\mathfrak{B}$ is a group and $\mathfrak{C}$ is not. But this difference implies, by Claim 24.3, that $\mathfrak{C}$ is not an elementary submodel of $\mathfrak{B}$.

The injection $h : \mathbb{Z} \to \mathbb{R}^+$ defined by $h(n) = 2^n$ is an embedding of $\mathfrak{B}$ into $\mathfrak{A}$. However, h is not an elementary embedding, since $\mathfrak{B} \models \forall v_0 \, (1 \neq v_0 + v_0)$, whereas $\mathfrak{A} \models \exists v_0 \, (h(1) = v_0 \cdot v_0)$.

It is not hard to see that the models $\mathfrak{B}$ and $\mathfrak{D}$ are isomorphic. So it may come as a surprise that $\mathfrak{D}$ is *not* an elementary submodel of $\mathfrak{B}$.

EXERCISE 24.2 (PG). Show that $\mathfrak{D}$ is not an elementary submodel of $\mathfrak{B}$.

EXAMPLE 24.2. Let $L_\leq$ be the language whose only nonlogical symbol is the binary relational symbol $\leq$, and consider the following models of $L_\leq$:

$$\mathfrak{A} = \langle \mathbb{Q} \cap (0, \infty), \leq \rangle; \, \mathfrak{B} = \langle \mathbb{Q}, \leq \rangle; \, \mathfrak{C} = \langle \mathbb{R}, \leq \rangle; \, \mathfrak{D} = \langle \mathbb{R}^+, \leq \rangle; \, \mathfrak{E} = \langle \mathbb{R} \cap [-1, \infty), \leq \rangle.$$

Since $\mathfrak{A}$, $\mathfrak{B}$, and $\mathfrak{C}$ are models of the complete theory DLONE,[2] it is perhaps easy to believe that $\mathfrak{A} \prec \mathfrak{B}$ and $\mathfrak{B} \prec \mathfrak{C}$. This is indeed the case, but a formal proof of this fact is nontrivial.

EXERCISE 24.3 (R). Suppose $\mathfrak{A} = \langle A, \leq_A \rangle$ and $\mathfrak{B} = \langle B, \leq_B \rangle$ are dense l.o.'s without endpoints. Show that $\mathfrak{A} \subseteq \mathfrak{B}$ implies $\mathfrak{A} \prec \mathfrak{B}$.

Hint: Assume that $a_0 <_A \cdots <_A a_{n-1}$ and $b_0 <_B \cdots <_B b_{n-1}$. Show that if φ is a formula of $L_\leq$ with free variables among $v_0, \ldots, v_{n-1}$, then

(+) $\qquad \mathfrak{A} \models \varphi[a_0, \ldots, a_{n-1}]$ if and only if $\mathfrak{B} \models \varphi[b_0, \ldots, b_{n-1}]$.

One can prove (+) by induction over the length of φ, using some of the ideas of the proof of Theorem 4.16.[3]

Exercise 24.3 implies in particular that $\mathfrak{A} \prec \mathfrak{B}$, $\mathfrak{B} \prec \mathfrak{C}$, and $\mathfrak{A} \prec \mathfrak{C}$. The latter can also be deduced using transitivity of the relation $\prec$.

---

[1] See the "Notation" section for a description of L_G.
[2] This theory was discussed in Section 6.1. The letters stand for Dense Linear Order with No Endpoints.
[3] If $\mathfrak{A} = \mathfrak{B}$, then property (+) expresses the fact that the elements of A are *indiscernible*. Sets of indiscernible elements play an important role in model theory.

EXERCISE 24.4 (G). Assume that $\mathfrak{A}$, $\mathfrak{B}$, and $\mathfrak{C}$ are models of L. Show that if h is an elementary embedding of $\mathfrak{A}$ into $\mathfrak{B}$ and j is an elementary embedding of $\mathfrak{B}$ into $\mathfrak{C}$, then $j \circ h$ is an elementary embedding of $\mathfrak{A}$ into $\mathfrak{C}$. Conclude that if $\mathfrak{A} \prec \mathfrak{B}$ and $\mathfrak{B} \prec \mathfrak{C}$, then $\mathfrak{A} \prec \mathfrak{C}$.

Now consider the models $\mathfrak{C}$, $\mathfrak{D}$, and $\mathfrak{E}$ of Example 24.2. Since $\mathfrak{D}$ has a smallest element but $\mathfrak{C}$ doesn't, it follows from Claim 24.3 that $\mathfrak{D}$ is not an elementary submodel of $\mathfrak{C}$. Somewhat more interesting is the fact that $\mathfrak{D}$ is not an elementary submodel of $\mathfrak{E}$ either. To see this, note that $\mathfrak{D} \models \forall v_1 (v_1 \leq 0 \to v_1 = 0)$,[4] whereas $\mathfrak{E} \models \exists v_1 (v_1 \leq 0 \wedge \neg (v_1 = 0))$.

In more human terms, while $\mathfrak{D}$ does not have any elements to the left of 0, $\mathfrak{E}$ does. It is instructive to compare this situation with the difference between $\mathfrak{A}$ and $\mathfrak{B}$. It is also true that $\mathfrak{B}$ contains elements to the left of 0 while $\mathfrak{A}$ doesn't. But since A, the universe of the model $\mathfrak{A}$, is equal to $\mathbb{Q} \cap (0, \infty)$, we have $0 \notin A$, and for every $q \in A$, there does exist an $r \in A$ with $r < q$. Thus, from $\mathfrak{A}$'s point of view, all of its elements have the same properties as they have from $\mathfrak{B}$'s point of view. Of course, $\mathfrak{B}$ has far more radical ideas than $\mathfrak{A}$ about how far to the left one can be, but $\mathfrak{A}$ is not bothered by such extremes, since they are simply not expressible in terms of the parameters known to $\mathfrak{A}$. In other words, if one informally defines "$\mathfrak{A}$'s language" as all expressions of L with parameters from A, and "$\mathfrak{B}$'s language" as the expressions of L with parameters from its universe $\mathbb{Q}$, then the language of $\mathfrak{A}$ is related to the language of $\mathfrak{B}$ as Orwellian Newspeak is related to natural language.

EXAMPLE 24.3. Let X be a nonempty set, let $\mathcal{U}$ be an ultrafilter on X, let $\mathfrak{A}$ be a model for L, and let $^X\mathfrak{A}/\mathcal{U}$ be the corresponding ultrapower. For each $a \in A$, let $g^a \in {}^X\mathfrak{A}/\mathcal{U}$ be the function that takes the value a for all $\xi \in X$. Then the function $h : A \to {}^X A/\mathcal{U}$ defined by: $h(a) = g^a/\mathcal{U}$ is an injection of A into $^X\mathfrak{A}/\mathcal{U}$.

EXERCISE 24.5 (PG). Show that the function h defined above is an elementary embedding of $\mathfrak{A}$ into $^X\mathfrak{A}/\mathcal{U}$. *Hint:* Use Łoś' Theorem.

The following theorem is often called the *Tarski-Vaught criterion for elementary submodels*.

THEOREM 24.4. *Let $\mathfrak{A} = \langle A, \ldots \rangle$, $\mathfrak{B} = \langle B, \ldots \rangle$ be models for L such that $\mathfrak{A} \subseteq \mathfrak{B}$. Then the following are equivalent:*
(a) $\mathfrak{A} \prec \mathfrak{B}$;
(b) *For every formula ψ of L with free variables among $v_0, \ldots, v_{n-1}$, for every $\ell < n$, and arbitrary parameters $a_0, \ldots, a_{\ell-1}, a_{\ell+1}, \ldots, a_{n-1} \in A$, if*

(24.1) $$\mathfrak{B} \models \exists v_\ell \psi[a_0, \ldots, a_{\ell-1}, a_{\ell+1}, \ldots, a_{n-1}],$$

then there exists $a_\ell \in A$ such that

(24.2) $$\mathfrak{B} \models \psi[a_0, \ldots, a_{\ell-1}, a_\ell, a_{\ell+1}, \ldots, a_{n-1}].$$

PROOF. For the implication (a) $\to$ (b), assume that $\mathfrak{A} \prec \mathfrak{B}$ and

(24.3) $$\mathfrak{B} \models \exists v_\ell \psi[a_0, \ldots, a_{\ell-1}, a_{\ell+1}, \ldots, a_{n-1}].$$

By condition (v) of Definition 24.1,

(24.4) $$\mathfrak{A} \models \exists v_\ell \psi[a_0, \ldots, a_{\ell-1}, a_{\ell+1}, \ldots, a_{n-1}],$$

---

[4] Recall that this is shorthand for $\mathfrak{D} \models \forall v_1 (v_1 \leq v_0 \to v_1 = v_0)[0]$.

and hence there exists $a_\ell \in A$ such that

(24.5) $$\mathfrak{A} \models \psi[a_0, \ldots, a_{\ell-1}, a_\ell, a_{\ell+1}, \ldots, a_{n-1}].$$

Applying (v) to the formula ψ, we infer that equation (24.2) holds.

Now assume that $\mathfrak{A}$ is a submodel of $\mathfrak{B}$ and (b) holds. If φ is a formula of L with all free variables among $v_0, \ldots, v_{n-1}$, let $(v)_\varphi$ be the statement:

$(v)_\varphi$ "For all n-tuples $\langle a_0, \ldots, a_{n-1}\rangle \in A^n$,
$\mathfrak{A} \models \varphi[a_0, \ldots, a_{n-1}]$ if and only if $\mathfrak{B} \models \varphi[a_0, \ldots, a_{n-1}]$."

By induction over the length of φ, we show that $(v)_\varphi$ holds for every formula φ of L.

If φ is an atomic formula, then $(v)_\varphi$ follows from Exercise 24.1.

If $(v)_\varphi$ and $(v)_\psi$ hold, then $(v)_{\neg\varphi}$ and $(v)_{\varphi\wedge\psi}$ follow from the clauses in the definition of the satisfaction relation $\models$ that pertain to negation and conjunction.

Now suppose that φ is of the form $\exists v_\ell \psi$ and $(v)_\psi$ holds. If

(24.6) $$\mathfrak{A} \models \exists v_\ell \psi[a_0, \ldots, a_\ell, \ldots, a_{n-1}],$$

then

(24.7) $$\mathfrak{A} \models \psi[a_0, \ldots a_{\ell-1}, a'_\ell, a_{\ell+1}, \ldots, a_{n-1}]$$

for some $a'_\ell \in A$ (possibly, $a'_\ell \neq a_\ell$), and $(v)_\psi$ implies that also

(24.8) $$\mathfrak{B} \models \psi[a_0, \ldots a_{\ell-1}, a'_\ell, a_{\ell+1}, \ldots, a_{n-1}].$$

By the existential quantifier clause in the definition of $\models$,

(24.9) $$\mathfrak{B} \models \exists v_\ell \psi[a_0, \ldots, a_{\ell-1}, a_\ell, a_{\ell+1}, \ldots, a_{n-1}].$$

Now assume formula (24.9) holds. Since v_ℓ is not a free variable in $\exists v_\ell \psi$, this can be written as

(24.10) $$\mathfrak{B} \models \exists v_\ell \psi[a_0, \ldots, a_{\ell-1}, a_{\ell+1}, \ldots, a_{n-1}].$$

By (b), there exists $a'_\ell \in A$ such that

(24.11) $$\mathfrak{B} \models \psi[a_0, \ldots a_{\ell-1}, a'_\ell, a_{\ell+1}, \ldots, a_{n-1}].$$

By $(v)_\psi$, this implies

(24.12) $$\mathfrak{A} \models \psi[a_0, \ldots a_{\ell-1}, a'_\ell, a_{\ell+1}, \ldots, a_{n-1}],$$

and from the definition of $\models$ we conclude that

(24.13) $$\mathfrak{A} \models \exists v_\ell \varphi[a_0, \ldots a_{\ell-1}, a_\ell, a_{\ell+1}, \ldots, a_{n-1}].$$

$\square$

LEMMA 24.5. *Let* $\mathfrak{B} = \langle B, (R_i)_{i\in I}, (F_j)_{j\in J}, (C_k)_{k\in K}\rangle$ *be a model of L, let δ be a limit ordinal > 0, and let $\langle A_\xi : \xi < \delta\rangle$ be a sequence of subsets of B such that $A_\xi \subseteq A_\eta$ for all $\xi < \eta < \delta$, and each A_ξ is the universe of an elementary submodel of $\mathfrak{B}$ (i.e., $\mathfrak{A}_\xi \prec \mathfrak{B}$, where $\mathfrak{A}_\xi = \langle A_\xi, (A_\xi^{\tau_0(i)} \cap R_i)_{i\in I}, (A_\xi^{\tau_1(j)+1} \cap F_j)_{j\in J}, (C_k)_{k\in K}\rangle$). Then $\bigcup_{\xi<\delta} A_\xi$ is the universe of an elementary submodel of $\mathfrak{B}$.*

24.1. ELEMENTARY FACTS ABOUT ELEMENTARY SUBMODELS

PROOF. Let $A = \bigcup_{\xi < \delta} A_\xi$, and let

$$\mathfrak{A} = \langle A, (\bigcup_{\xi < \delta} A_\xi^{\tau_0(i)} \cap R_i)_{i \in I}, (\bigcup_{\xi < \delta} A_\xi^{\tau_1(j)+1} \cap F_j)_{j \in J}, (C_k)_{k \in K} \rangle.$$

We want to show that $\mathfrak{A} \prec \mathfrak{B}$.

It suffices to verify that condition (b) of Theorem 24.4 is satisfied. Let ψ be an arbitrary formula of L with free variables among $v_0, \ldots, v_{n-1}$, let $\ell < n$, and let $a_0, \ldots, a_{\ell-1}, a_{\ell+1}, \ldots, a_{n-1}$ be arbitrary elements of A such that

(24.14) $\qquad \mathfrak{B} \models \exists v_\ell \psi[a_0, \ldots, a_{\ell-1}, a_{\ell+1}, \ldots, a_{n-1}].$

We want to find $a_\ell \in A$ such that

(24.15) $\qquad \mathfrak{B} \models \psi[a_0, \ldots, a_{\ell-1}, a_\ell, a_{\ell+1}, \ldots, a_{n-1}].$

Since δ is a limit ordinal, we can find $\xi < \delta$ such that $\{a_0, \ldots, a_{\ell-1}, a_{\ell+1}, \ldots, a_{n-1}\} \subseteq A_\xi$. Since $\mathfrak{A}_\xi \prec \mathfrak{B}$, condition (b) of Theorem 24.4 holds for $\mathfrak{A}_\xi$ and $\mathfrak{B}$. Thus there exists $a_\ell \in A_\xi$ such that

(24.16) $\qquad \mathfrak{B} \models \psi[a_0, \ldots, a_{\ell-1}, a_\ell, a_{\ell+1}, \ldots, a_{n-1}].$

Since $A_\xi \subseteq A$, this a_ℓ is also an element of A, and we are done. $\square$

Now it is your turn to practice applications of Theorem 24.4.

EXERCISE 24.6 (PG). Let $\mathfrak{B}$ and $\langle \mathfrak{A}_\xi : \xi < \delta \rangle$ be as in the assumptions of Lemma 24.5. Show that $\mathfrak{A}_\xi \prec \mathfrak{A}_\eta$ for all $\xi < \eta < \delta$.

EXERCISE 24.7 (PG). (a) Theorem 24.4 can be treated as a criterion that tells us when the identity mapping is an elementary embedding. Formulate and prove a generalization of Theorem 24.4 to a criterion that tells us when an arbitrary embedding of a model $\mathfrak{A}$ into a model $\mathfrak{B}$ is elementary.

(b) Use the criterion found in point (a) to prove the following:

CLAIM 24.6. *If $h : A \to B$ is an isomorphism of models $\mathfrak{A} = \langle A, \ldots \rangle$ and $\mathfrak{B} = \langle B, \ldots \rangle$, then h is an elementary embedding of $\mathfrak{A}$ into $\mathfrak{B}$.*

REMARK 24.7. (a) The models $\mathfrak{D} \subseteq \mathfrak{B}$ of Example 24.1 and $\mathfrak{D} \subseteq \mathfrak{E}$ of Example 24.2 are isomorphic, and yet the smaller models are not elementary submodels of the larger ones. This does not contradict Claim 24.6 though, since the inclusion maps are different from the isomorphisms.

(b) Theorem 6.4 asserts that isomorphic models are elementarily equivalent. This is an immediate consequence of Claims 24.3 and 24.6.

By Corollary 21.31, if L is a countable language, and if $\mathfrak{A} = \langle A, \ldots \rangle$ is an uncountable model of L, then $\mathfrak{A}$ has many countable submodels; in fact, the universes of these models form a club subset of $[A]^{\leq \aleph_0}$. Are at least some of these countable submodels elementary?

To study this question, let us first assume that $\mathfrak{A}$ has the following property:

(SK) $\qquad$ *For each $n \in \omega \setminus \{0\}$, each formula φ of L with free variables among $v_0, \ldots, v_n$, and each $\ell \leq n$ there exists a functional symbol $f_{\varphi, n, \ell}$ in L such that for all $a_0, \ldots, a_{\ell-1}, a_{\ell+1}, \ldots, a_n \in A$:*
$\mathfrak{A} \models \exists v_\ell \varphi(a_0, \ldots, a_{\ell-1}, v_\ell, a_{\ell+1}, \ldots, a_n)$ *implies*
$\mathfrak{A} \models \varphi(a_0, \ldots, a_{\ell-1}, f_{\varphi, n, \ell}(a_0, \ldots a_{\ell-1}, a_{\ell+1}, \ldots, a_n), a_{\ell+1}, \ldots, a_n).$

If $\mathfrak{A}$ satisfies condition SK, then we call $\mathfrak{A}$ a *Skolemized model*. Let $Form_L^n$ denote the set of formulas of L with free variables among $v_0, \ldots, v_n$. The set $\{F_{\varphi,n,\ell} : 0 < n < \omega, \ell \leq n, \varphi \in Form_L^n\}$ of interpretations of the function symbols $f_{\varphi,n,\ell}$ is called a set of *Skolem functions*[5] for L.

LEMMA 24.8. *Suppose $\mathfrak{B}$ is a Skolemized model and $\mathfrak{A}$ is a nonempty submodel of $\mathfrak{B}$. Then $\mathfrak{A} \prec \mathfrak{B}$.*

PROOF. If $\mathfrak{B} = \langle B, (R_i)_{i \in I}, (F_j)_{j \in J}, (C_k)_{k \in K} \rangle$ is any model of L and $\mathfrak{A}$ is a submodel of $\mathfrak{B}$, then the universe of $\mathfrak{A}$ is closed under all functions F_j. Thus, if $\mathfrak{B}$ is Skolemized, then $\mathfrak{A}$ and $\mathfrak{B}$ satisfies condition (b) of Theorem 24.4, and hence $\mathfrak{A} \prec \mathfrak{B}$. □

So, apparently, at least some uncountable models have lots of elementary submodels. But how common is it for a model $\mathfrak{A}$ to satisfy condition SK? Unfortunately, models of the most popular first-order languages never satisfy condition SK. If L has a Skolemized model, there must be a functional symbol $f_{\varphi,n,\ell}$ in L for every $\varphi \in Form_L^n$ and $\ell \leq n$. For example, the language of set theory L_S contains no functional symbols whatsoever, and hence L_S has no Skolemized models. For a similar reason, no model of L_G is Skolemized, not even the one-element group $\mathfrak{G}_1$. For $\mathfrak{G}_1$ to be Skolemized, we would need, among other things, a function symbol $f_{\varphi,3,3}$ of arity 3 to represent the Skolem function for the formula "$\exists v_3(v_3 = (v_0 * v_1) * v_2)$," and no such symbol is available in L_G.

Isn't that silly? After all, the constant e seems to be such a wonderful "all purpose" Skolem function for the one-element group.

Let us try to give a precise meaning to this idea. Recall from Chapter 6 that if L^+ is a *richer* first-order language than L, i.e., if the nonlogical symbols of L^+ are $\{r_i : i \in I^+\} \cup \{f_j : j \in J^+\} \cup \{c_k : k \in K^+\}$, with $I \subseteq I^+$, $J \subseteq J^+$, and $K \subseteq K^+$; and if $\mathfrak{B}^+ = \langle B, (R_i)_{i \in I^+}, (F_j)_{j \in J^+}, (C_k)_{k \in K^+} \rangle$ is a model of L^+, then the *L-reduct* of $\mathfrak{B}^+$ is the model $\mathfrak{B} = \langle B, (R_i)_{i \in I}, (F_j)_{j \in J}, (C_k)_{k \in K} \rangle$. The notion of an elementary submodel behaves nicely with respect to reducts.

CLAIM 24.9. *Suppose L^+ is a richer language than L, $\mathfrak{A}^+$ and $\mathfrak{B}^+$ are models for L^+, and $\mathfrak{A}$, $\mathfrak{B}$ are the corresponding L-reducts.*
   (a) *If φ is a formula of L with free variables among $v_0, \ldots, v_n$ and $a_0, \ldots, a_n$ are elements of the universe of $\mathfrak{A}$, then $\mathfrak{A} \models \varphi[a_0, \ldots, a_n]$ if and only if $\mathfrak{A}^+ \models \varphi[a_0, \ldots, a_n]$.*
   (b) *If $\mathfrak{A}^+ \prec \mathfrak{B}^+$, then also $\mathfrak{A} \prec \mathfrak{B}$.*

EXERCISE 24.8 (PG). Prove Claim 24.9.

Now let us consider a language L_G^+ that, in addition to the nonlogical symbols e and $*$ of L_G, has functional symbols f_j of arity j for every $j \in \omega \setminus \{0\}$. The language L_G^+ has exactly one one-element model $\mathfrak{G}_1^+ = \langle \{\bar{e}\}, \bar{*}, (F_j)_{j \in \omega \setminus \{0\}} \rangle$, where $F_j(\bar{e}, \ldots, \bar{e}) = \bar{e}$ for all $j \in \omega \setminus \{0\}$. Clearly, $\mathfrak{G}_1$ is the L_G-reduct $\mathfrak{G}_1^+$.

EXERCISE 24.9 (G). Convince yourself that $\mathfrak{G}_1^+$ satisfies SK.

---

[5]Of course, if φ does not have any free variables except v_0, then the Skolem function $F_{\varphi,0,0}$ is really a constant; and thus $f_{\varphi,\ell,\ell}$ should be a constant symbol rather than a functional symbol. We choose to sidestep this technicality by requiring $n > 0$ and allowing any number of dummy variables.

24.1. ELEMENTARY FACTS ABOUT ELEMENTARY SUBMODELS

We have shown that although $\mathfrak{G}_1$ is not a Skolemized model of L_G, it is the reduct of a Skolemized model of a richer (but still countable) language L_G^+. As it turns out, $\mathfrak{G}_1$ is by no means unique in this respect.

THEOREM 24.10. *Let L be a first-order language with nonlogical symbols $\{r_i : i \in I\} \cup \{f_j : j \in J\} \cup \{c_k : k \in K\}$, and let $\mathfrak{A} = \langle A, (R_i)_{i \in I}, (F_j)_{j \in J}, (C_k)_{k \in K}\rangle$ be a model for L with nonempty universe. Then there exist $J^+ \supseteq J$ and a Skolemized model $\mathfrak{A}^+$ for the language L^+ with nonlogical symbols $\{r_i : i \in I\} \cup \{f_j : j \in J^+\} \cup \{c_k : k \in K\}$ such that $\mathfrak{A}$ is the L-reduct of $\mathfrak{A}^+$. Moreover, J^+ can be chosen in such a way that $|J^+| = |I| + |J| + |K| + \aleph_0$.*

PROOF. Let $\mathfrak{B}$ be a model with nonempty universe of a first-order language L_m with nonlogical symbols $\{r_i : i \in I_m\} \cup \{f_j : j \in J_m\} \cup \{c_k : k \in K_m\}$. We construct a language $sk(L_m)$ and a model $sk(\mathfrak{B})$ as follows: For every $n \in \omega \backslash \{0\}$, $\ell \leq n$, and $\varphi \in Form_{L_m}^n$ we put a new functional symbol $f_{\varphi,n,\ell}$ of arity n into the set of nonlogical symbols of $sk(L_m)$.

EXERCISE 24.10 (PG). Convince yourself that if κ is an infinite cardinal such that L_m has at most κ nonlogical symbols, then $sk(L_m)$ also has at most κ nonlogical symbols.

Next we pick an element b^* of the universe B of $\mathfrak{B}$, and we construct $sk(\mathfrak{B})$ as follows:

(1) The universe of $sk(\mathfrak{B})$ is the same set B as the universe of $\mathfrak{B}$;
(2) The interpretations of the nonlogical symbols of L_m in $sk(\mathfrak{B})$ are the same as in $\mathfrak{B}$;
(3) For each new functional symbol $f_{\varphi,n,\ell}$ we define an interpretation $F_{\varphi,n,\ell} : B^n \to B$ as follows: If $\mathfrak{B} \models \exists v_\ell \, \varphi[b_0, \dots, b_{\ell-1}, b_{\ell+1}, \dots, b_n]$, then we pick $b_\ell \in B$ such that $\mathfrak{B} \models \varphi[b_0, \dots, b_{\ell-1}, b_\ell, b_{\ell+1}, \dots, b_n]$ and let $F_{\varphi,n,\ell}(b_0, \dots b_{\ell-1}, b_{\ell+1}, \dots, b_n) = b_\ell$.
If $\mathfrak{B} \models \neg \exists v_\ell \, \varphi[b_0, \dots, b_{\ell-1}, b_{\ell+1}, \dots, b_n]$,
we let $F_{\varphi,n,\ell}(b_0, \dots b_{\ell-1}, b_{\ell+1}, \dots, b_n) = b^*$.

The attentive reader will have noticed that, in general, for a given $\mathfrak{B}$, there will be many possible choices for $sk(\mathfrak{B})$. Since we are assuming the Axiom of Choice, this is not a problem; for the purpose of this proof it suffices to assume that we have associated $sk(L_m)$ and $sk(\mathfrak{B})$ with every relevant language L_m and model $\mathfrak{B}$.

Now let L and $\mathfrak{A}$ be as in the assumptions of Theorem 24.10. One would like to define $L^+ = sk(L)$ and $\mathfrak{A}^+ = sk(\mathfrak{A})$. We have made sure that for all formulas φ of L there are functional symbols $f_{\varphi,n,\ell}$ whose interpretations witness SK:

EXERCISE 24.11 (G). Convince yourself that if $n \in \omega \backslash \{0\}$, $\ell \leq n$, $\varphi \in Form_L^n$, and $a_0, \dots, a_n \in A$, then $\mathfrak{A}^+ \models \exists v_\ell \varphi(a_0, \dots, a_{\ell-1}, v_\ell, a_{\ell+1}, \dots, a_n)$ implies

$$\mathfrak{A}^+ \models \varphi(a_0, \dots, a_{\ell-1}, f_{\varphi,n,\ell}(a_0, \dots a_{\ell-1}, a_{\ell+1}, \dots, a_n), a_{\ell+1}, \dots, a_n).$$

Hint: Use Claim 24.9(a).

Unfortunately, the language of $\mathfrak{A}^+$ is L^+, and we need functional symbols $f_{\varphi,n,\ell}$ for every formula φ of L^+, not just of L!

There is a standard remedy for this kind of problem: Let $L_0 = L$, $\mathfrak{A}_0 = \mathfrak{A}$, and construct recursively $L_m, \mathfrak{A}_m$ by letting $L_{m+1} = sk(L_m)$ and $\mathfrak{A}_{m+1} = sk(\mathfrak{A}_m)$. For each $m \in \omega$, the relational and constant symbols of L_m are the same as for L. The set of functional symbols of L_m can be written as $\{f_j : j \in J_m\}$, where $J_m \subseteq J_{m+1}$

for each $m \in \omega$. Let $J^+ = \bigcup_{m \in \omega} J_m$, and let L^+ be the language with relational symbols $\{r_i : i \in I\}$, functional symbols $\{f_j : j \in J^+\}$, and constant symbols $\{c_k : k \in K\}$. Let $\mathfrak{A}^+ = \langle A, (R_i)_{i \in I}, (F_j)_{j \in J^+}, (C_k)_{k \in K}\rangle$, where the interpretation F_j of the functional symbol f_j in $\mathfrak{A}^+$ is the same as in $\mathfrak{A}_{m(j)}$, with $m(j)$ being the smallest $m \in \omega$ such that $f_j \in J_m$.

EXERCISE 24.12 (PG). Convince yourself that L^+ and $\mathfrak{A}^+$ are as required. □

Theorem 24.10 has a number of interesting consequences:

THEOREM 24.11. *Let κ be a regular uncountable cardinal, and let*

$$\mathfrak{B} = \langle B, (R_i)_{i \in I}, (F_j)_{j \in J}, (C_k)_{k \in K}\rangle$$

be a model for a language L with nonempty universe and fewer than κ nonlogical symbols. Then the set

$$\{A \in [B]^{<\kappa} : A \text{ is the universe of an elementary submodel of } \mathfrak{B}\}$$

is club in $[B]^{<\kappa}$.

PROOF. Let κ, $\mathfrak{B}$ be as in the assumptions. By Theorem 24.10, there exists a language L^+ with fewer than κ nonlogical symbols and a Skolemized model $\mathfrak{B}^+$ of L^+ such that $\mathfrak{B}$ is the L-reduct of $\mathfrak{B}^+$. By Corollary 21.31, the family $\mathcal{A} = \{A \in [B]^{<\kappa} : A \text{ is the universe of a submodel of } \mathfrak{B}\}$ is club in $[B]^{<\kappa}$. By Lemma 24.8, each $A \in \mathcal{A}$ is the universe of an elementary submodel of $\mathfrak{B}^+$. By Claim 24.9(b), each $A \in \mathcal{A}$ also is the universe of an elementary submodel of $\mathfrak{B}$. □

Now we have all the ingredients to prove Theorem 6.8.

THEOREM 24.12 (The (Downward) Löwenheim-Skolem Theorem). *Let L be a first-order language that contains only countably many nonlogical symbols, and let T be a theory in L. If T has any model at all, then T has a countable model.*

PROOF. Let L, T be as in the assumptions, and let $\mathfrak{B} = \langle B, \ldots \rangle$ be a model of L such that $\mathfrak{B} \models T$. If B is countable, we are done. If B is uncountable, then by Theorem 24.11 there exists $A \in [B]^{<\aleph_1}$ such that A is the universe of an elementary submodel $\mathfrak{A}$ of $\mathfrak{B}$. Since $\mathfrak{A}$ is an elementary submodel, it follows from Claim 24.3 that $\mathfrak{A} \models T$. □

EXERCISE 24.13 (PG). Find a generalization of Theorem 24.12 to uncountable languages, and prove it.

Corollary 21.25 also has an analogue for elementary submodels.

COROLLARY 24.13. *Let $\mathfrak{B} = \langle B, \ldots\rangle$ be a model of a first-order language L with nonlogical symbols $\{r_i : i \in I\} \cup \{f_j : j \in J\} \cup \{c_k : k \in K\}$, and let $A \subseteq B$. Then there exists an elementary submodel $\mathfrak{A}^+ = \langle A^+, \ldots\rangle$ of $\mathfrak{B}$ such that $A \subseteq A^+$ and $|A^+| \leq \max\{|A|, |I| + |J| + |K|, \aleph_0\}$.*

PROOF. Let $\lambda = \max\{|A|, |I| + |J| + |K|, \aleph_0\}$, and let $\kappa = \lambda^+$. By Theorem 24.11, the family $\mathcal{C} = \{A^+ \in [B]^{<\kappa} : A^+ \text{ is the universe of an elementary submodel of } \mathfrak{B}\}$ is club in $[B]^{<\kappa}$. By Example 21.9, the family $\check{A} = \{A^+ \in [B]^{<\kappa} : A \subseteq A^+\}$ is club in $[B]^{<\kappa}$. Thus, $\mathcal{C} \cap \check{A} \neq \emptyset$. Any element A^+ of $\mathcal{C} \cap \check{A}$ is as required. □

24.2. Applications of elementary submodels in set theory

Naturally, we are most interested in elementary submodels of models of set theory. All the models we are going to consider in this section will be standard models, i.e., models of the form $\langle M, \bar{\in}\rangle$, where M is a set and $\bar{\in}$ stands for the standard interpretation of the binary relational symbol $\in$, that is, $\bar{\in}$ denotes the "real" membership relation restricted to elements of M. Most of the time, we will refer to a model $\langle M, \bar{\in}\rangle$ simply as "the model M."

Of course, there is a major problem with models of ZFC: By Gödel's Second Incompleteness Theorem, ZFC does not prove their existence. When we speak of "models of set theory" in this section, we don't necessarily mean models of ZFC, but models of a "sufficiently large fragment of ZFC."

Let us illustrate this notion with an example. In Chapter 12 we defined for every infinite cardinal λ the class H_λ of all sets hereditarily of cardinality less than λ; i.e., $H_\lambda = \{x : |TC(x)| < \lambda\}$, where $TC(x)$ denotes the transitive closure of x. We showed that each H_λ is a transitive set (i.e., H_λ is not a proper class, and if $y \in H_\lambda$ and $x \in y$, then $x \in H_\lambda$), and that the model $\langle H_\lambda, \bar{\in}\rangle$ satisfies quite a few of the axioms of ZFC. Thus, a suitably chosen H_λ may be a model for a "sufficiently large fragment of ZFC." But what does "suitably chosen" mean? Our first application of elementary submodels in set theory will be an alternative proof of the Δ-System Lemma. In the proof, we will fix an uncountable family of finite sets B, and show that B contains an uncountable Δ-system[6] $A \subseteq B$. We will pick λ such that $|TC(B)| < \lambda$, i.e., $B \in H_\lambda$. Then we will argue, using an elementary submodel of H_λ, that

(24.17) $\qquad H_\lambda \models \exists A \subseteq B \,(A \text{ is an uncountable } \Delta\text{-system})$.

But are H_λ's beliefs about the existence of an uncountable Δ-subsystem of B correct? What we really want is the following:[7]

(24.18) $\qquad \mathbf{V} \models \exists A \subseteq B \,(A \text{ is an uncountable } \Delta\text{-system})$.

Does (24.17) imply (24.18)? Let us first show how to answer this question the hard way. If (24.17) holds, then there exists $A \in H_\lambda$ such that

(24.19) $\qquad\qquad\qquad H_\lambda \models A \subseteq B$;

(24.20) $\qquad\qquad\qquad H_\lambda \models A \text{ is a } \Delta\text{-system}$;

and

(24.21) $\qquad\qquad\qquad H_\lambda \models A \text{ is uncountable}$.

We want to show that (24.19), (24.20), and (24.21) remain true if H_λ is replaced by $\mathbf{V}$. By Claim 12.19(c), (24.19) implies that $\mathbf{V} \models A \subseteq B$. Now consider (24.20). This can be translated into:

(24.22) $\qquad \exists r \in H_\lambda \forall a, b \in A \cap H_\lambda \,(H_\lambda \models a \neq b \rightarrow a \cap b = r)$.

Since H_λ is transitive, it is not hard to see that $H_\lambda \models a \neq b \rightarrow a \cap b = r$ if and only if the implication $a \neq b \rightarrow a \cap b = r$ is really true. Similarly, transitivity of

---

[6] Actually, we will show something stronger, but let us keep this illustration simple.

[7] The satisfaction relation for proper classes was defined in Chapter 12. Essentially, the expression "$\mathbf{V} \models \varphi$" is nothing more than a suggestive synonym for "φ."

H_λ implies that $A \cap H_\lambda = A$. Thus, (24.22) implies:

(24.23) $\qquad \exists r \, \forall a, b \in A \, (a \neq b \to a \cap b = r)$.

In other words, A really is a Δ-system.

EXERCISE 24.14 (G). Show that if $f \in H_\lambda$ is such that

(24.24) $\qquad H_\lambda \models f$ is an injection from A into ω,

then f really is an injection from A into ω.

Note that in the last step of transforming (24.22) into (24.23) we took advantage of the fact that if there exists a kernel r for A in H_λ, then obviously such r exists in $\mathbf{V}$. Formula (24.21) is more tricky in this respect. It translates into

(24.25) $\qquad \neg \exists f \in H_\lambda \, (H_\lambda \models f$ is an injection from A into $\omega)$,

which by Exercise 24.14 translates into:

(24.26) $\qquad \neg \exists f \in H_\lambda (f$ is an injection from A into $\omega)$.

The nonexistence of a certain object in H_λ does not automatically preclude the existence of such an object in $\mathbf{V}$. However, in the case of f we are lucky: An injection from A into ω is a subset of $A \times \omega$, and it is not hard to see that H_λ is closed under subsets and Cartesian products of finitely many factors. Thus, any injection from A into ω must be an element of H_λ, and we have proved that (24.17) implies (24.18). A similar argument can be used to show that (24.18) implies (24.17); but this is not needed for the present purpose.

Let us analyze our reasoning at a more abstract level. In Chapter 12, we associated with every formula φ of L_S and every class $\mathbf{X}$ a formula $\varphi_{/\mathbf{X}}$ (called the *relativization of φ to $\mathbf{X}$*) that is obtained by replacing each quantifier "$\exists v_i$" occurring in φ by "$\exists v_i \in \mathbf{X}$" and each quantifier "$\forall v_i$" occurring in φ by "$\forall v_i \in \mathbf{X}$." Then we defined the relation $\mathbf{X} \models \varphi$ as a fancy way of writing $\varphi_{/\mathbf{X}}$, and remarked (Exercise 12.9) that if $\mathbf{X}$ happens to be a set, then $\varphi_{/\mathbf{X}}$ holds if and only if $\langle \mathbf{X}, \bar{\in} \rangle \models \varphi$ in the usual sense (i.e., in the sense of the definition given in the "Notation" section). We called a formula $\varphi \in Form_{L_s}^{n-1}$ *absolute for* $\mathbf{X}$ if $\varphi(x_0, \ldots, x_{n-1}) \leftrightarrow \varphi_{/\mathbf{X}}(x_0, \ldots, x_{n-1})$ for all $x_0, \ldots, x_{n-1} \in \mathbf{X}$. Our proof that (24.17) implies (24.18) is nothing else but one half of the proof that the formula "v_0 contains an uncountable Δ-system" is absolute for H_λ whenever λ is an infinite cardinal.

Now let us go back to the abstract situation. Suppose we are given a formula $\varphi \in Form_{L_s}^{n-1}$ and sets $a_0, \ldots, a_{n-1}$, and we want to prove, with the help of elementary submodels, that $\varphi(a_0, \ldots, a_{n-1})$ holds. All we need is an H_λ such that $a_0, \ldots, a_{n-1} \in H_\lambda$, the formula φ is absolute for H_λ, and $H_\lambda \models \varphi[a_0, \ldots, a_{n-1}]$. In many cases, absoluteness of φ for H_λ can be verified by arguments similar to the reasoning in our illustration. However, most of the time even this is not necessary. The next theorem provides a convenient shortcut.

THEOREM 24.14 (Reflection Principle). *Let φ be a formula of L_S with free variables among $v_0, \ldots, v_{n-1}$. Define $\mathbf{C}_\varphi = \{\alpha : \forall a_0, \ldots, a_{n-1} \in V_\alpha \, (\mathbf{V} \models \varphi[a_0, \ldots, a_{n-1}] \leftrightarrow V_\alpha \models \varphi[a_0, \ldots, a_{n-1}])\}$. Then $\mathbf{C}_\varphi$ contains a closed unbounded class of ordinals.*

How can Theorem 24.14 help us in choosing suitable H_λ's for our arguments?

LEMMA 24.15. *Let $\mathbf{E} = \{\lambda \in \mathbf{Card} : V_\lambda = H_\lambda\}$. Then $\mathbf{E}$ is a closed unbounded class of cardinals.*

24.2. APPLICATIONS OF ELEMENTARY SUBMODELS IN SET THEORY

PROOF. If $\lambda = \sup_{\alpha < cf(\lambda)} \lambda_\alpha$ is a limit cardinal, then

(24.27) $$V_\lambda = \bigcup_{\alpha < cf(\lambda)} V_{\lambda_\alpha};$$

and

(24.28) $$H_\lambda = \bigcup_{\alpha < cf(\lambda)} H_{\lambda_\alpha}.$$

It follows immediately from (24.27) and (24.28) that $\mathbf{E}$ is a closed class. To show that $\mathbf{E}$ is unbounded in $\mathbf{ON}$, let λ_0 be an arbitrary cardinal. Define recursively for $n \in \omega$:

$$\lambda_{2n+1} = \min\{\alpha \geq \lambda_{2n} : H_{\lambda_{2n}} \subseteq V_\alpha\}; \quad \lambda_{2n+2} = \min\{\kappa \geq \lambda_{2n+1} : V_{\lambda_{2n+1}} \subseteq H_\kappa\}.$$

Let $\lambda_\omega = \sup_{n \in \omega} \lambda_n$. Then $\lambda_\omega \geq \lambda$. By (24.27) and (24.28), $\lambda_\omega \in \mathbf{E}$. We have shown that $\mathbf{E}$ is unbounded in $\mathbf{ON}$. □

Now suppose we are given a formula $\varphi \in Form_{LS}^{n-1}$ and sets $a_0, \ldots, a_{n-1}$, and we want to prove, with the help of elementary submodels, that $\varphi(a_0, \ldots, a_{n-1})$ holds. Since every closed unbounded class of cardinals is in particular a closed unbounded class of ordinals, and since the intersection of two closed unbounded classes of ordinals is a closed unbounded class of ordinals,[8] Theorem 24.14 and Lemma 24.15 imply that there exists a cardinal λ such that $H_\lambda = V_\lambda$, and $H_\lambda \models \varphi[a_0, \ldots, a_{n-1}]$ if and only if $\mathbf{V} \models \varphi[a_0, \ldots, a_{n-1}]$. In most cases, we may as well start our argument with picking λ as above, and then restrict the proof to showing that $H_\lambda \models \varphi[a_0, \ldots, a_{n-1}]$.[9] Why did we write "in most cases"? Don't Theorem 24.14 and Lemma 24.15 imply that there is always a suitable λ around? True enough, but you have to be careful about what you can and what you cannot assume about λ. Note that Theorem 24.14 guarantees that $\mathbf{C}_\varphi$ contains arbitrary large cardinals, even cardinals of arbitrary large cofinality, but Theorem 24.14 does *not* guarantee that $\mathbf{C}_\varphi$ contains any regular cardinals. As long as you only need to assume that λ is sufficiently large, or even that $cf(\lambda) > \omega$, you may safely use Theorem 24.14 and Lemma 24.15. But if for some reason your argument uses the assumption that λ is a regular uncountable cardinal, you are back to square one and you must either verify absoluteness of $\varphi_{H_\lambda}[a_0, \ldots, a_{n-1}]$ directly, or modify your argument.

Let us summarize what we have said so far by giving an "official" definition of the phrase "sufficiently large λ" that will appear at the beginning of several arguments later in this chapter: If $a_0, \ldots, a_{n-1}$ are given and if we want to prove a formula $\varphi(a_0, \ldots, a_{n-1})$ with the help of elementary submodels, then λ is "sufficiently large" if $a_0, \ldots, a_{n-1} \in H_\lambda$, the formula φ is absolute for H_λ, and H_λ satisfies all the (finitely many) axioms of ZFC that will be used (implicitly or explicitly) in the proof.[10] Theorem 24.14 and Lemma 24.15 imply that such λ exists. However, note that our approach does not guarantee that if λ is suitable and $\kappa > \lambda$,

---

[8]This was shown in Exercise 21.19.
[9]Of course, we might as well show that $V_\lambda \models \varphi[a_0, \ldots, a_{n-1}]$, but since the structure of H_λ is easier to describe, set theorists usually prefer working with H_λ.
[10]This is what we mean when we speak of a "sufficiently large fragment of ZFC."

then κ is also suitable. We shall always put the phrase "sufficiently large" in scare-quotes to alert you that it should not be taken too literally.

While we are at it, let us mention another important application of Theorem 24.14. Recall from Chapter 12 the definitions of the L_α-hierarchy and of the class **L**.

EXERCISE 24.15 (PG). Show that the class $\mathbf{D} = \{\alpha \in \mathbf{ON} : L_\alpha = \mathbf{L} \cap V_\alpha\}$ is closed and unbounded in **ON**. *Hint:* Modify the argument of the proof of Lemma 24.15.

The following result appeared in Volume 1 as Lemma 12.35.

COROLLARY 24.16. *Let $\alpha \in \mathbf{ON}$ and $x, a_0, \ldots, a_{k-1} \in L_\alpha$, and let $\varphi \in Form_{L_S}^{n-1}$. Then there exists $\beta > \alpha$ such that for every $z \in x$:*

$$\langle L_\beta, \bar{\in} \rangle \models \varphi[z, a_0, \ldots, a_{k-1}] \quad \textit{iff} \quad \mathbf{L} \models \varphi[z, a_0, \ldots, a_{k-1}].$$

PROOF. Let $\alpha \in \mathbf{ON}$, and let $\varphi, x, a_0, \ldots, a_{k-1}$ be as in the assumptions. Let ψ be a representation of **L**, i.e., let ψ be a formula with one free variable such that $\mathbf{L} = \{y : \psi(y)\}$. By Theorem 24.14 and Exercise 21.19, there exists $\beta > \alpha$ such that for all $z \in V_\beta$:

(24.29) $$\mathbf{V} \models \psi[z] \text{ iff } V_\beta \models \psi[z];$$

and

(24.30) $$\mathbf{V} \models \varphi_{/\mathbf{L}}[z, a_0, \ldots, a_{n-1}] \text{ iff } V_\beta \models \varphi_{/\mathbf{L}}[z, a_0, \ldots, a_{n-1}].$$

The left hand side of (24.30) is equivalent to

(24.31) $$\mathbf{L} \models \varphi[z, a_0, \ldots, a_{n-1}];$$

and by (24.29), the right hand side of (24.30) is equivalent to

(24.32) $$V_\beta \cap \mathbf{L} \models \varphi[z, a_0, \ldots, a_{n-1}].$$

By Exercise 24.15, we can choose β as above in such a way that $V_\beta \cap \mathbf{L} = L_\beta$, which allows us to rewrite (24.32) as:

(24.33) $$\langle L_\beta, \bar{\in} \rangle \models \varphi[z, a_0, \ldots, a_{n-1}].$$

Since (24.31) is equivalent to (24.33) for all $z \in L_\beta$, and since $x \subseteq L_\beta$ by transitivity of L_β, Corollary 24.16 follows. □

Now let us prove Theorem 24.14.

PROOF OF THEOREM 24.14. Let φ be as in the assumption. Let $\varphi_0, \ldots, \varphi_s$ be a list of all subformulas of φ (including, of course, φ itself). For each φ_j on this list that is of the form $\exists v_{i_j} \varphi_\ell$ for some $\ell \leq s$, we define a functional class $\mathbf{G}_j$ as follows: Fix $m_j > i_j$ such that the free variables of φ_ℓ are among $v_0, \ldots, v_{m_j}$. The domain of $\mathbf{G}_j$ will be the class of all m_j-tuples $\langle a_0, \ldots, a_{i_j-1}, a_{i_j+1}, \ldots, a_{m_j} \rangle$ (we will use the shorthand $\vec{a}$ to denote such tuples). If $\exists v_{i_j} \varphi_\ell(\vec{a})$, then we define $\mathbf{G}_j(\vec{a}) = \min\{\alpha : \exists a_{i_j} \in V_\alpha \, \varphi_\ell(a_0, \ldots, a_{i_j-1}, a_{i_j}, a_{j_i+1}, \ldots, a_{m_j})\}$. If $\neg \exists v_{i_j} \varphi_\ell(\vec{a})$, then we define $\mathbf{G}_j(\vec{a}) = 0$.

EXERCISE 24.16 (PG). Convince yourself that $\mathbf{G}_j$ really is a functional class, i.e., convince yourself that there is a formula ψ_j of L_S such that $\mathbf{G}_j = \{\langle \vec{a}, \alpha \rangle : \psi_j(\vec{a}, \alpha)\}$.

24.2. APPLICATIONS OF ELEMENTARY SUBMODELS IN SET THEORY

For every ordinal γ, the restriction $\mathbf{G}_j \upharpoonright (V_\gamma)^{m_j}$ is a function. By the Axiom of Replacement, the range of the above restriction is a set, and hence there exists β such that $rng(\mathbf{G}_j \upharpoonright (V_\gamma)^{m_j}) \subseteq V_\beta$. Thus, we can define a functional class $\mathbf{F}_j : \mathbf{ON} \to \mathbf{ON}$ as follows: $\mathbf{F}_j(\zeta) = \max\{\zeta, \min\{\eta : rng(\mathbf{G}_j \upharpoonright (V_\zeta)^{m_j}) \subseteq V_\eta\}\}$.

EXERCISE 24.17 (G). Convince yourself that the functional class $\mathbf{F}_j$ defined above is normal in the sense of the definition preceding Claim 21.23.

By Claim 21.23, the class $Fix(\mathbf{F}_j)$ defined as $\{\alpha \in \mathbf{ON} : \mathbf{F}_j(\alpha) = \alpha\}$ is closed and unbounded in $\mathbf{ON}$. Define $\mathbf{D} = Fix(\mathbf{F}_0) \cap Fix(\mathbf{F}_1) \cap \cdots \cap Fix(\mathbf{F}_s)$. By Exercise 21.19, the class $\mathbf{D}$ is also closed and unbounded in $\mathbf{ON}$. It remains to show that $\mathbf{D} \subseteq \mathbf{C}_\varphi$.

EXERCISE 24.18 (PG). Show that for all $\alpha \in \mathbf{D}$ and all $a_0, \ldots, a_{n-1} \in V_\alpha$,
$$\mathbf{V} \models \varphi[a_0, \ldots, a_{n-1}] \text{ iff } \mathbf{V}_\alpha \models \varphi[a_0, \ldots, a_{n-1}].$$

Hint: Use an argument similar to the proof of Theorem 24.4. □

Theorem 24.14 implies that for every finite subtheory T of ZFC, there is an abundance of models of T. Indeed, suppose $T = \{\varphi_0, \ldots, \varphi_m\}$, and let φ be the sentence $\varphi_0 \wedge \cdots \wedge \varphi_m$. By Theorem 24.14, there is a proper class of ordinals α such that $V_\alpha \models \varphi$, and hence, $V_\alpha \models T$.

One is naturally tempted to indulge in the following reasoning: Let us enumerate the axioms of ZFC by $\{\varphi_0, \varphi_1, \varphi_2, \ldots\}$, let $T_n = \{\varphi_0, \varphi_1, \ldots, \varphi_n\}$, and let $\mathbf{C}_n$ denote the class $\{\alpha : V_\alpha \models T_n\}$. By Exercise 21.20, $\mathbf{C} = \bigcap_{n \in \omega} \mathbf{C}_n$ is a proper class, hence nonempty. Let $\alpha = \min \mathbf{C}$. Then $V_\alpha \models T_n$ for every n; hence $V_\alpha \models$ ZFC. So, there exists a set model of ZFC, and we have contradicted Gödel's Second Incompleteness Theorem!

What is wrong here? In Exercise 21.20, we alerted you that forming the intersection $\bigcap_{i \in I} \mathbf{D}_i$ of a set of proper classes is a problematic operation—since there is no such thing as a nonempty set of proper classes! In some cases, the use of the symbol $\bigcap_{i \in I} \mathbf{D}_i$ may still be justified though. If $\mathbf{D}$ is a class of ordered pairs, and $\mathbf{D}_i = \{x : \langle i, x \rangle \in \mathbf{C}\}$, then $\bigcap_{i \in I} \mathbf{D}_i$ can be understood as shorthand for the class $\{x : \forall i \in I \, (\langle i, x \rangle \in \mathbf{D})\}$. However, for the classes $\mathbf{C}_n$ considered above there is no class $\mathbf{C}$ such that $\langle n, \alpha \rangle \in \mathbf{C}$ if and only if $V_\alpha \models \varphi_n$ for all $n \in \omega$ and $\alpha \in \mathbf{ON}$. Any representation of such a class $\mathbf{C}$ would have to be an infinitely long formula. Thus the expression $\bigcap_{n \in \omega} \mathbf{C}_\mathbf{n}$ is meaningless, and we get no contradiction.

EXERCISE 24.19 (PG). Convince yourself that we have just proved the following:

COROLLARY 24.17. *The theory ZFC is not finitely axiomatizable.*[11]

It is worth comparing these limitations of Theorem 24.14 with the amount of reflection one gets from large cardinals. Let κ be a strongly inaccessible cardinal. In Chapter 12 we showed that $V_\kappa \models$ ZFC. By Theorem 24.11, the set $\mathcal{A} = \{A \in [V_\kappa]^{<\kappa} : \langle A, \bar{\in} \rangle \prec \langle V_\kappa, \bar{\in} \rangle\}$ is club in $[V_\kappa]^{<\kappa}$. Note that each $A \in \mathcal{A}$ is in particular a model of ZFC. Moreover, there are also lots of elementary submodels of $\langle V_\kappa, \bar{\in} \rangle$ of the form $\langle V_\beta, \bar{\in} \rangle$.

THEOREM 24.18. *Let κ be a strongly inaccessible cardinal. Then the set $B = \{\beta < \kappa : \langle V_\beta, \bar{\in} \rangle \prec \langle V_\kappa, \bar{\in} \rangle\}$ is club in κ.*

---

[11]I.e., there is no finite theory T in L_S such that for all sentences φ of L_S, $T \vdash \varphi$ iff ZFC $\vdash \varphi$.

PROOF. Let κ, B be as above. Then κ is a club subset of $[\kappa]^{<\kappa}$ (see Example 21.10). Let $\mathcal{A}$ be as in the discussion preceding Theorem 24.18. Then $B = \mathcal{A} \cap \kappa$, and since the intersection of two club sets is club, the theorem follows. □

REMARK 24.19. There is nothing special about the language L_S here. For example, if $R \subseteq V_\kappa^n$ for some $n \in \omega \setminus \{0\}$, then we can treat the structure $\langle V_\kappa, \bar{\in}, R \rangle$ as a model of a language L_S^+ whose only nonlogical symbols are $\in$ and a relational symbol r of arity n. The same argument as in the proof of Theorem 24.18 shows that the set $B_R = \{\beta < \kappa : \langle V_\beta, \bar{\in}, R \cap (V_\beta)^n \rangle \prec \langle V_\kappa, \bar{\in}, R \rangle\}$ is *club* in κ.

Note that Theorem 24.18 implies that the first ordinal α such that $V_\alpha \models$ ZFC, if it exists, is not a strongly inaccessible cardinal. Since all *club* sets contain ordinals of countable cofinality, Theorem 24.18 also implies that it is possible that $V_\alpha \models$ ZFC even though α has countable cofinality (compare this with Remark 12.20).

Finally, let us look at Theorem 24.18 from the point of view of consistency strength. For $n \in \omega$, define recursively a theory ZFC^n in L_S as follows: $\text{ZFC}^0 =$ ZFC, $\text{ZFC}^{n+1} = \text{ZFC}^n \cup \{CON(\text{ZFC}^n)\}$, where $CON(T)$ can be interpreted as "There exists a model of the theory T."

EXERCISE 24.20 (PG). Let B be as in Theorem 24.18, and let $(\beta_n)_{n\in\omega}$ be an enumeration of the first ω elements of B in increasing order. Show that for all $n \in \omega$, $V_{\beta_n} \models \text{ZFC}^n$. Conclude that for each $n \in \omega$, ZFC + "∃ a strongly inaccessible cardinal" ⊢ ZFC^n.

Now let us assume H_λ is a model of a "sufficiently large" fragment T of ZFC, and let us assume that M is a countable elementary submodel of H_λ. What does M look like?

Since $H_\lambda \models \exists x \forall y \, (y \notin x)$, also $M \models \exists x \forall y \, (y \notin x)$. Hence, there exists $a_0 \in M$ such that $M \models \forall y \, (y \notin a_0)$. Since M is an elementary submodel of H_λ, we also have $H_\lambda \models \forall y \, (y \notin a_0)$, and since H_λ is transitive, a_0 must be the empty set. Thus we have shown that $\emptyset \in M$. A similar argument could now be used to show that $\{\emptyset\} \in M$, and by induction one could show that $n \in M$ for every $n \in \omega$. This is important information that will be used later on, but it is only a special case of the following more general observation. Recall that $a \in A$ is *definable over a model* $\mathfrak{A} = \langle A, \ldots \rangle$ *with parameters* $a_0, \ldots, a_{k-1}$, if there exists a formula φ with free variables $v_0, \ldots, v_k$ such that $\mathfrak{A} \models \varphi[a_0, \ldots, a_{k-1}, a]$ and $\mathfrak{A} \models \forall x \, (\varphi(a_0, \ldots, a_{k-1}, x) \to x = a)$. If $k = 0$, then we simply say that a is *definable over* $\mathfrak{A}$.

LEMMA 24.20. *Suppose* $\mathfrak{A} = \langle A, \ldots, \rangle$ *and* $\mathfrak{B} = \langle B, \ldots, \rangle$ *are models of a first order language L such that* $\mathfrak{B} \prec \mathfrak{A}$. *If a is definable over* $\mathfrak{A}$ *with parameters* $a_0, \ldots, a_n \in B$, *then* $a \in B$.

EXERCISE 24.21 (G). Prove Lemma 24.20.

One immediate consequence of Lemma 24.20 is that elementary submodels of H_λ are closed under finite subsets.

EXERCISE 24.22 (G). Suppose λ is "sufficiently large" and $M \prec H_\lambda$. Show that if $x \subseteq M$ and x is finite, then $x \in M$. *Hint:* If $x = \{a_0, \ldots, a_n\}$, consider a suitable formula with parameters $a_0, \ldots, a_n$.

EXERCISE 24.23 (PG). Show that if λ is "sufficiently large" and $M \prec H_\lambda$, then $V_\omega \subseteq M$. *Hint:* Use Lemma 24.20 and induction over the wellfounded relation $\bar{\in}$.

Lemma 24.20 implies that the definable ordinals $\omega, \omega_1, \omega_2, \ldots, \omega_\omega, \omega_{\omega_1}, \ldots$ can all be assumed to be members of M. Of course, if M is countable, not all countable ordinals can be elements of M. So, what does $M \cap \omega_1$ look like in this case? To settle this question, we need a technical lemma.

LEMMA 24.21. *Suppose $M \prec H_\lambda$, κ is a cardinal less than λ, and $\kappa \subseteq M$. Then for every $x \in M$, if $H_\lambda \models |x| = \kappa$, then $x \subseteq M$.*

PROOF. First note that although in general it is not the case that $\kappa \subseteq M$ implies $\kappa \in M$,[12] our assumptions do imply that $\kappa \in M$: If there exists $x \in M$ such that $H_\lambda \models |x| = \kappa$, then in particular, $H_\lambda \models \exists \mu (|x| = \mu)$. By elementarity, $M \models \exists \mu (|x| = \mu)$; and hence there exists $\mu \in M$ such that $M \models |x| = \mu$. Again by elementarity, $H_\lambda \models |x| = \mu$, hence $\mu = \kappa$, and thus $\kappa \in M$.

Now $M \models \exists f \in {}^\kappa x \ f[\kappa] = x$; hence there exists $f \in M$ such that $M \models f \in {}^\kappa x \wedge f[\kappa] = x$. But then $H_\lambda \models f \in {}^\kappa x \wedge f[\kappa] = x$; hence f really maps κ onto x. If $y \in x$, then there is some $\xi \in \kappa$ such that $f(\xi) = y$. But $\kappa \subseteq M$ implies that $\xi \in M$, and hence $M \models f(\xi) = \tilde{y}$ for some $\tilde{y}$. Since $H_\lambda \models f(\xi) = y$, this $\tilde{y}$ must be y. □

COROLLARY 24.22. *If $M \prec H_\lambda$ and x is a countable set such that $x \in M$, then $x \subseteq M$.*

PROOF. This follows immediately from Exercise 24.23 and Lemma 24.21. □

Now suppose $\alpha \in M \cap \omega_1$. Then α is a countable set of ordinals; hence by Corollary 24.22, $\alpha \subseteq M$. In other words, $\beta \in M$ for every $\beta \in \alpha$. A similar argument applies if $\alpha < \kappa^+$ and $\kappa \subseteq M$. We have proved the following:

CLAIM 24.23. *Suppose $M \prec H_\lambda$.*
(a) *If M is countable, then $M \cap \omega_1 = \delta$ for some countable ordinal δ.*
(b) *If $|M| = \kappa$ and $\kappa \subseteq M$, then $M \cap \kappa^+ = \delta$ for some ordinal $\delta < \kappa^+$.*

Now we are ready to present some applications of elementary submodels. Our first example will be a solution to Exercise 16.2.

LEMMA 24.24. *Let B be an uncountable family of finite subsets of ω_1. Then there exists a countable subset $N \subseteq \omega_1$ such that every $b \in B$ which is not contained in N is a member of an uncountable Δ-system $A \subseteq B$ with kernel $r \subseteq N$.*

PROOF. Let B be as in the assumptions. Fix a "sufficiently large" λ, and let M be a countable elementary submodel of H_λ such that $B \in M$ (recall that the existence of such M follows from Corollary 24.13). Let $N = M \cap \omega_1$. By Claim 24.23, N is equal to a countable ordinal δ. We show that N is as required. Consider $b \in B$ such that b is not contained in N. Let $r = b \cap N$. By Exercise 24.22, b is not an element of M, but r is. Now let α be an arbitrary countable ordinal in M. Then $\alpha < \delta$, and thus

(24.34) $$H_\lambda \models r \subseteq b \wedge b \setminus r \neq \emptyset \wedge \min(b \setminus r) > \alpha.$$

In particular,

(24.35) $$H_\lambda \models \exists x \in B \, (r \subseteq x \wedge x \setminus r \neq \emptyset \wedge \min(x \setminus r) > \alpha).$$

---

[12] For example, if $M = V_\kappa$ then $\kappa \subseteq M$ but $\kappa \notin M$.

The parameters α and r of (24.35) are all in M; hence by elementarity,

(24.36) $\qquad M \models \exists x \in B\,(r \subseteq x \wedge x\backslash r \neq \emptyset \wedge \min(x\backslash r) > \alpha).$

Since α was assumed to be an arbitrary element of $M \cap \omega_1$, we have

(24.37) $\qquad M \models \forall \alpha < \omega_1 \exists x \in B\,(r \subseteq x \wedge x\backslash r \neq \emptyset \wedge \min(x\backslash r) > \alpha).$

By elementarity,

(24.38) $\qquad H_\lambda \models \forall \alpha < \omega_1 \exists x \in B\,(r \subseteq x \wedge x\backslash r \neq \emptyset \wedge \min(x\backslash r) > \alpha).$

Since $H_\lambda \supseteq \omega_1$, (24.38) allows us to recursively construct a transfinite sequence $(\alpha_\xi)_{\xi<\omega_1}$ of countable ordinals and a sequence $(b_\xi)_{\xi<\omega_1}$ such that r is a proper subset of b_ξ and $\delta < \alpha_\xi < \min(b_\xi\backslash r) \leq \max b_\xi < \alpha_\eta$ for all $\xi < \eta < \omega_1$. It is easy to see that the set $\{b_\xi : \xi < \omega_1\}$ is a δ-system with kernel r, as desired. □

EXERCISE 24.24 (G). Prove the following generalization of Lemma 24.24:

LEMMA 24.25. *Let κ be an uncountable cardinal, and let B be a family of finite subsets of κ^+ with $|B| = \kappa^+$. Then there exists a subset $N \subseteq \kappa^+$ of cardinality κ such that every $b \in B$ which is not contained in N is a member of a Δ-system $A \in [B]^{\kappa^+}$ with kernel $r \subseteq N$.*

In the proof of Lemma 24.24, the assumption that B consists of subsets of ω_1 played a crucial role. Can we get away without using this assumption?

LEMMA 24.26. *Let B be an uncountable family of finite sets. Then there exists a countable set N such that every $b \in B$ which is not contained in N is a member of an uncountable Δ-system $A \subseteq B$ with kernel $r \subseteq N$.*

PROOF. Let B be as in the assumptions. Fix a "sufficiently large" λ, and let M be a countable elementary submodel of H_λ such that $B \in M$. Let $N = M$. We show that N is as required. Consider $b \in B$ such that b is not contained in M, and let $r = b \cap N$. By Exercise 24.22, b is not an element of M, but r is.

In the proof of Lemma 24.24, we proceeded by considering countable ordinals in M. This won't help us here, since b may not be a set of ordinals. We need to consider something more general: Let D be an *arbitrary countable set* such that $D \in M$ and $r \subseteq D$. By Corollary 24.22, $D \subseteq M$, and thus $b \cap D = r$ and $b \backslash D \neq \emptyset$. It follows that

(24.39) $\qquad M \models \forall D \supseteq r\,(|D| \leq \aleph_0 \to \exists a \in B\,(a \neq r \wedge a \cap D = r)).$

By elementarity,

(24.40) $\qquad H_\lambda \models \forall D \supseteq r\,(|D| \leq \aleph_0 \to \exists a \in B\,(a \neq r \wedge a \cap D = r)).$

EXERCISE 24.25 (G). Complete the proof of Lemma 24.26. □

Now let us consider a situation where the kernels of Δ-systems may be infinite.

THEOREM 24.27. *Let μ and ν be infinite cardinals such that $\nu^\mu = \nu$. Let B be a family of sets such that $|B| > \nu$ and $|b| \leq \mu$ for all $b \in B$. Then there exists a set N of cardinality ν such that every $b \in B$ that is not contained in N is a member of a Δ-system $A \subseteq B$ with kernel $r \subseteq N$ such that $|A| > \nu$.*

EXERCISE 24.26 (G). Convince yourself that Theorem 24.27 implies Theorem 16.3 for the case when both κ and λ are successor cardinals. *Hint:* Let $\kappa = \mu^+$ and $\lambda = \nu^+$.

24.2. APPLICATIONS OF ELEMENTARY SUBMODELS IN SET THEORY

PROOF OF THEOREM 24.27. Let μ, ν, B be as in the assumptions. Fix a "sufficiently large" λ, and let M be an elementary submodel of H_λ such that $\mu, \nu, B \in M$. We'd like to show that $N = M$ is as required. Thus, in particular, M must be of cardinality ν. Moreover, if $b \in B$ is such that b is not contained in M and $r = b \cap N$, then b should not be an element of M, but r should be. How can we make sure that this is the case? The first of these two requirements is relatively easy to handle: Let us choose M in such a way that $\nu \subseteq M$. By Corollary 24.13, it is possible to find $M \prec H_\lambda$ of cardinality ν such that $\nu \cup \{\nu, B\} \subset M$.[13] Note that the assumptions of Theorem 24.2 imply $\mu < \nu$, and hence we have $\varrho \subseteq M$ and $\varrho \in M$ for every $\varrho \leq \mu$. If M is as above and $b \in M$ is a set of cardinality $\varrho \leq \mu$, then by elementarity, $M \models |b| = \varrho$. By Lemma 24.21, $b \subseteq M$. In other words, if $b \in B$ is not a subset of M, then $b \notin M$.

Making sure that $r = b \cap M$ is an element of M is trickier. In contrast to the proofs of Lemmas 24.24– 24.26, r does not need to be finite, and Exercise 24.22 no longer applies. We need a new concept. Let κ be an infinite cardinal. We say that M is *closed under subsets of size less than* κ if $[M]^{<\kappa} \subseteq M$.

EXAMPLE 24.4. Exercise 24.22 implies that every elementary submodel of H_λ is closed under subsets of size less than $\aleph_0$. If λ is regular, then H_λ itself is closed under subsets of size less than λ.

Now let $\kappa = \mu^+$, and suppose we have an elementary submodel $M \prec H_\lambda$ such that $\nu \cup \{\nu, B\} \subseteq M$ and M is closed under subsets of size less than κ. Let b, r be as above. Then r is a subset of M of size $\leq \mu < \kappa$, and hence $r \in M$. Now we can reason as in the proof of Lemma 24.26: Let D be an arbitrary set of cardinality $\leq \nu$ such that $D \in M$ and $r \subseteq D$. Since $\nu \subseteq M$, Lemma 24.21 implies that $D \subseteq M$, and thus $b \cap D = r$ and $b \setminus D \neq \emptyset$. Hence

(24.41) $\qquad M \models \forall D \supseteq r \, (|D| \leq \nu \to \exists a \in B \, (a \neq r \wedge a \cap D = r)).$

By elementarity,

(24.42) $\qquad H_\lambda \models \forall D \supseteq r \, (|D| \leq \nu \to \exists a \in B \, (a \neq r \wedge a \cap D = r)).$

As in Exercise 24.25, we can recursively construct transfinite sequences $(D_\xi)_{\xi<\nu^+}$ of sets of size ν and $(b_\xi)_{\xi<\nu^+}$ of elements of B such that $b_\xi \cap D_\xi = r$, $b_\xi \setminus D_\xi \neq \emptyset$, $b_\xi \subseteq D_\eta$, and $M \subseteq D_\xi \subseteq D_\eta$ for all $\xi < \eta < \nu^+$. The set $\{b_\xi : \xi < \nu^+\}$ will be the desired Δ-system.

But does there exist a suitable M? To complete the proof of Theorem 24.27, we need one more lemma.

LEMMA 24.28. *Let κ be a regular cardinal, let ν be a cardinal such that $\nu^{<\kappa} = \nu$, and let λ be a cardinal $> \nu$. Then for every $A \in [H_\lambda]^{\leq \nu}$ there exists $M \prec H_\lambda$ such that $A \subseteq M$, $|M| = \nu$, and M is closed under subsets of size less than κ.*

PROOF. Let κ, ν, λ, A be as in the assumptions. By recursion over $\alpha \leq \nu$ we construct a sequence $(M_\alpha)_{\alpha \leq \nu}$ such that for all $\alpha < \beta < \nu$:

(i) $A \subseteq M_0$;
(ii) $[M_\alpha]^{<\kappa} \subseteq M_\beta$;
(iii) $|M_\alpha| = \nu$;
(iv) $M_\alpha \prec H_\lambda$.

---

[13]Actually, at this point we only need $\mu \subseteq M$, but at a later stage of the argument we will use the inclusion $\nu \subseteq M$.

The construction is straightforward: Use Corollary 24.13 to construct M_0 such that (i), (iii), and (iv) hold. Having constructed M_α, note that since $\nu = \nu^{<\kappa}$, (iii) implies that $|[M_\alpha]^{<\kappa}| = \nu$. Thus, we can apply Corollary 24.13 with $[M_\alpha]^{<\kappa}$ in the role of A to find $M_{\alpha+1}$. Now (ii) implies that for $\alpha < \beta$:

(ii)° $\quad M_\alpha \subseteq M_\beta$.

Thus at limit stages $\delta \leq \nu$, we can define $M_\delta = \bigcup_{\alpha<\delta} M_\alpha$. By Lemma 24.5, $M_\delta \prec H_\lambda$.

Now let us show that M_ν is closed under subsets of size less than κ. Note that the assumption $\nu^{<\kappa} = \nu$ implies that $cf(\nu) \geq \kappa$. Thus, if $x \in [M_\nu]^{<\kappa}$, then there exists $\alpha < \nu$ such that x is already a subset of M_α. In other words, $x \in [M_\alpha]^{<\kappa}$, and (ii) applied to $\beta = \nu$ yields $x \in M_\nu$, as desired. $\square\square$

Next let us present two examples of applications of elementary submodels in topology. In this type of applications, one usually considers a topological space X and an elementary submodel M of a suitable H_λ with $X \in M$, and then one derives properties of X by considering M's version of X. But what is "M's version of X"? Recall that a topological space is formally defined as a pair $\langle X, \tau \rangle$. The most natural candidate for "M's version of X" would be the pair $\langle X \cap M, \tau \cap M \rangle$. Unfortunately, in most cases, $\tau \cap M$ is not a topology, since in general, arbitrary unions of elements of $\tau \cap M$ are not in M! However, $\tau \cap M$ is still a base for a topology τ^M on X. Depending on the particular application, "M's version of X" could be either $\langle X, \tau^M \rangle$ or its subspace $\langle X \cap M, \tau^M \rangle$.

A *network* for a topological space X is a collection $\mathcal{N}$ of subsets of X such that for every $x \in X$ and every open neighborhood U of x there exists $N \in \mathcal{N}$ with $x \in N \subseteq U$. The *network weight* of X is defined as follows:

$$nw(X) = \min\{|\mathcal{N}| : \mathcal{N} \text{ is a network for } X\} + \aleph_0.$$

Note that a network for X consisting of open sets is the same thing as a base for X. It follows that $nw(X) \leq w(X)$ for every topological space X. The inequality may be strict: For example, the space $\langle \omega+1, \tau^{\mathcal{F}} \rangle$ of Example 26.8(b) has uncountable weight, but countable network weight, since the set of singletons is obviously a network for $\langle \omega+1, \tau^{\mathcal{F}} \rangle$. However, for compact Hausdorff spaces the network weight is always the same as the weight.

LEMMA 24.29 (Arhangel'skii). *Let $\langle X, \tau_1 \rangle$ be a Hausdorff space and assume $nw(X) = \kappa$. Then there exists a topology $\tau_0 \subseteq \tau_1$ on X such that $\langle X, \tau_0 \rangle$ is Hausdorff and has weight $\leq \kappa$.*

COROLLARY 24.30 (Arhangel'skii). *For every compact Hausdorff space X, $nw(X) = w(X)$.*

PROOF. If $\langle X, \tau_1 \rangle$ is compact Hausdorff and τ_0 is as in Lemma 24.29, then $\tau_0 = \tau_1$ (see Theorem 26.3). $\square$

PROOF OF LEMMA 24.29. Let X, τ_1, κ be as in the assumptions, and let $\mathcal{N}$ be a network for $\langle X, \tau_1 \rangle$ of size κ. Choose a "sufficiently large" λ, and let $M \prec H_\lambda$ be such that $\mathcal{N} \cup \{\mathcal{N}, X, \tau_1\} \subseteq M$ and $|M| = \kappa$. Let τ_0 be the topology on X generated by the base $M \cap \tau_1$. Clearly, $\tau_0 \subseteq \tau_1$, and by definition, $\langle X, \tau_0 \rangle$ has weight $\leq \kappa$. Let us show that $\langle X, \tau_0 \rangle$ is Hausdorff. Consider two distinct points $x_0, x_1 \in X$ (not necessarily in $X \cap M$). We want to show that there are disjoint $W_0, W_1 \in \tau_0$ such that $x_0 \in W_0$ and $x_1 \in W_1$. Since τ_1 is a Hausdorff topology, there exist disjoint open sets $U_0, U_1 \in \tau_1$ such that $x_i \in U_i$ for $i \in \{0,1\}$. Unfortunately, nothing we have said so far guarantees that U_0 and U_1 are in τ_0. But here comes the trick: By the definition of a network, there exist $N_0, N_1 \in \mathcal{N}$ such that $x_i \in N_i \subseteq U_i$ for $i \in \{0,1\}$. Thus

(24.43) $\qquad H_\lambda \models \exists V_0, V_1 \in \tau_1 \, (V_0 \cap V_1 = \emptyset \land N_0 \subseteq V_0 \land N_1 \subseteq V_1).$

Since $\mathcal{N} \subseteq M$, the parameters N_0 and N_1 of the last formula are in M, and elementarity implies:

(24.44) $\qquad M \models \exists V_0, V_1 \in \tau_1 \, (V_0 \cap V_1 = \emptyset \land N_0 \subseteq V_0 \land N_1 \subseteq V_1).$

Thus, there are disjoint open $W_0, W_1 \in M \cap \tau_1$ (and hence in τ_0) such that $N_0 \subseteq W_0$ and $N_1 \subseteq W_1$. Since $x_i \in N_i$ for $i \in \{0,1\}$, we have shown that x_0 and x_1 are separated in τ_0. □

THEOREM 24.31 (Arhangel'skii). *If X is a Lindelöf, first countable Hausdorff space, then $|X| \leq 2^{\aleph_0}$.*

PROOF. Let $\langle X, \tau \rangle$ be as in the assumptions; choose a "sufficiently large" λ, and let $M \prec H_\lambda$ be such that $X, \tau \in M$, $|M| = 2^{\aleph_0}$, and M is closed under countable subsets. Such M exists by Lemma 24.28. Let $Y = X \cap M$. Obviously, $|Y| \leq 2^{\aleph_0}$.

CLAIM 24.32. *For every $y \in Y$, M contains a local base $\mathcal{B}_y$ at y.*

PROOF. Let $y \in Y$. Then

(24.45) $\qquad H_\lambda \models \exists \mathcal{B}_y \, (\mathcal{B}_y$ is a countable local base at y).

By elementarity,

(24.46) $\qquad M \models \exists \mathcal{B}_y \, (\mathcal{B}_y$ is a countable local base at y).

Thus, there exists $\mathcal{B}_y \in M$ such that $\mathcal{B}_y$ is a countable local base at y. By Corollary 24.22, $\mathcal{B}_y \subseteq M$. □

CLAIM 24.33. *Y is a closed subspace of X.*

PROOF. By first countability, if $x \in cl\, Y$, then x is the limit of a converging sequence $(y_n)_{n \in \omega}$ of elements of Y. One can treat this sequence as a countable subset of $\omega \times Y$. Since M is closed under countable subsets, the sequence $(y_n)_{n \in \omega}$ is also an element of M. By Hausdorffness, convergent sequences have unique limits, and thus x is definable over M with parameters $X, \tau, (y_n)_{n \in \omega}$. Hence Lemma 24.20 implies that $x \in M$. □

Now let us wrap up the argument by showing that $X = Y$. Suppose not, and let $x \in X \setminus Y$. For every $y \in Y$, let $U_y \in \mathcal{B}_y$ be such that $x \notin U_y$. Such U_y exists since X is Hausdorff and $\mathcal{B}_y$ is a local base at y. The family $\mathcal{U} = \{U_y : y \in Y\}$ is an open cover of Y by elements of M such that $\bigcup \mathcal{U}$ misses x, and hence is not a cover of X. So far there is no contradiction since $\mathcal{U}$ does not need to be an element of M. But Y is a closed subspace of a Lindelöf space, hence is Lindelöf itself. Thus there exists a countable $\mathcal{U}' \subseteq \mathcal{U}$ such that $\bigcup \mathcal{U}' \supseteq Y$. Since M is closed under countable subsets, $\mathcal{U}'$ is an element of M. Moreover, since $\mathcal{U}'$ covers every element of $X \cap M$, we have:

(24.47) $\qquad M \models \mathcal{U}'$ is an open cover of X.

But since $x \notin \bigcup \mathcal{U}'$,

(24.48) $\qquad\qquad H_\lambda \models \mathcal{U}'$ is not an open cover of X.

This contradicts the choice of M as an elementary submodel of H_λ. $\qquad\square$

For dessert, let us show that some of the large cardinals you have encountered in this text can be characterized in terms of elementary embeddings. Let us say that a cardinal κ has *the Reflection Property*[14] if for every $R \subseteq V_\kappa$ there exists $\alpha < \kappa$ such that $\langle V_\alpha, \bar{\in}, R\rangle \prec \langle V_\kappa, \bar{\in}, R\rangle$. By Remark 24.19, every strongly inaccessible cardinal has the Reflection Property. The converse is also true.

THEOREM 24.34. *If κ has the Reflection Property, then κ is strongly inaccessible.*

PROOF. Assume that κ has the Reflection Property. Since no $\langle V_n, \bar{\in}\rangle$ is an elementary submodel of $\langle V_m, \bar{\in}\rangle$ for $n < m \leq \omega$, we know that $\kappa > \omega$. Now let us show that κ is a regular cardinal. If not, there are $\mu < \kappa$ and a function $F : \mu \to \kappa$ with range unbounded in κ. Let $R = \{\mu\} \cup F$, and let $\alpha < \kappa$ be such that $\langle V_\alpha, \bar{\in}, R\rangle \prec \langle V_\kappa, \bar{\in}, R\rangle$. Since

(24.49) $\qquad\qquad \langle V_\kappa, \bar{\in}, R\rangle \models \mu$ is the only ordinal in R,

elementarity implies that

(24.50) $\qquad\qquad \langle V_\alpha, \bar{\in}, R\rangle \models \mu$ is the only ordinal in R.

In particular, $\mu \in V_\alpha$. Since V_α is transitive, we also have $\mu \subseteq V_\alpha$. But this implies that for each $\xi \in \mu$ there exists η such that $\langle \xi, \eta\rangle \in F \cap V_\alpha$. Thus $F \subseteq V_\alpha$, which contradicts the assumption that the range of F is unbounded in κ.

Finally, let us show that κ is a strong limit cardinal. If not, there is $\lambda < \kappa$ such that $2^\lambda \geq \kappa$. Fix a surjection $G : \mathcal{P}(\lambda) \to \kappa$, and let $R = \{\lambda + 1\} \cup G$. Let $\alpha < \kappa$ be such that $\langle V_\alpha, \bar{\in}, R\rangle \prec \langle V_\kappa, \bar{\in}, R\rangle$. Again, this implies that $\lambda + 1 \in V_\alpha$, and by definition of the cumulative hierarchy, $\mathcal{P}(\lambda) \in V_\alpha$.

EXERCISE 24.27 (G). Show that $\kappa \subseteq V_\alpha$, and explain why this yields a contradiction. $\qquad\square$

Now let us consider the "dual" of the Reflection Property. Let us say that κ has the *Extension Property* if for every $R \subseteq V_\kappa$ there exist a transitive set $X \neq V_\kappa$ and $S \subseteq X$ such that $\langle V_\kappa, \bar{\in}, R\rangle \prec \langle X, \bar{\in}, S\rangle$. What does the Extension Property tell us about κ?

THEOREM 24.35. *An ordinal κ has the Extension Property if and only if it is a weakly compact cardinal.*

We shall not give a complete proof of Theorem 24.35 here, but only the most interesting part of this proof.

LEMMA 24.36. *Suppose κ is a strongly inaccessible cardinal with the Extension Property. Then κ is weakly compact.*

---

[14]Be apprised that if you see the phrase "Reflection Property" elsewhere in the literature, it probably has a broader meaning than in this text.

PROOF. Let κ be as in the assumption. By Corollary 15.24, it suffices to show that κ has the Tree Property. Let $\langle T, \leq_T \rangle$ be a κ-tree. We want to show that this tree has a cofinal branch. Without loss of generality, we may assume that $T \subseteq V_\kappa$. Let R denote the strict p.o. relation $<_T$. Then $R \subseteq V_\kappa \times V_\kappa$, hence $R \subseteq V_\kappa$. By the Extension Property, there exist a transitive set $X \neq V_\kappa$ and $S \subseteq X$ such that $\langle V_\kappa, \bar{\in}, R \rangle \prec \langle X, \bar{\in}, S \rangle$. Let us fix such X and S for the remainder of this proof.

EXERCISE 24.28 (G). Use elementarity to show that $S \cap V_\kappa = R$. Hint: Treat $\langle V_\kappa, \bar{\in}, R \rangle$ and $\langle X, \bar{\in}, S \rangle$ as models of a language L_S^+ as described in Remark 24.19, and consider the formula $r(x)$.

By assumption, $X \setminus V_\kappa \neq \emptyset$. Let us show that, in particular, $\kappa \in X$. In Chapter 12, we defined for each set x an ordinal $rank(x) = \min\{\alpha : x \in V_{\alpha+1}\}$. Since $rank(x) \in \kappa \subseteq V_\kappa$ for all $x \in V_\kappa$, we have

(24.51) $\qquad V_\kappa \models \forall x \exists \alpha \in \mathbf{ON} \, (rank(x) = \alpha)$.

By elementarity, X has the same property. Now consider $x \in X \setminus V_\kappa$, and let $\alpha \in X$ be such that

(24.52) $\qquad X \models \alpha \in \mathbf{ON} \wedge rank(x) = \alpha$.

By Claim 12.19(b), the formula "$\alpha \in \mathbf{ON}$" is absolute for transitive models.[15] Hence α really is an ordinal, and by the definition of the $rank$ function,

(24.53) $\qquad X \models x \in V_{\alpha+1}$.

EXERCISE 24.29 (PG). Show by induction over α that each of the formulas "$x \in V_\alpha$" and "$rank(x) = \alpha$" is absolute for transitive models.

By Exercise 24.29, formula (24.53) implies that x really is an element of $V_{\alpha+1}$. Since $x \notin V_\kappa$ and $V_{\alpha+1} \subseteq V_\kappa$ for all $\alpha < \kappa$, it follows that $\alpha \geq \kappa$. Thus, we have found an element α of X that either is equal to κ or contains κ as an element, and transitivity of X implies that $\kappa \in X$.

Now consider the relation R. For every $\alpha < \kappa$, the α-th level $T(\alpha)$ of T can be defined as the set of all t such that $\langle \hat{t}, R \rangle$ is a strict wellorder of order type α, where $\hat{t} = \{s : \langle s, t \rangle \in R\}$. The expression "$t \in T(\alpha)$" can be written in the language L_S^+ of Remark 24.19 as:[16] "there exists a bijection f with domain α such that for all s, ($s \in rng(f)$ iff $r(s,t)$) and for all $\xi, \eta \in \alpha$ ($\xi \in \eta$ iff $r(f(\xi), f(\eta))$)." Our assumption that T is a κ-tree implies that $|T(\alpha)| < \kappa$ for all $\alpha < \kappa$. Thus, each $T(\alpha)$ is actually an element of V_κ. Since the definition of the $T(\alpha)$'s can be expressed in L_S^+, the model $\langle X, \bar{\in}, S \rangle$ also has its version of the $T(\alpha)$'s, and elementarity implies that if $\alpha < \kappa$ and x are such that

(24.54) $\qquad \langle V_\kappa, \bar{\in}, R \rangle \models x = T(\alpha)$,

then also

(24.55) $\qquad \langle X, \bar{\in}, S \rangle \models x = T(\alpha)$.

Since both V_κ and X are transitive, it follows that for any s,

(24.56) $\qquad \langle V_\kappa, \bar{\in}, R \rangle \models s \in T(\alpha)$ iff $\langle X, \bar{\in}, S \rangle \models s \in T(\alpha)$.

---

[15] I.e., if $\mathbf{Y}$ is a transitive class and $\alpha \in \mathbf{Y}$, then $\mathbf{Y} \models \alpha \in \mathbf{ON}$ if and only if $\mathbf{V} \models \alpha \in \mathbf{ON}$.

[16] Admittedly, we are going to use a few abbreviations rather than L_S^+ in its pristine form.

Since $ht(T) = \kappa$, and since $\mathbf{ON} \cap V_\kappa = \kappa$, we have
(24.57) $$\langle V_\kappa, \bar{\in}, R \rangle \models \forall \alpha \in \mathbf{ON}\, (T(\alpha) \neq \emptyset).$$
By elementarity we get:
(24.58) $$\langle X, \bar{\in}, S \rangle \models \forall \alpha \in \mathbf{ON}\, (T(\alpha) \neq \emptyset).$$
In particular,
(24.59) $$\langle X, \bar{\in}, S \rangle \models T(\kappa) \neq \emptyset.$$
Let $t_\kappa \in X$ be such that
(24.60) $$\langle X, \bar{\in}, S \rangle \models t_\kappa \in T(\kappa).$$
Note that
(24.61) $$\langle V_\kappa, \bar{\in}, R \rangle \models \forall \beta \in \mathbf{ON}\, \forall t \in T(\beta) \forall \alpha < \beta\, \exists! s \in T(\alpha)\, (r(s,t)).$$
By elementarity,
(24.62) $$\langle X, \bar{\in}, S \rangle \models \forall \beta \in \mathbf{ON}\, \forall t \in T(\beta) \forall \alpha < \beta\, \exists! s \in T(\alpha)\, (r(s,t)).$$
In particular, for each $\alpha < \kappa$ there exists exactly one s_α such that
(24.63) $$\langle X, \bar{\in}, S \rangle \models s_\alpha \in T(\alpha) \wedge r(s_\alpha, t_\kappa).$$
By (24.56), if $\langle X, \bar{\in}, S \rangle \models s_\alpha \in T(\alpha)$, then s_α really is an element of $T(\alpha)$; in particular, $s_\alpha \in V_\kappa$. Moreover, since $\hat{t}$ is wellordered by S, for all $\alpha < \beta < \kappa$ we have
(24.64) $$\langle X, \bar{\in}, S \rangle \models r(s_\alpha, s_\beta), \quad (i.e., \langle s_\alpha, s_\beta \rangle \in S).$$
By Exercise 24.28, formula (24.64) implies that $\langle s_\alpha, s_\beta \rangle \in R$. It follows that the set $\{s_\alpha : \alpha < \kappa\}$ is a cofinal branch of T, as desired. $\square$

Now let us consider $V_{\kappa+1}$, where κ is a cardinal. By Theorem 24.35, $\kappa + 1$ does not have the Extension Property. But perhaps $\kappa + 1$ could "almost" have the Extension Property in the following sense:

(∗)   *There exists an elementary embedding $j : V_{\kappa+1} \to M$ of $V_{\kappa+1}$ into a transitive set $M \neq V_{\kappa+1}$ such that $V_\kappa \subseteq M$ and $j \restriction V_\kappa$ is the identity.*

Perhaps we could even have:

(∗∗)   *There exists an elementary embedding $j : V_{\kappa+1} \to M$ of $V_{\kappa+1}$ into a transitive set M that contains $V_{\kappa+1}$ as a proper subset such that $j \restriction V_\kappa$ is the identity.*

It turns out that (∗) and (∗∗) give alternative characterizations of measurable cardinals.

THEOREM 24.37. *For every cardinal κ the following are equivalent:*

(a)   *κ has Property (∗);*
(b)   *κ has Property (∗∗);*
(c)   *κ is measurable.*

PROOF. First suppose κ has Property (∗), and let j, M be an elementary embedding and a transitive set that witness (∗). We are going to show that $V_{\kappa+1}$ is a proper subset of M, which implies that j and M also witness (∗∗).

Since

(24.65) $$V_{\kappa+1} \models \kappa \text{ is the largest ordinal},$$

elementarity of j implies that

(24.66) $\qquad M \models j(\kappa)$ is the largest ordinal.

By Claim 12.19(b), the formula "$x \in \mathbf{ON}$" is absolute for transitive models, and thus $j(\kappa)$ really is an ordinal.

By assumption, $j(\alpha) = \alpha$ for all ordinals $\alpha < \kappa$. Since j is one-to-one, $j(\kappa)$ must be either equal to or larger than κ. We will show that the former is inconsistent with our assumptions. If $X \in V_{\kappa+1}$, then $X \subseteq V_\kappa$, and thus

(24.67) $\qquad M \models j(X) \subseteq V_{j(\kappa)}.$

Absoluteness of V_α implies that

(24.68) $\qquad \mathbf{V} \models j(X) \subseteq V_{j(\kappa)}$ for every X in the domain of j.

Now consider $x \in V_\kappa$. By elementarity,

(24.69) $\qquad (M \models j(x) \in j(X))$ iff $x \in X$.

Since the formula "$x \in y$" is absolute and $j \upharpoonright V_\kappa$ is the identity, (24.69) implies that

(24.70) $\qquad x \in j(X)$ iff $x \in X$.

We have thus shown that

(24.71) $\qquad j(X) \cap V_\kappa = X$ for every X in the domain of j (i.e., in $V_{\kappa+1}$).

If $j(\kappa)$ were equal to κ, then (24.67) and (24.71) together would imply that j is the identity, which contradicts our choice of j. We have shown that

(24.72) $\qquad j(\kappa) > \kappa.$

Since M is transitive, $j(\kappa) \subseteq M$; in particular, $\kappa \in M$. Note that

(24.73) $\qquad V_{\kappa+1} \models \forall \alpha \in \mathbf{ON}\, \exists x\, (x = V_\alpha).$

By elementarity,

(24.74) $\qquad M \models \forall \alpha \in \mathbf{ON}\, \exists x\, (x = V_\alpha).$

In particular,

(24.75) $\qquad M \models \exists x\, (x = V_\kappa).$

By Exercise 24.29, $V_\kappa \in M$, and if $X \in V_{\kappa+1}$, then (24.71) implies that X is the intersection of two elements of M. Since $V_{\kappa+1}$ is closed under intersections, so is M, and thus $X \in M$. Hence $V_{\kappa+1} \subseteq M$, and we have shown that j witnesses $(**)$.

Now let us contemplate how j acts on subsets X of κ. If

(24.76) $\qquad V_{\kappa+1} \models X \subseteq \kappa,$

then

(24.77) $\qquad M \models j(X) \subseteq j(\kappa).$

By (24.71), $j(X)$ contains the same elements of κ as X, but it may also contain elements $\geq \kappa$. In particular, if

(24.78) $\qquad V_{\kappa+1} \models X$ is an unbounded subset of κ,

then

(24.79) $\qquad M \models j(X)$ is an unbounded subset of $j(\kappa)$.

A strong version of the converse of (24.79) is also true:

EXERCISE 24.30 (PG). Show that if X is a bounded subset of κ, then $j(X) = X$. *Hint:* Consider $\alpha < \kappa$ such that $X \subseteq \alpha$.

In particular, $j(X)$ may or may not contain the ordinal κ itself. Let us consider the family
$$\mathcal{U} = \{X \subseteq \kappa : \kappa \in j(X)\}.$$
If $X, Y \subseteq \kappa$ are complements of each other, i.e., if
(24.80) $$V_{\kappa+1} \models X \cup Y = \kappa \wedge X \cap Y = \emptyset,$$
then
(24.81) $$M \models j(X) \cup j(Y) = j(\kappa) \wedge j(X) \cap j(Y) = \emptyset.$$
In other words, exactly one of the sets X, Y is in $\mathcal{U}$. One might now suspect that $\mathcal{U}$ is an ultrafilter on κ, and this is indeed the case.

EXERCISE 24.31 (G). Show that $\mathcal{U}$ is closed under finite intersections and under supersets.

By Exercise 24.30, if X is a bounded subset of κ, then $\kappa \notin j(X)$, and hence $X \notin \mathcal{U}$. In particular, $\mathcal{U}$ is a nonprincipal ultrafilter on κ. So far we have shown nothing unusual about $\mathcal{U}$. But here comes the punchline that proves measurability of κ:

CLAIM 24.38. *The ultrafilter $\mathcal{U}$ defined above is κ-complete.*

PROOF. Let $\lambda < \kappa$, and let $\{X_\xi : \xi < \lambda\}$ be a family of elements of $\mathcal{U}$. By the definition of $\mathcal{U}$, for each $\xi < \lambda$:
(24.82) $$X_\xi \subseteq \kappa \wedge \kappa \in j(X_\xi).$$
Define:
$$Z = \{\langle \xi, \alpha \rangle : \xi < \lambda \wedge \alpha \in X_\xi\}.$$
Note that $Z \in V_{\kappa+1}$. The intersection $Y = \bigcap_{\xi < \lambda} X_\xi$ can be defined as follows:
$$Y = \{\alpha < \kappa : \forall \xi < \lambda \langle \xi, \alpha \rangle \in Z\}.$$
By elementarity,
$$j(Y) = \{\alpha < j(\kappa) : \forall \xi < j(\lambda) \langle \xi, \alpha \rangle \in j(Z)\}.$$
Note that $j(Z)$ consists of pairs of ordinals such that the first coordinate is less than $j(\lambda)$ and the second coordinate is less than $j(\kappa)$. Thus $j(Z)$ may pick up a whole lot of pairs with new second coordinates that were not in Z. In particular, (24.82) implies that
(24.83) $$\forall \xi < \lambda (\langle \xi, \kappa \rangle \in j(Z)).$$
But since $j(\lambda) = \lambda$, the sets of first coordinates of Z and $j(Z)$ are the same. Hence (24.83) implies that $\kappa \in j(Y)$, and thus $Y \in \mathcal{U}$. □

Now assume that κ is a measurable cardinal. We are going to show that κ has Property (∗). Let $\mathcal{U}$ be a nonprincipal, κ-complete ultrafilter on κ, and let $\langle {}^\kappa V_{\kappa+1}/\mathcal{U}, E \rangle$ be the corresponding ultrapower of the model $\langle V_{\kappa+1}, \bar{\in} \rangle$, as defined in Section 13.2. By Theorem 13.14, the relation E is wellfounded. It is not hard to verify that the relation E is also extensional in the sense of Definition 4.38. Thus, by Theorem 4.39, there exists a (unique) isomorphism $F : {}^\kappa V_{\kappa+1}/\mathcal{U} \to M$

24.2. APPLICATIONS OF ELEMENTARY SUBMODELS IN SET THEORY

of ${}^\kappa V_{\kappa+1}/\mathcal{U}$ onto a transitive set M (the Mostowski collapse of ${}^\kappa V_{\kappa+1}/\mathcal{U}$). Let $h : V_{\kappa+1} \to {}^\kappa V_{\kappa+1}/\mathcal{U}$ be the elementary embedding of $\langle V_{\kappa+1}, \in \rangle$ into $\langle {}^\kappa V_{\kappa+1}/\mathcal{U}, E \rangle$ defined in Example 24.3, and define $j : V_{\kappa+1} \to M$ as $j = F \circ h$.

We show that j and M witness that κ has property $(*)$. By Claim 24.6 and Exercise 24.4, j is an elementary embedding.

CLAIM 24.39. *For all $X \in V_\kappa$, $j(X) = X$.*

PROOF. For each $z \in V_{\kappa+1}$, let g^z denote the function in ${}^\kappa V_{\kappa+1}$ that takes the value z for all α in its domain. By the definition of h, Claim 24.39 can be reformulated as follows:

"For all $X \in V_\kappa$, $F(g^X_{/\mathcal{U}}) = X$."

We will use this reformulation later in the proof.

Now let us prove the claim by induction over the wellfounded relation $\bar{\in}$. By elementarity, $j(\emptyset) = \emptyset$. Now suppose $X \in V_\kappa$ and $j(x) = x$ for all $x \in X$. We want to show that $j(X) = X$. By elementarity, $j(x) \in j(X)$ for all $x \in X$. Hence, it suffices to show that if $y \in j(X)$, then $y = x$ for some $x \in X$. So, let $y \in j(X)$. By transitivity of M, there exists $g \in {}^\kappa V_{\kappa+1}$ such that $F(g_{/\mathcal{U}}) = y$. Fix such a g.

Note that $j(X) = F(g^X_{/\mathcal{U}})$. Since F is an isomorphism, the choice of g implies that

(24.84) $${}^\kappa V_{\kappa+1}/\mathcal{U} \models g_{/\mathcal{U}} \in g^X_{/\mathcal{U}}.$$

This means that the set $B = \{\beta \in \kappa : g(\beta) \in g^X(\beta)\}$ is in $\mathcal{U}$. In view of the definition of g^X we have $B = \{\beta \in \kappa : g(\beta) \in X\}$, and hence $B = \bigcup_{x \in X} \{\beta \in \kappa : g(\beta) = x\}$. Since κ is measurable and hence strongly inaccessible (Theorem 23.14), by Exercise 12.29, $V_\kappa = H_\kappa$. Thus $|X| < \kappa$, and κ-completeness of $\mathcal{U}$ implies that for some $x \in X$ the set $\{\beta \in \kappa : g(\beta) = x\}$ is in $\mathcal{U}$, i.e., $g \sim_\mathcal{U} g^x$. By the inductive assumption, this implies $F(g_{/\mathcal{U}}) = x$, and we have proved Claim 24.39. □

Next we will show that $j(\kappa)$ is an ordinal greater than κ. Since $V_{\kappa+1}$ does not contain ordinals $> \kappa$, this implies that $M \setminus V_{\kappa+1} \neq \emptyset$, which takes care of the other requirement in $(*)$.

CLAIM 24.40. $j(\kappa) > \kappa$.

PROOF. By the previous claim, $j(\alpha) = \alpha$ (i.e., $F(g^\alpha_{/\mathcal{U}}) = \alpha$) for all $\alpha < \kappa$. Thus it suffices to show that there exists $g^{id} \in {}^\kappa V_{\kappa+1}$ such that for all $\alpha < \kappa$:

(24.85) $${}^\kappa V_{\kappa+1}/\mathcal{U} \models g^\alpha_{/\mathcal{U}} \in g^{id}_{/\mathcal{U}} \in g^\kappa_{/\mathcal{U}}.$$

Define:
$$g^{id}(\beta) = \beta \text{ for all } \beta < \kappa.$$

Then $\{\beta : g^{id}(\beta) \in g^\kappa(\beta)\} = \kappa \in \mathcal{U}$, and for each $\alpha < \kappa$,
$$\{\beta : g^\alpha(\beta) \in g^{id}(\beta)\} = \kappa \setminus (\alpha+1) \in \mathcal{U}.$$

By definition of the ultrapower, (24.85) follows. □□

If g^{id} is defined as in the proof of Claim 24.40, is $F(g^{id}_{/\mathcal{U}})$ actually *equal* to κ? The answer depends on the properties of $\mathcal{U}$. If there exists a regressive function $g : \kappa \to \kappa$ such that $g^{-1}\{\alpha\} \notin \mathcal{U}$ for each $\alpha < \kappa$, then

(24.86) $$\text{For all } \alpha < \kappa, \quad {}^\kappa V_{\kappa+1}/\mathcal{U} \models g^\alpha_{/\mathcal{U}} \in g_{/\mathcal{U}}.$$

Moreover, by regressiveness of g,

(24.87) $$\{\beta \in \kappa : g(\beta) \in g^{id}(\beta)\} = \kappa \setminus \{0\} \in \mathcal{U},$$

and hence

(24.88) $$^{\kappa}V_{\kappa+1}/\mathcal{U} \models g_{/\mathcal{U}} \in g^{id}_{/\mathcal{U}}.$$

Thus $F(g_{/\mathcal{U}})$ is between all the α's that are less than κ and $F(g^{id}_{/\mathcal{U}})$, and hence $F(g^{id}_{/\mathcal{U}})$ cannot be κ.

On the other hand, if for every regressive function $g : \kappa \to \kappa$ there exists $A \in \mathcal{U}$ such that $g \restriction A$ is constant, then for every $g_{/\mathcal{U}}$ such that

(24.89) $$^{\kappa}V_{\kappa+1}/\mathcal{U} \models g_{/\mathcal{U}} \in g^{id}_{/\mathcal{U}}$$

there exists $\alpha < \kappa$ such that

(24.90) $$^{\kappa}V_{\kappa+1}/\mathcal{U} \models g_{/\mathcal{U}} = g^{\alpha}_{/\mathcal{U}},$$

and hence $F(g^{id}_{/\mathcal{U}})$ is equal to κ.

Let us call an ultrafilter $\mathcal{U}$ on κ *normal* if for every $X \in \mathcal{U}$ and for every regressive function $g : X \to \kappa$ there exists $A \in \mathcal{U}$ such that $g \restriction A$ is constant. Note that we have proved the following:

THEOREM 24.41. *Let $\mathcal{U}$ be a nonprincipal, κ-complete ultrafilter on a measurable cardinal κ, let M be the Mostowski collapse of the ultrapower $^{\kappa}V_{\kappa+1}/\mathcal{U}$, let $F : {}^{\kappa}V_{\kappa+1}/\mathcal{U} \to M$ be the collapsing function, and let $g^{id} : \kappa \to \kappa$ be defined by $g^{id}(\beta) = \beta$ for all $\beta < \kappa$. Then $F(g^{id}_{/\mathcal{U}}) = \kappa$ if and only if $\mathcal{U}$ is normal.*

In Chapter 23, we also considered "normal" ultrafilters, but they were defined in a different way. It turns out that the two definitions of normal ultrafilters are equivalent.

EXERCISE 24.32 (PG). Let κ be an infinite cardinal. Show that $\mathcal{U}$ is a normal ultrafilter on κ iff $\mathcal{U}$ is closed under diagonal intersections, i.e., iff for every sequence $\langle X_\alpha : \alpha \in \kappa \rangle$ of elements of $\mathcal{U}$ the diagonal intersection $\Delta_{\alpha < \kappa} X_\alpha \in \mathcal{U}$. *Hint:* For the "if"-direction, imitate the proof of Fodor's Lemma (Theorem 21.12).

By Exercise 23.8(a), if there exists a uniform, normal ultrafilter on κ, then κ is measurable. The converse is also true.

THEOREM 24.42. *Let κ be a measurable cardinal. Then there exists a uniform, normal ultrafilter on κ.*

PROOF. Let $\mathcal{U}$ be any κ-complete nonprincipal ultrafilter on κ, and let $F : {}^{\kappa}V_{\kappa+1}/\mathcal{U} \to M$ be as in Theorem 24.41. Then there exists *some* $g : \kappa \to \kappa$ with $F(g_{/\mathcal{U}}) = \kappa$. Pick such g and define:

$$\mathcal{V} = \{X \subseteq \kappa : g^{-1}X \in \mathcal{U}\}.$$

EXERCISE 24.33 (PG). Show that $\mathcal{V}$ is a uniform, normal ultrafilter on κ. □

Let us conclude this chapter with an interesting consequence of Property $(**)$.

THEOREM 24.43. *Suppose κ is a measurable cardinal and $2^\lambda = \lambda^+$ for each infinite cardinal $\lambda < \kappa$. Then $2^\kappa = \kappa^+$.*

PROOF. Let κ be as in the assumptions, and let j, M be as in $(**)$. Since

(24.91) $\quad V_{\kappa+1} \models \forall \lambda < \kappa \, (\lambda \text{ is an infinite cardinal} \to 2^\lambda = \lambda^+),$

by elementarity,

(24.92) $\quad M \models \forall \lambda < j(\kappa) \, (\lambda \text{ is an infinite cardinal} \to 2^\lambda = \lambda^+).$

How realistic are M's notions of infinite cardinals?

EXERCISE 24.34 (G). Convince yourself that if λ really is an infinite cardinal, then $M \models \lambda$ is an infinite cardinal. *Hint:* Argue as in Exercise 24.14.

By Exercise 24.34,

(24.93) $\quad\quad\quad\quad\quad M \models \kappa$ is an infinite cardinal.

Hence (24.92) implies in particular that

(24.94) $\quad\quad\quad\quad\quad M \models 2^\kappa = \kappa^+.$

This means, there exist $f \in M$ and an ordinal $\beta \in M$ such that

(24.95) $\quad\quad\quad\quad\quad M \models f$ is a bijection from $\mathcal{P}(\kappa)$ onto β,

and for every ordinal α such that $\kappa < \alpha < \beta$,

(24.96) $\quad\quad\quad\quad\quad M \models \alpha$ is not a cardinal.

By Claim 12.19(j), there exists a set $x \in M$ such that f really is a bijection from x onto β and

(24.97) $\quad\quad\quad\quad\quad M \models x = \mathcal{P}(\kappa).$

Since M witnesses $(**)$ and $\mathcal{P}(\kappa) \subseteq V_\kappa$, also $\mathcal{P}(\kappa) \subseteq M$. Moreover, since M is transitive, it is not hard to see that $y \in x$ if and only if y really is a subset of κ. In other words, x really is equal to $\mathcal{P}(\kappa)$. Thus, f really is a bijection from $\mathcal{P}(\kappa)$ onto β, and we have shown that $2^\kappa = |\beta|$. By (24.96) and Exercise 24.34, none of the ordinals between κ and β is a cardinal. Thus $\beta = \kappa^+$, and we have proved Theorem 24.43. □

Mathographical Remarks

Theorem 24.35 appears as Theorem 4.5 with a complete proof in *The Higher Infinite*, by Akihiro Kanamori, Springer Verlag, 1994. The latter is also the ideal source for learning more about characterizations of large cardinals in terms of elementary embeddings. If you want to learn more about applications of elementary submodels in topology, we recommend Alan Dow's survey article *An introduction to applications of elementary submodels to topology*, Topology Proceedings 13(1) (1988), 17–72.

CHAPTER 25

Boolean Algebras

In Section 13.3, we introduced Boolean algebras and outlined their connections to totally disconnected compact Hausdorff spaces and partial orders. The present chapter contains a more detailed treatment of selected topics in the theory of Boolean algebras. Throughout this chapter, letters a, b, c, d are reserved for elements of a Boolean algebra. If no confusion is likely, we shall suppress the phrase "Let a, b be elements of a Boolean algebra $\mathfrak{B} = \langle B, +, \cdot, -, 0, 1 \rangle$."

We start with a connection between Boolean algebras and a certain class of algebraic rings. The *symmetric difference* between two elements of a Boolean algebra is defined as:
$$a \Delta b = a \cdot -b + -a \cdot b.$$
In a set algebra, this definition coincides with the usual one.

EXERCISE 25.1 (G). Show that $a \Delta b = (-a) \Delta (-b)$ in every Boolean algebra.

LEMMA 25.1. *Let a, b, c be elements of a Boolean algebra. Then*

(i) $a = b$ if and only if $a \Delta b = 0$;
(ii) $a \Delta b = b \Delta a$;
(iii) $a \Delta (b \Delta c) = (a \Delta b) \Delta c$;
(iv) $a \cdot (b \Delta c) = (a \cdot b) \Delta (a \cdot c)$.

PROOF. (i) We have

$$
\begin{array}{rll}
a = b & \text{iff} & a \leq b \text{ and } b \leq a \quad \text{(Lemma 13.19)} \\
& \text{iff} & a \cdot -b = 0 \text{ and } -a \cdot b = 0 \quad \text{(Exercise 13.41(iii))} \\
& \text{iff} & a \cdot -b + b \cdot -a = 0 \\
& \text{iff} & a \Delta b = 0.
\end{array}
$$

Point (ii) is obvious.

EXERCISE 25.2 (PG). Prove points (iii) and (iv). □

DEFINITION 25.2. Let $\mathfrak{R} = \langle R, \oplus, \cdot, 0, 1 \rangle$ be an algebraic ring with identity, $p \in R$. If $p^2 = p$, we say that p is *idempotent*. $\mathfrak{R}$ is a *Boolean ring* if all elements of R are idempotent.

Note that we do not require that $0 \neq 1$. The one-element ring is a Boolean ring.

Let $\mathfrak{B} = \langle B, +, \cdot, -, 0, 1 \rangle$ be a Boolean algebra. We define:
$$r(\mathfrak{B}) = \langle B, \Delta, \cdot, 0, 1 \rangle.$$
Lemma 25.1 implies that $r(\mathfrak{B})$ is a Boolean ring.

Now let $\mathfrak{R} = \langle R, \oplus, \cdot, 0, 1\rangle$ be a Boolean ring. We want to show that $\mathfrak{R}$ is a commutative ring. Idempotence and distributivity imply that

$$a \oplus b = (a \oplus b) \cdot (a \oplus b) = a \cdot a \oplus a \cdot b \oplus b \cdot a \oplus b \cdot b = a \oplus a \cdot b \oplus b \cdot a \oplus b.$$

. Subtracting $a \oplus b$ from both sides of this equation and using commutativity of $\oplus$, we obtain:

(25.1) $$a \cdot b \oplus b \cdot a = 0.$$

For $a = b$, equation (25.1) yields:

(25.2) $$a \oplus a = 0.$$

It follows that every element of a Boolean ring is of order ≤ 2. Equations (25.1) and (25.2) combined imply that every Boolean ring is commutative.

For $\mathfrak{R}$ as above and $a, b \in R$ we define $a + b = a \oplus b \oplus (a \cdot b)$ and $-a = a \oplus 1$, and we let $b(\mathfrak{R}) = \langle R, +, \cdot, -, 0, 1\rangle$.

EXERCISE 25.3 (PG). Show that $b(\mathfrak{R})$ as defined above is a Boolean algebra.

THEOREM 25.3. *Every Boolean algebra $\mathfrak{B}$ is isomorphic to $b(r(\mathfrak{B}))$. Every Boolean ring $\mathfrak{R}$ is isomorphic to $r(b(\mathfrak{R}))$.*

EXERCISE 25.4 (PG). Prove Theorem 25.3.

EXERCISE 25.5 (G). Let $\mathfrak{B} = \langle B, +, \cdot, 0, 1\rangle$ be a Boolean algebra, $I \subseteq B$. Show that I is an ideal in the Boolean algebra $\mathfrak{B}$ (i.e., an ideal in the p.o. $\langle B\backslash\{1\}, \leq_B\rangle$) if and only if I is an ideal in the Boolean ring $r(\mathfrak{B})$.

Let $\mathfrak{A}$ and $\mathfrak{B}$ be Boolean algebras, $\pi : A \to B$ a homomorphism. We define the *kernel* of π as follows:

$$ker\,(\pi) = \{a \in A : \pi(a) = 0\}.$$

EXERCISE 25.6 (G). (a) Convince yourself that π is also a homomorphism from the algebraic ring $r(\mathfrak{A})$ into $r(\mathfrak{B})$, and conclude that its kernel $ker\,(\pi)$ is an ideal in the Boolean algebra $\mathfrak{A}$.

(b) Prove directly (i.e., without reference to algebraic rings) that the kernel of every homomorphism of Boolean algebras is an ideal in the domain of the homomorphism.

If I is an ideal in a Boolean algebra $\mathfrak{B}$, then the *quotient algebra induced by I* is the algebra $\mathfrak{B}/I = b(r(\mathfrak{B})/I)$, where $r(\mathfrak{B})/I$ is the quotient ring. More explicitly, $\mathfrak{B}/I = \langle B/I, +_{/I}, \cdot_{/I}, -_{/I}, I, I^*\rangle$, where B/I is the set of equivalence classes of the relation $\sim_I$ defined by $a \sim_I b$ if and only if $a\Delta b \in I$; $+_{/I}, \cdot_{/I}$, and $-_{/I}$ are the operations on B/I induced by the corresponding operations in $\mathfrak{B}$, and $I^* = \{-a : a \in I\}$ is the dual filter of I. Every quotient algebra $\mathfrak{B}/I$ is a homomorphic image of $\mathfrak{B}$; this is witnessed by the *canonical homomorphism* $\pi_I : B \to B/I$ which is defined by:

$$\pi_I(b) = \{a \in B : a \sim_I b\}.$$

An element a of a Boolean algebra $\mathfrak{B}$ is called an *atom* if $a \neq 0$ and $\forall b \in B\,(b \leq a \to (b = 0 \vee b = a))$. The set of all atoms of B is denoted by $At(B)$. We say that a Boolean algebra B is *atomic*, if $At(B)$ is a dense subset of B, i.e., if for each $b \in B^+$ there is an $a \in At(B)$ such that $a \leq b$. If $At(B) = \emptyset$, then B is *atomless*.

EXERCISE 25.7 (G). Let B be a Boolean algebra, and consider the function $\varphi : B \to \mathcal{P}(At(B))$ defined by the formula $\varphi(b) = \{a \in At(B) : a \leq b\}$.

(a) Show that φ is a homomorphism from B into the set algebra $\mathcal{P}(At(B))$. Find the kernel of φ, and show that φ is an injection if and only if B is atomic.

(b) Show that if B is a complete atomic Boolean algebra, we have $B \cong \mathcal{P}(At(B))$.

EXERCISE 25.8 (PG). (a) Let X be a compact, totally disconnected Hausdorff space. Show that a clopen set a is an atom in the Boolean algebra $Clop(X)$ if and only if $a = \{x\}$ for some isolated point $x \in X$.

(b) Conclude that a Boolean algebra B is atomic if and only if the set of isolated points is dense in $st(B)$, and that B is atomless if and only if $st(B)$ has no isolated points.

(c) Let D be the Cantor set defined in Exercise 13.44. Show that $Clop(D)$ is a countable atomless Boolean algebra.

(d) Show that every countable, atomless Boolean algebra is isomorphic to $Clop(D)$. Hint: Do a back-and-forth construction as in the proof of Theorem 4.16.

The last exercise has an important consequence.

COROLLARY 25.4. *Let A be a complete atomless Boolean algebra that contains a countable dense subset. Then A is isomorphic to the algebra $RO(D)$ of regular open subsets of the Cantor set.*

PROOF. $Clop(D)\backslash\{\emptyset\}$ is dense in the complete Boolean algebra $RO(D)$. Thus, by Exercise 25.8(c), $RO(D)$ contains a dense countable atomless subalgebra. Now Exercise 25.8(d) and Corollary 13.35 imply that, up to isomorphism, $RO(D)$ is the only complete Boolean algebra with this property. □

As we remarked in Exercise 13.44, D is homeomorphic to $D(2)^\omega$. More generally, for any infinite cardinal κ we set $D_\kappa = D(2)^\kappa$ and we let $Fr(\kappa)$ denote the Boolean algebra $Clop(D_\kappa)$.

EXERCISE 25.9 (G). (a) Show that for every infinite cardinal κ, the Boolean algebra D_κ satisfies the c.c.c. Hint: Use Theorem 16.4.

(b) Show that if λ is an infinite cardinal and B is a dense subalgebra of a Boolean algebra A, then B satisfies the λ-c.c. if and only if A does.

For any infinite cardinal κ we have $Fr(\kappa) \subseteq RO(D_\kappa)$. Instead of $RO(Fr(\kappa))$ we shall write $cFr(\kappa)$. Since $D(2)^\kappa$ has a base consisting of clopen sets, $Fr(\kappa)$ is dense in $cFr(\kappa)$. Thus, $cFr(\kappa)$ is the completion of $Fr(\kappa)$. Moreover, Exercise 25.9 implies that $cFr(\kappa)$ satisfies the c.c.c.

EXERCISE 25.10 (PG). Let $\kappa \geq \omega$, let $\mathcal{B}_\kappa$ denote the Boolean algebra of all Borel subsets of D_κ, and let $\mathcal{M}_\kappa$ denote the ideal of all $A \in \mathcal{B}_\kappa$ that are meager. Show that $cFr(\kappa) \cong \mathcal{B}_\kappa / \mathcal{M}_\kappa$.

LEMMA 25.5. *Let κ be an infinite cardinal, B a Boolean algebra with the κ-c.c., and E a dense subset of B. Then $|B| \leq |E|^{<\kappa}$.*

PROOF. Consider a function $\pi : B \to [E]^{<\kappa}$ such that for each $b \in B$, the function value $\pi(b)$ is a set X of pairwise disjoint elements of E maximal with respect to the property that $e \leq_B b$ for all $e \in A$.[1] It is not hard to verify that π must be injective, and thus witnesses the inequality $|B| \leq |E|^{<\kappa}$. □

---

[1] Such a set is often called a *maximal antichain in E below b*.

Let us consider $cFr(\kappa)$ for $\kappa \geq \omega$. The Boolean algebra $cFr(\kappa)$ satisfies the c.c.c., and $Fr(\kappa)$ is dense in $cFr(\kappa)$. By Lemma 25.5, this implies that $|cFr(\kappa)| \leq \kappa^{\aleph_0}$. Thus for every infinite cardinal κ with $\kappa^{\aleph_0} = \kappa$ there exists a complete Boolean algebra of cardinality κ. Now we are going to show that this is also a necessary condition; i.e., for each infinite complete Boolean algebra B the equality $|B| = |B|^{\aleph_0}$ holds. First we show this for a special class of complete Boolean algebras. In order to define this class, we need to introduce one more bit of notation.

Let $\mathfrak{B} = \langle B, +, \cdot, -, 0, 1 \rangle$ be a Boolean algebra, $a \in B$. We define $B|a = \{b \in B : b \leq a\}$. The structure

$$\mathfrak{B}|a = \langle B|a, +_a, \cdot_a, -_a, 0_a, 1_a \rangle,$$

where $x +_a y = x + y$, $x \cdot_a y = x \cdot y$, $-_a x = -x \cdot a$, $0_a = 0$, and $1_a = a$, is also a Boolean algebra. $\mathfrak{B}|a$ is called the *factor algebra* (or simply *factor*) of $\mathfrak{B}$ relative to a.

EXERCISE 25.11 (PG). Prove the following "Cantor-Schröder-Bernstein Theorem" for complete Boolean algebras: If A and B are complete Boolean algebras such that there exist $a \in A$ and $b \in B$ with $A \cong B|b$ and $B \cong A|a$, then $A \cong B$.

Let B be an infinite Boolean algebra with $|B| = \kappa$. We say that B is *weakly κ-homogeneous* (or simply *weakly homogeneous*), if for each $a \in B^+$ the cardinality of $B|a$ is equal to κ. Note that each wekly homogeneous Boolean algebra is atomless.

LEMMA 25.6. *Let B be a complete, weakly homogeneous Boolean algebra with $|B| \geq \aleph_0$. Then $|B| = |B|^{\aleph_0}$.*

PROOF. We need a preliminary observation.

EXERCISE 25.12 (PG). Suppose B is an infinite Boolean algebra. Then there exists an infinite set of pairwise disjoint, non-zero elements of B. *Hint:* Use König's Lemma.

Let B be as in the assumption and let $\kappa = |B|$. By 25.12, we can choose a set of pairwise disjoint elements $\{a_i : i \in \omega\} \subseteq B^+$. For every sequence $(b_i)_{i \in \omega}$ such that $b_i \leq a_i$ for all $i \in \omega$, let $s((b_i)_{i \in \omega}) = \sum_{i \in \omega} b_i$. Note that $(b_i)_{i \in \omega} \neq (b'_i)_{i \in \omega}$ implies $s((b_i)_{i \in \omega}) \neq s((b'_i)_{i \in \omega})$. Thus s is an injection of the set $\prod_{i \in \omega} B|a_i$ into B. By weak κ-homogeneity of B, the existence of such an injection implies that $|B| \geq \kappa^{\aleph_0}$, and the lemma follows. □

The prime examples of Boolean algebras that are not weakly homogeneous are atomic Boolean algebras. Fortunately, for them the analogue of Lemma 25.6 is easy to prove.

EXERCISE 25.13 (G). Show that if B is an infinite, complete, atomic Boolean algebra, then $|B|^{\aleph_0} = |B|$. *Hint:* Use Exercise 25.7.

Exercise 25.13 allows us to reduce our problem to the case of atomless Boolean algebras. To do this in an elegant fashion, let us introduce one more notion.

Let $\mathfrak{A} = \langle A, +_A, \cdot_A, -_A, 0_A, 1_A \rangle$ and $\mathfrak{B} = \langle B, +_B, \cdot_B, -_B, 0_B, 1_B \rangle$ be two Boolean algebras. We define the *product algebra* $\mathfrak{A} \times \mathfrak{B} = \langle A \times B, +, \cdot, -, 0, 1 \rangle$ as

follows:
$$\langle a_0, b_0\rangle + \langle a_1, b_1\rangle = \langle a_0 +_A a_1, b_0 +_B b_1\rangle;$$
$$\langle a_0, b_0\rangle \cdot \langle a_1, b_1\rangle = \langle a_0 \cdot_A a_1, b_0 \cdot_B b_1\rangle;$$
$$-\langle a_0, b_0\rangle = \langle -_A a_0, -_B b_0\rangle;$$
$$0 = \langle 0_A, 0_B\rangle;$$
$$1 = \langle 1_A, 1_B\rangle.$$

LEMMA 25.7. *Let B be a Boolean algebra, $a \in B$. Then*
$$B \cong (B|a) \times (B|-a).$$

EXERCISE 25.14 (G). Prove Lemma 25.7. *Hint:* Consider the map $\pi : B \to (B|a) \times (B|-a)$ defined by $\pi(b) = \langle b \cdot a, b \cdot -a\rangle$.

THEOREM 25.8. *Let B be a complete, infinite Boolean algebra. Then $|B| = |B|^{\aleph_0}$.*

PROOF. Let B be as in the assumption, and let At be the set of all atoms of B. Let $a = \sum At$. Then $B|-a$ is atomless.

EXERCISE 25.15 (G). (a) Let a be as above. Show that at least one of the equalities $|B|a| = |B|$, $|B|-a| = |B|$ holds.

(b) Derive Theorem 25.8 from the following lemma:

LEMMA 25.9 (Pierce). *Let B be an atomless, complete, infinite Boolean algebra. Then $|B| = |B|^{\aleph_0}$.*

PROOF. Let B be as in the assumptions. Consider the set
$$J = \{0\} \cup \{a \in B^+ : B|a \text{ is weakly homogeneous.}\}.$$

EXERCISE 25.16 (R). (a) Show that J is a dense ideal in B.

(b) Conclude that there exists a maximal pairwise disjoint subset of B that is contained in $J\setminus\{0\}$.

(c) Derive Lemma 25.9. *Hint:* Use an argument similar to the one in the proof of Lemma 25.6. □□

Let B be a Boolean algebra, $X \subseteq B$. Then $\langle X\rangle$ denotes the smallest subalgebra of B that contains all elements of X. If B is a subalgebra of A and $a \in A$, we write $B(a)$ instead of $\langle B \cup \{a\}\rangle$.

Now let A be a Boolean algebra, B a subalgebra of A, and $a \in A \setminus B$. Consider the set
$$M = \{b_0 \cdot a + b_1 \cdot -a : b_0, b_1 \in B\}.$$
Clearly, $0, 1 \in M$. We show that M is closed under $+$, $\cdot$, and $-$: Let $b_0, b_1, b_2, b_3 \in B$. We have:
$$(b_0 \cdot a + b_1 \cdot -a) + (b_2 \cdot a + b_3 \cdot -a) = (b_0 + b_2) \cdot a + (b_1 + b_3) \cdot -a;$$
$$(b_0 \cdot a + b_1 \cdot -a) \cdot (b_2 \cdot a + b_3 \cdot -a) = (b_0 \cdot b_2) \cdot a + (b_1 \cdot b_3) \cdot -a;$$
$$-(b_0 \cdot a + b_1 \cdot -a) = -b_0 \cdot a + -b_1 \cdot -a.$$

Thus M is a subalgebra of B. Since $M \subseteq B(a)$, we have $M = B(a)$.

Let B and C be Boolean algebras with $B \subseteq C$. We say that C is a *finite extension* of B if there exists a finite set $\{a_0, \ldots, a_{n-1}\}$ such that $C = \langle B \cup \{a_0, \ldots, a_{n-1}\}\rangle$. We call C is a *simple extension* of B if there exists a such that $C = \langle B \cup \{a\} \rangle$.

Let A, B, and C be Boolean algebras with $A \subseteq B$, and let $\pi : A \to C$ be a homomorphism from A into C. We are going to explore under what conditions π can be extended to a homomorphism from B into C.

Let A be a Boolean algebra. A pair $\langle S, T \rangle$ of subsets of A will be called a *cut* if $S \leq T$, i.e., if $s \leq t$ for all $s \in S$ and $t \in T$. Note that the word "cut" is used here in a different meaning from that of Section 13.3. An element $a \in A$ *fills* the cut if $S \leq a \leq T$. A cut $\langle S, T \rangle$ is called a *gap* if there is no $a \in A$ that fills this cut.

Let B be a subalgebra of A and let $a \in A$. We define $S_a = \{b \in B : b \leq a\}$ and $T_a = \{b \in B : a \leq b\}$. Clearly, $\langle S_a, T_a \rangle$ is a cut. Moreover, if $f : B \to C$ is a homomorphism from $\mathfrak{B}$ into $\mathfrak{C}$, then $\langle f[S_a], f[T_a] \rangle$ is a cut in C.

LEMMA 25.10. *Let A, B, C, a, f be as above. If some $c \in C$ fills the cut $\langle f[S_a], f[T_a] \rangle$, then f can be extended to a homomorphism $f^+ : B(a) \to C$.*

PROOF. We define $f^+(b_0 \cdot a + b_1 \cdot -a) = f(b_0) \cdot c + f(b_1) \cdot -c$.

EXERCISE 25.17 (G). Show that the above definition of f^+ does not depend on the choice of representations of the elements of $B(a)$.

Now let us show that f^+ preserves the operations $+$ and $-$. We have:

$$f^+((b_0 \cdot a + b_1 \cdot -a) + (b_2 \cdot a + b_3 \cdot -a)) = f^+((b_0 + b_2) \cdot a + (b_1 + b_3) \cdot -a)$$
$$= f(b_0 + b_2) \cdot a + f(b_1 + b_3) \cdot -a = f^+(b_0 \cdot a + b_1 \cdot -a) + f^+(b_2 \cdot a + b_3 \cdot -a).$$

Moreover,

$$-(b_0 \cdot a + b_1 \cdot -a) = -(b_0 \cdot a) \cdot -(b_1 \cdot -a) = -b_0 \cdot -b_1 + -b_0 \cdot a + -b_1 \cdot -a$$
$$= (-b_0 \cdot -b_1) \cdot a + (-b_0 \cdot -b_1) \cdot -a + -b_0 \cdot a + -b_1 \cdot -a = -b_0 \cdot a + -b_1 \cdot -a.$$

Hence,

$$f^+(-(b_0 \cdot a + b_1 \cdot -a)) = f(-b_0) \cdot a + f(-b_1) \cdot -a$$
$$= -f(b_0) \cdot a + -f(b_1) \cdot -a = -(f(b_0) \cdot a + f(b_1) \cdot -a) = -f^+(b_0 \cdot a + b_1 \cdot -a).$$

Since $f^+(1_B) = 1_C$, Exercise 13.40 and duality imply that f^+ is a homomorphism of Boolean algebras. □

EXERCISE 25.18 (G). Let A, B, C be Boolean algebras with $B \subset A$, let $a \in A \setminus B$, and let $f : B \to C$ be an embedding of Boolean algebras. Show that if there is $c \in C$ such that
   (i) c fills the cut $\langle f[S_a], f[T_a] \rangle$,
   (ii) $\forall b \in B \, (b \cdot a \neq 0 \to f(b) \cdot c \neq 0)$, and
   (iii) $\forall b \in B \, (b \cdot -a \neq 0 \to f(b) \cdot -c \neq 0)$,
then f can be extended to an embedding $f^+ : B(a) \to C$.

THEOREM 25.11 (Sikorski's Extension Theorem). *Let A, B, C be Boolean algebras such that B is a subalgebra of A and C is complete. Moreover, let $f : B \to C$ be a homomorphism. Then f can be extended to a homomorphism $f^+ : A \to C$.*

PROOF. Let $\mathcal{F}$ be the family of all functions f^* such that $dom(f^*)$ is a subalgebra of A, $B \subseteq dom(f^*)$, and f^* is a homomorphism from $dom(f^*)$ into C. Then $\langle \mathcal{F}, \subseteq \rangle$ is a p.o. in which the union $\bigcup K$ of any chain K is an upper bound of K. Zorn's Lemma implies that $\mathcal{F}$ has a maximal element f^+. Lemma 25.10 implies that f^+ is as required. □

EXAMPLE 25.1. A homomorphic image of a complete Boolean algebra does not need to be complete. To see this, consider $B = \mathcal{P}(\omega)$, $A = \mathcal{P}(\omega)/Fin$, and let $\pi : B \to A$ be the canonical homomorphism. The image $\langle \pi[S], \pi[T] \rangle$ of each Hausdorff gap $\langle S, T \rangle$ is a gap in A, and thus Theorem 20.2 implies that A is not complete. Note that the above example also shows that a homomorphic image of a Boolean algebra B does not need to be isomorphic to a subalgebra of B.

EXAMPLE 25.2. The following exercise shows that a subalgebra of a Boolean algebra B does not need to be a homomorphic image of B.

EXERCISE 25.19 (G). Show that $Finco\,(\omega)$ is not a homomorphic image of $\mathcal{P}(\omega)$.

THEOREM 25.12. Let $\mathfrak{A} = \langle A, +_A, \cdot_A, -_A, 0_A, 1_A \rangle$ and $\mathfrak{B} = \langle B, +_B, \cdot_B, -_B, 0_B, 1_B \rangle$ be Boolean algebras, $\mathfrak{A}^\bullet = \langle A, \leq_A \rangle$ and $\mathfrak{B}^\bullet = \langle B, \leq_B \rangle$ the associated p.o.'s. Then $\mathfrak{A} \cong \mathfrak{B}$ if and only if $\mathfrak{A}^\bullet \cong \mathfrak{B}^\bullet$.

PROOF. Since the relation $\leq$ is definable in Boolean algebras, the "only if" direction is immediate.

For the "if" direction, assume that $f : \mathfrak{A}^\bullet \to \mathfrak{B}^\bullet$ is an order isomorphism. Since 0_A is the minimum element of $\mathfrak{A}^\bullet$ and 0_B is the minimum element of $\mathfrak{B}^\bullet$, we must have $f(0_A) = 0_B$. For a similar reason, $f(1_A) = 1_B$. For each $a, b \in A$, one can define $a + b$ as the least upper bound of $\{a, b\}$ in $\mathfrak{A}^\bullet$, and $a \cdot b$ as the greatest lower bound of $\{a, b\}$ in $\mathfrak{A}^\bullet$. Similarly, $-a$ is the maximum element $c \in A$ such that the greatest lower bound of $\{a, c\}$ is 0_A. Thus all constants and operations in $\mathfrak{A}$ are uniquely determined by the structure $\mathfrak{A}^\bullet$, and it follows that f is also an isomorphism of Boolean algebras. □

THEOREM 25.13. Let A and B be complete Boolean algebras, let $X \subseteq A$ and $Y \subseteq B$ be dense in the associated p.o.'s $\langle A \backslash \{0_A\}, \leq_A \rangle$ and $\langle B \backslash \{0_B\}, \leq_B \rangle$, and let $h : X \to Y$ be an order isomorphism. Then there exists a unique extension $h^+ : A \to B$ of h to an isomorphism from the Boolean algebra A onto the Boolean algebra B.

PROOF. By Theorem 25.12 it suffices to show that there exists a unique extension of h to an order isomorphism h^+ from $\langle A, \leq_A \rangle$ onto $\langle B, \leq_B \rangle$.

For $a \in A$ we define:

$$h^+(a) = \sup \{h(x) : x \in X, x \leq a\}.$$

EXERCISE 25.20 (G). Show that h^+ is an order-preserving extension of h such that $h^+(0_A) = 0_B$.

Let $a, b \in A$, $a \neq b$. Without loss of generality, assume that $a \cdot -b \neq 0_A$. Then there exists $x \in X$ such that $x \leq a \cdot -b$. Thus $0_B \neq h^+(x) \leq h^+(a)$, but $h^+(x) \cdot h^+(b) = 0_B$. Hence, $h^+(a) \neq h^+(b)$ and we have shown that h^+ is one-to-one.

Finally, let $b \in B$. Then $b = \sup \{y \in Y : y \leq b\}$. Consider $a \in A$ defined as

$$a = \sup \{h^{-1}(y) : y \in Y, y \leq b\}.$$

Then $h^+(a) = b$, and we have shown that $h^+[A] = B$. Thus, h^+ is an isomorphism from A onto B.

We leave the proof of uniqueness to the reader. □

EXAMPLE 25.3. Let $\langle L, \leq \rangle$ be a linear order with a minimum element 0_L. We extend L to $L \cup \{\infty\}$, where ∞ is an object that is not an element of L, and we extend $\leq$ to $L \cup \{\infty\}$ by requiring that $a < \infty$ for all $a \in L$. For $a, b \in L \cup \{\infty\}$ let

$$[a, b) = \{c \in L : a \leq c < b\}.$$

We call $[a, b)$ a *half-open interval*. In particular, $[0_L, \infty) = L$.

Let $s = \langle a_0, \ldots, a_{2n-1} \rangle$ be an increasing finite sequence of elements of $L \cup \{\infty\}$. We define:

$$\pi(s) = \bigcup_{i<n} [a_{2i}, a_{2i+1}).$$

In particular, if s is the empty sequence $\langle \rangle$, then $\pi(s) = \emptyset$. Let $Intalg\,(L)$ be the set of all $\pi(s)$ as above. Then $Intalg\,(L)$ is closed under unions, intersections, and complements. Thus $Intalg\,(L)$ is a Boolean subalgebra of $\mathcal{P}(L)$. We call $Intalg\,(L)$ the *interval algebra* of L.

EXERCISE 25.21 (G). (a) Show that $Intalg\,(L)$ is closed under unions, intersections, and complements.

(b) Let $\mathbb{Q}^+$ be the set of nonnegative rational numbers with their natural order. Show that $Intalg\,(\mathbb{Q}^+)$ is a countable, atomless Boolean algebra.

(c) Show that $Intalg\,(\omega) \cong Finco\,(\omega)$.

(d) Show that if $\langle L, \leq \rangle$ is a linear order with a minimum element 0, and $L_1 \subseteq L$ is such that $0 \in L_1$, then $Intalg\,(L_1)$ is a subalgebra of $Intalg\,(L)$.

(e) Conclude that every countable interval algebra can be embedded in the Boolean algebra $Intalg\,(\mathbb{Q}^+)$.

Now we are going to explore which Boolean algebras are isomorphic to interval algebras. Let B be a Boolean algebra. A subset A of B is called a *ramification system* if the following conditions are satisfied:

(i) $1 \in A$, $0 \notin A$;
(ii) $\forall a, b \in A (a \cdot b = 0 \vee a \leq b \vee b \leq a)$.

A is a *generating* ramification system (abbreviated gR), if it satisfies (i), (ii), and

(iii) $\langle A \rangle = B$.

Not every Boolean algebra contains a generating ramification system.

EXAMPLE 25.4. We show that $\mathcal{P}(\omega)$ does not contain a gR. Assume towards a contradiction that $A \subseteq \mathcal{P}(\omega)$ satisfies (i)–(iii). Then $|A| > \omega$. Moreover, if $a, b \in A$ have nonempty intersection, then a and b are comparable. Therefore, every antichain in $\langle A, \subseteq \rangle$ consists of pairwise disjoint sets and thus is countable.

EXERCISE 25.22 (G). Use the Erdős-Duschnik-Miller Theorem and Ramsey's Theorem to show that $\langle A, \subseteq \rangle$ contains an infinite chain L of order type ω or ω^*.

Let L be as in Exercise 25.22. If L has order type ω, let $L = \{a_i : i < \omega\}$ with $a_i \subseteq a_{i+1}$ for all $i < \omega$. Let $x = \bigcup_{i<\omega}(a_{2i+1} \setminus a_{2i})$. Then $x \notin \langle A \rangle$. The case where L has order type ω^* is analogous.

EXERCISE 25.23 (G). Show that every denumerable Boolean algebra B contains a gR. *Hint:* Let $(b_i)_{i<\omega}$ be an enumeration of the elements of B and define recursively for each $s \in {}^{<\omega}2$ an $a_s \in B$ such that $a_\emptyset = 1$, $a_{s \frown 0} = a_s \cdot b_n$, and $a_{s \frown 1} = a_s \cdot -b_n$ for all $s \in {}^{<\omega}2$.

Now we are going to look at the kind of ramification systems that generate interval algebras. Let B be a Boolean algebra and let $A \subseteq B$. Then A is a *linear ramification system* if A is a chain such that $0 \notin A$ and $1 \in A$. We say that A is a *linear generating ramification system (lgR)*, if A is a linear ramification sytem with $\langle A \rangle = B$.

EXAMPLE 25.5. Let $\langle L, \leq \rangle$ be a l.o. with smallest element 0. Then the family $\{L\} \cup \{[0, a) : a \in L \setminus \{0\}\}$ is an lgR for $Intalg(L)$.

The result of Exercise 25.23 can be strengthened as follows:

LEMMA 25.14. *Each countable Boolean algebra B has an lgR.*

PROOF. Let $(b_i)_{i<\omega}$ be an enumeration of B. We define recursively a sequence $(C_i)_{i<\omega}$ of chains in B such that for all $i < \omega$:

(i) $0 \notin C_0$, $1 \in C_0$;
(ii) $\langle C_i \rangle = \langle \{b_k : k < i\} \rangle$.

We let $C_0 = \{1\}$. Suppose C_n has been defined, and $C_n = \{c_0^n, \ldots, c_{j(n)-1}^n\}$, where $c_k^n < c_{k+1}^n$ for $k < j(n) - 2$.

We let $c_{-1}^n = 0$, and define for $k < j(n)$:

$$d_k^{n+1} = c_{k-1}^n + c_k^n \cdot b_n.$$

Let

$$C_{n+1} = C_n \cup (\{d_k^{n+1} : k < j(n)\} \setminus \{0\}).$$

It follows from the construction that C_{n+1} is a chain. Note that

$$b_n = \sum_{k<j(n)} d_k^{n+1} \cdot -c_{k-1}^n,$$

hence $b_n \in \langle C_{n+1} \rangle$, and thus $\langle C_{n+1} \rangle = \langle \{b_k : k \leq n\} \rangle$. It follows that $\bigcup_{i<\omega} C_i$ is an lgR. □

THEOREM 25.15. *Let B be a Boolean algebra, A an lgR for B. For each $b \in B$ there exist uniquely determined $n < \omega$ and $a_0, \ldots, a_{2n+1} \in A \cup \{0\}$ such that $a_0 < \cdots < a_{2n+1}$ and $b = \sum_{i \leq n} -a_{2i} \cdot a_{2i+1}$.*

EXERCISE 25.24 (G). Prove Theorem 25.15.

COROLLARY 25.16. *Let B be a Boolean algebra.*
(a) *If A is an lgR for B, then $B \cong Intalg(\langle A, \geq_B \rangle)$.*
(b) *B is isomorphic to an interval algebra if and only if B has an lgR.*
(c) *Every countable Boolean algebra is isomorphic to an interval algebra.*
(d) *Every countable Boolean algebra can be embedded in the countable atomless Boolean algebra $Intalg(\mathbb{Q}^+)$.*

EXERCISE 25.25 (R). (a) Show that

$$Intalg(\omega) \cong Intalg(\omega + \omega^*) \text{ and } Intalg(\omega_1 + \omega) \cong Intalg(\omega_1).$$

(b) Is $Intalg(\omega_1)$ isomorphic to $Intalg(\omega_1 + \omega^*)$ or to $Intalg(\omega_1 + \omega_1^*)$?

R. L. Vaught developed a procedure that allows us to test pairs of countable Boolean algebras for isomorphism. Now we are going to describe this procedure.

DEFINITION 25.17. A binary relation R on a set of Boolean algebras is a *Vaught relation* if it satisfies the following conditions:

(V1) R is symmetric, i.e., if $A R B$, then $B R A$;
(V2) If $A R B$ and A is trivial, then B is also trivial;
(V3) (The back-and-forth–property) If $A R B$ and $a \in A$, then there exists $b \in B$ such that $(A|a) R (B|b)$ and $(A| - a) R (B| - b)$.

For example, being isomorphic is a Vaught relation.

THEOREM 25.18 (Vaught). *Let A and B be countable Boolean algebras. If there exists a Vaught relation R such that $A R B$, then A and B are isomorphic.*

PROOF. We need a technical concept. Let $\bar{b} = (b_i)_{i<n}$ be a sequence of elements of a Boolean algebra. We say that $(b_i)_{i<n}$ is a *partition of unity* if the b_i's are pairwise disjoint and $\sum_{i<n} b_i = 1$.

Let $n \leq k \leq \omega$, and let $\bar{a} = (a_i)_{i<k}$ be a sequence of elements of a Boolean algebra. For $e \in {}^n 2$ we define:

$$p_e(\bar{a}) = \prod_{i<n} c_i,$$

where

(25.3) $$c_i = \begin{cases} a_i, & \text{if } e(i) = 0; \\ -a_i, & \text{otherwise.} \end{cases}$$

Then $\{p_e(\bar{a}) : e \in {}^n 2\}$ is a partition of unity.

Now suppose R is a Vaught relation such that $A R B$, and let $(a_{2n})_{n<\omega}$ and $(b_{2n+1})_{n<\omega}$ be enumerations, possibly with repetitions, of A and B respectively. We construct recursively $a_1, a_3, a_5 \ldots \in A$ and $b_0, b_2, b_4, \ldots \in B$ such that for each $n < \omega$:

(R)$_n$ $(A|p_e^A(a_0, a_1, \ldots, a_{n-1}) R (B|p_e^B(b_0, b_1, \ldots, b_{n-1}))$ for each $e \in {}^n 2$, and
(I)$_n$ There exists an isomorphism $f_n : \langle\{a_i : i < n\}\rangle \to \langle\{b_i : i < n\}\rangle$ with $f(a_i) = b_i$ for all $i < n$.

Note that the isomorphism f_n in (I)$_n$ is uniquely determined. Moreover, $A = \bigcup_{n<\omega} \langle\{a_i : i < n\}\rangle$ and $B = \bigcup_{n<\omega} \langle\{b_i : i < n\}\rangle$. It follows that $\bigcup_{n<\omega} f_n$ will be an isomorphism from A onto B.

Now let us carry out the construction. For $n = 0$ we have ${}^0 2 = \{\emptyset\}$. Since $p_\emptyset^A(\emptyset) = 1_A$ and $p_\emptyset^B(\emptyset) = 1_B$, (R)$_0$ and (I)$_0$ hold.

Suppose n is even and $b_0, a_1, b_2, \cdots, a_{n-1}$ have been constructed in such a way that (R)$_n$ and (I)$_n$ hold. For each $e \in {}^n 2$ we choose $r_e \in B|(p_e^B(b_0, b_1, \ldots, b_{n-1}))$ such that $(A|(p_e(a_0, a_1, \ldots, a_{n-1}) \cdot a_n) R (B|r_e)$ and

$$(A|(p_e^A(a_0, a_1, \ldots, a_{n-1}) \cdot -a_n) R (B|(p_e(b_0, b_1, \ldots, b_{n-1}) \cdot -r_e)).$$

By (V2), such r_e exists.

We define $b_n = \sum\{r_e : e \in {}^n 2\}$. For odd n, we reverse the roles of A and B in the above construction.

EXERCISE 25.26 (G). Convince yourself that Properties (R)$_n$ and (I)$_n$ hold after the n-th step of this construction. □

Let B be a Boolean algebra. We call B *superatomic* if each of its homomorphic images is atomic.

LEMMA 25.19. *A Boolean algebra B is superatomic if and only if each homomorphic image of B other than the trivial Boolean algebra has an atom.*

PROOF. The implication from the left to the right is clear.

To prove the converse, assume that B is not superatomic. Then there exist a homomorphic image A of B and $a \in A \setminus \{0\}$ such that $A|a$ is atomless. But $A|a$ is a homomorphic image of A and hence of B. Since $a \neq 0$, the Boolean algebra $A|a$ is nontrivial and contains no atom. □

THEOREM 25.20. *A Boolean Algebra B is superatomic if and only if each subalgebra of B is atomic.*

PROOF. Assume that B is a superatomic Boolean algebra, and let B' be a subalgebra of B. Without loss of generality we assume that B is nontrivial, i.e., $|B| > 1$. We show that B' is atomic.

Let I be an ideal in B, maximal in the family of all ideals J in B such that $J \cap B' = \{0\}$. Let $\pi : B \to B/I$ be the canonical homomorphism induced by I. Then $\pi \lceil B'$ is an embedding of B' into B/I. Let $a \in B' \setminus \{0\}$. Then $\pi(a) \neq 0$, and since B/I is atomic, there exists an atom $b' \in B/I$ such that $b' \leq \pi(a)$. Then $\pi^{-1}\{0\} \cup \pi^{-1}\{b'\}$ is an ideal in B that contains the ideal I as a proper subset. It follows from the choice of I that $\pi^{-1}\{b'\} \cap B' \neq \emptyset$. Hence there exists a $b \in B'$ such that $\pi(b) = b'$; hence B' is atomic.

To prove the converse, let B be a Boolean algebra such that all of its subalgebras are atomic. Suppose towards a contradiction that B is not superatomic. Let $\pi : B \to A$ be a homomorphism onto an atomless Boolean algebra A. Consider a countable atomless subalgebra A' of A. Let $\{a_i : i \in \omega\}$ be a gR for A' such that $a_0 = 1_{A'}$. Recursively we choose, for each $i \in \omega$, a $b_i \in B$ such that:

(i) $b_0 = 1_B$ and $\pi(b_i) = a_i$ for all i;
(ii) For all $i, j \in \omega$, if $a_i \leq_{A'} a_j$, then $b_i \leq_B b_j$;
(iii) For all $i \in \omega$ if $a_i \cdot_{A'} a_j = 0_{A'}$, then $b_i \cdot_B b_j = 0_B$.

Then the subalgebra of B generated by $\{b_i : i \in \omega\}$ is isomorphic to A' and thus is atomless. □

EXERCISE 25.27 (PG). (a) Show that $\mathit{Intalg}\,(\omega^{\cdot 2})$ is superatomic.

(b) Show that the class of superatomic Boolean algebras is closed under finite products, subalgebras, and homomorphic images.

Now let us investigate the countable superatomic algebras. An important tool in this investigation will be the *Cantor-Bendixson derivative*.

Let A be a Boolean algebra. The ideal generated by the atoms of A will be denoted by $I(A)$. Note that $I(A) = A$ if and only if A is finite. For each ordinal α we define recursively an ideal $I_\alpha(A)$, a Boolean algebra A_α, and a homomorphism $\pi_\alpha : A \to A_\alpha$ as follows: Let

$$I_0(A) = \{0\}.$$

If $I_\alpha(A)$ has already been defined, then we define $A_\alpha = A/I_\alpha(A)$ and we let $\pi_\alpha : A \to A_\alpha$ be the corresponding canonical homomorphism. Now suppose $\alpha > 0$, and $I_\beta(A)$ has already been defined for all $\beta < \alpha$. If $\alpha = \beta + 1$ for some β, then we let

$$I_\alpha(A) = \pi_\beta^{-1} I(A_\beta).$$

If α is a limit ordinal, then we let
$$I_\alpha(A) = \bigcup_{\beta < \alpha} I_\beta(A).$$

For $\alpha \leq \beta$ we have $I_\alpha(A) \subseteq I_\beta(A)$. Thus there exists a smallest α such that $I_\alpha(A) = I_\beta(A)$ for all $\beta > \alpha$.

Now suppose that A is superatomic. In this case, each A_α is atomic, and hence there exists α such that $I_\alpha(A) = A$. The smallest such α must be a equal to $\beta + 1$, where β is the smallest ordinal such that A_β is finite. Let n be the number of atoms in A_β. We refer to the ordered pair $\langle \beta, n \rangle$ as the *Cantor-Bendixson characteristics* of A. In particular, we call β the *Cantor-Bendixson height* of A (or simply the *height* of A) and write $h(A)$ for β. Note that $n(A)$ is the number of atoms of $A_{h(A)}$, and that the equality $n(A) = 0$ can only hold if A is the trivial Boolean algebra.

EXAMPLE 25.6. The Cantor-Bendixson characteristics of the trivial Boolean algebra are $\langle 0, 0 \rangle$. The characteristics of the interval algebra $Intalg(\omega)$ are $\langle 1, 1 \rangle$. More generally, if $\langle \alpha, n \rangle$ is an arbitrary pair with $\alpha \in \mathbf{ON}$ and $n \in \omega \setminus \{0\}$, then $\langle \alpha, n \rangle$ are the Cantor-Bendixson characteristics of the superatomic Boolean algebra $Intalg(\omega^{\cdot \alpha} \cdot n)$.

LEMMA 25.21. *Let A and B be superatomic Boolean algebras.*
  (i)   *If $A \subseteq B$, then $h(A) \leq h(B)$;*
  (ii) *If $h(A) = h(B)$, then $n(A \times B) = n(A) + n(B)$;*
  (iii) *If $h(A) < h(B)$, then $n(A \times B) = n(B)$;*
  (iv) *If $h(A) = \alpha$, $\beta < \alpha$, $n < \omega$, then there exists $a \in A$ with $h(A|a) = \beta$ and $n(A|a) = n$.*

EXERCISE 25.28 (PG). Prove Lemma 25.21.

THEOREM 25.22. *Let A and B be countable superatomic Boolean algebras with the same Cantor-Bendixson characteristics $\langle \alpha, n \rangle$. Then $A \cong B$.*

PROOF. We prove the theorem by induction over α.

If $\alpha = 0$, then A and B are finite Boolean algebras with the same number of atoms, and hence isomorphic.

Let $\alpha > 0$. First note that we can restrict ourselves to the case where $n = 1$: Suppose $a_0, \ldots, a_{n-1} \in A$ and $b_0, \ldots, b_{n-1} \in B$ are such that $\{a_0, \ldots, a_{n-1}\}$ is a partition of 1_A, $\{b_0, ..., b_{n-1}\}$ is a partition of 1_B, and a_i/I_α, b_i/I_α are atoms of A_α, respectively B_α. Then each of the algebras $A|a_i$ and $B|b_i$ has Cantor-Bendixson characteristics $\langle \alpha, 1 \rangle$, and the isomorphisms between $A|a_i$ and $B|b_i$ can be combined into an isomorphism from A onto B.

Now let $\mathcal{A} = \{A|a : a \in A\}$, $\mathcal{B} = \{B|b : b \in B\}$, and define a relation R by

$C R D$ iff $C, D \in \mathcal{A} \cup \mathcal{B}$, and C, D have the same
Cantor-Bendixson characteristics.

We show that R is a Vaught relation: Clearly, (V1) and (V2) are satisfied. Now suppose $C R D$. If the height of C is smaller than α, then $C \cong D$ by the inductive assumption, and (V3) follows. Suppose $h(C) = \alpha$. Then C has Cantor-Bendixson characteristics $\langle \alpha, 1 \rangle$. Let $c \in C$. It follows from (i) and (ii) of Lemma 25.21 that $h(C|c) < \alpha$ or $h(C|-c) < \alpha$. Without loss of generality, assume $h(C|c) < \alpha$. Then Lemma 25.21(iv) implies that there exists $d \in D$ with $h(C|c) = h(D|d)$ and

$n(C|c) = n(D|d)$. By (ii) and (iii) of Lemma 25.21, $C|(-c)$ and $D|(-d)$ have the same Cantor-Bendixson characteristics. □

Let B be a Boolean algebra, $a, b \in B$. We say that a and b are *disjoint* if $a \cdot b = 0$. Similarly, we say that $A \subseteq B^+$ is a *disjoint subset* of B if $c \cdot d = 0$ for all $c, d \in A$ with $c \neq d$. The *cellularity* of B is defined as:

$$c(B) = \sup\{|A| : A \text{ is a disjoint subset of } B\}.$$

Thus, if $|B|$ satisfies the κ-c.c., then $c(B) \leq \kappa$.

If C is a subalgebra of B, then $c(C) \leq c(B)$. We let $c(a)$ stand for $c(B|a)$.

EXAMPLE 25.7. Clearly, $c(\mathcal{P}(\omega)) = \aleph_0$. Let $A = \mathcal{P}(\omega)/Fin$, where Fin is the ideal of finite subsets of ω, and assume that X is an almost disjoint family of infinite subsets of ω such that $|X| = 2^{\aleph_0}$. Since the image of X under the canonical homomorphism from B onto A is a disjoint subset of A of the same cardinality as X, we see that $c(A) = 2^\omega$. Similarly, for each $a \in A^+$, $c(a) = 2^\omega$.

We say that *the cellularity of B is attained*, if there is a disjoint family $A \subseteq B$ such that $|A| = c(B)$.

If $c(B)$ is a successor cardinal, then the cellularity of B is of course attained. Theorem 25.23 below asserts that the same is true if $c(B)$ is singular. If $c(B)$ is a regular limit cardinal, then $c(B)$ may or may not be attained.

EXAMPLE 25.8. Let κ be a weakly inaccessible cardinal, and let $\mathbb{L}$ be the set of all finite functions with domain contained in $\kappa \times \omega$ and such that if $p \in \mathbb{L}$ and $\langle \alpha, n \rangle \in dom(p)$, then $p(\alpha, n) \in \alpha$. The p.o. $\langle \mathbb{L}, \supseteq \rangle$ is called *Lévy's partial order* or the *Lévy collapse*.

EXERCISE 25.29 (PG). (a) Show that Lévy's p.o. is separative, and thus dense in B^+ for some (uniquely determined) complete Boolean algebra B.

(b) Let B be as in point (a). Convince yourself that for each $\alpha \in \kappa \setminus \{0\}$ the set $\{\langle\langle \alpha, 0 \rangle, \beta \rangle : \beta < \alpha\}$ is a disjoint subset of B. Conclude that the cellularity of B is equal to κ.

(c) Show that there is no disjoint subset $A \subseteq B^+$ of size κ. Conclude that the cellularity of B is not attained. *Hint:* If a counterexample A exists, then there exists one that is a subset of $\mathbb{L}$. Apply the Δ-System Lemma to the domains of elements of such an A to derive a contradiction.

THEOREM 25.23 (Erdős-Tarski). *Let B be a Boolean algebra such that $c(B)$ is singular. Then $c(B)$ is attained.*

PROOF. Suppose $c(B) = \lambda$, $cf\,\lambda = \kappa < \lambda$, and let $(\lambda_\alpha)_{\alpha < \kappa}$ be a strictly increasing sequence of cardinals such that $\sup\{\lambda_\alpha : \alpha < \kappa\} = \lambda$. We distinguish three cases:

Case 1: There exists an $a \in B$ such that $c(b) = \lambda$ for every $b \in B$ with $0 < b \leq a$.

Then one can construct a disjoint family A with $|A| = \lambda$ as follows: Choose a disjoint family $\{b_\alpha : \alpha < \kappa\}$ in $B|a$ and for $\alpha < \kappa$ a disjoint family $A_\alpha = \{a_\beta^\alpha : \beta < \lambda_\alpha\}$ in $B|b_\alpha$. Then $A = \bigcup_{\alpha<\kappa} A_\alpha$ is a disjoint family of cardinality λ.

Suppose Case 1 does not apply. We define:

$$S = \{a \in B^+ : c(a) < \lambda\}.$$

EXERCISE 25.30 (G). Show that there exists a maximal disjoint subset $C \subseteq B^+$ such that $C \subseteq S$.

Fix a set C as in Exercise 25.30.

Case 2: $\sup\{c(a) : a \in C\} = \lambda$.

We construct recursively a sequence $(a_\alpha)_{\alpha<\kappa}$ of distinct elements of C such that $c(a_\alpha) > \lambda_\alpha$ for each $\alpha < \kappa$. This is possible since $c(a) < \lambda$ for each $a \in C$.

For each $\alpha < \kappa$ we choose a disjoint family $A_\alpha \subseteq B|a_\alpha$ such that $|A_\alpha| \geq \lambda_\alpha$. Then $A = \bigcup_{\alpha<\kappa} A_\alpha$ is as required.

Case 3: $\sup\{c(a) : a \in C\} = \mu < \lambda$.

We prove that in this case $|C| = \lambda$. Assume towards a contradiction that $|C| = \mu' < \lambda$. Let $C = \{c_\alpha : \alpha < \mu'\}$, and denote $\max\{\mu, \mu'\}$ by ν. Let D be a disjoint family in B with $|D| > \nu$. For $\alpha < \mu'$ consider

$$D_\alpha = \{c_\alpha \cdot a : a \in D, c_\alpha \cdot a \neq 0\},$$

and let $E = \bigcup\{D_\alpha : \alpha < \mu'\}$. On the one hand, E is a disjoint family with $|E| \geq |D| > \nu$. But on the other hand, $|E| = \sum_{\alpha<\mu'} |D_\alpha| \leq \mu' \cdot \mu < \nu$, and we get a contradiction. □

Let B be a Boolean algebra, $A, X \subseteq B^+$. We say that X is a *disjoint refinement* of A if X is a disjoint family and there exists a bijection $f : A \to X$ such that $f(a) \leq a$ for all $a \in A$.

EXERCISE 25.31 (G). (a) Let $A \subseteq (\mathcal{P}(\omega)/Fin)^+$ be such that $|A| \leq \aleph_0$. Show that A has a disjoint refinement.

(b) Does every countable $A \subseteq \mathcal{P}(\omega)\setminus\{\emptyset\}$ have a disjoint refinement?

The next theorem generalizes the result of Exercise 25.31(a).

THEOREM 25.24 (Balcar and Vojtáš). *Let B be a Boolean algebra, κ an infinite cardinal with $\kappa < c(a)$ for each $a \in B^+$. Then every $A \in [B^+]^{\leq\kappa}$ has a disjoint refinement.*

PROOF. Without loss of generality we may assume that A is infinite. Let $(a_\alpha)_{\alpha<\lambda}$ be an enumeration of A. For $Y \subseteq B$ and $a \in B$ define $Y(a) = \{y \in Y : y \cdot a \neq 0\}$.

CLAIM 25.25. *There exists a disjoint family $Y \subseteq B$ such that $|Y(a_\alpha)| = \kappa^+$ for each $\alpha < \lambda$.*

If Y is as in Claim 25.25, then one can recursively construct a sequence $(y_\alpha)_{\alpha<\lambda}$ of pairwise distinct elements of Y so that $y_\alpha \cdot a_\alpha \neq 0$. For all $\alpha < \lambda$ let $x_\alpha = y_\alpha \cdot a_\alpha$ and $X = \{x_\alpha : \alpha < \lambda\}$. The function $\pi : A \to X$ defined by $\pi(a_\alpha) = x_\alpha$ witnesses that X is a refinement of A.

PROOF OF CLAIM 25.25. We construct an increasing sequence $(Y_\alpha)_{\alpha<\lambda}$ of disjoint families such that

(i) $\forall \alpha < \lambda \, |Y_{\alpha+1}(a_\alpha)| = \kappa^+$ and
(ii) $\forall \alpha, \beta < \lambda \, (Y_\alpha(a_\beta) = \{\emptyset\} \lor |Y_\alpha(a_\beta)| = \kappa^+)$.

Note that if (i) and (ii) hold, then $Y = \bigcup_{\alpha<\lambda} Y_\alpha$ is as required.

We let $Y_0 = \emptyset$. If α is a limit ordinal $< \lambda$, then we let $Y_\alpha = \bigcup_{\beta<\alpha} Y_\beta$. If $\alpha = \beta + 1$ for some β, then we inspect $Y_\beta(a_\beta)$. If $|Y_\beta(a_\beta)| = \kappa^+$, then we let

$Y_\alpha = Y_\beta$. If $Y_\beta(a_\alpha) = \{\emptyset\}$, then we pick a disjoint family Z of size κ^+ such that $\forall z \in Z\, (0 < z \leq a_\beta)$ (such a family exists since $c(a_\beta) > \kappa$), and we define

$$Y_\alpha = Y_\beta \cup (Z \setminus \bigcup \{Z(a_\gamma) : \gamma < \lambda \wedge |Z(a_\gamma)| \leq \kappa\}).$$

Note that since $\lambda \leq \kappa$, the implication $|Z(a_\gamma)| = \kappa^+ \to |Y_\alpha(a_\gamma)| = \kappa^+$ holds. In particular, $|Y_\alpha(a_\beta)| = \kappa^+$. □□

Let B be a Boolean algebra, $X, Y \subseteq B$, and $z \in B$. We write $X \perp Y$ if $x \perp y$ (i.e., $x \cdot y = 0$) for all $x \in X$ and $y \in Y$. Instead of $\{z\} \perp Y$ we simply write $z \perp Y$. Similarly, we write $X \leq Y$ if $x \leq y$ for all $x \in X$ and $y \in Y$. An element b of B *separates* X and Y if $X \leq b$ and $Y \perp b$, or vice versa. Note that if some b separates the sets X and Y, then $X \perp Y$. We say that B has the *countable separation property* if for all $X, Y \in [B]^{\leq \aleph_0}$ such that $X \perp Y$ there exists $b \in B$ that separates X and Y.

EXERCISE 25.32 (G). Let B be a Boolean algebra. Show that the following are equivalent:

(i) B has the countable separation property;
(ii) For all disjoint sets $X, Y \in [B]^{\leq \aleph_0}$ such that $X \perp Y$ there exists $b \in B$ separating X and Y;
(iii) $\forall X, Y \in [B]^{\leq \aleph_0}\, (X \leq Y \to \exists b \in B\, (X \leq b \leq Y))$.

LEMMA 25.26. *The countable separation property is preserved under homomorphic images.*

PROOF. Let B be a Boolean algebra with the countable separation property, and let $\pi : B \to A$ be a homomorphism from B onto A. First note that if $\{a_i : i < \omega\}$ is a disjoint family in A and $c_i \in \pi^{-1}(a_i)$ for each $i < \omega$, then the family $\{b_i : i < \omega\}$, where $b_i = c_i \cdot -\sum_{k<i} c_k$, is disjoint in B and has the property that $\pi(b_i) = a_i$ for each $i < \omega$. Thus, if $X, Y \in [A]^{\leq \aleph_0}$ are disjoint families with $X \perp Y$, we can choose disjoint families $U, V \in [B]^{\leq \aleph_0}$ with $f[U] = X$, $f[V] = Y$, and $U \perp V$. Then some $a \in B$ separates U and V, and hence $f(a)$ separates X and Y. □

EXERCISE 25.33 (G). We say that a Boolean algebra B has the $\aleph_1$-*separation property* if for all $X, Y \in [B]^{\leq \aleph_1}$ such that $X \perp Y$ there exists $a \in B$ that separates X and Y. Show that the $\aleph_1$-separation property is not preserved under homomorphic images. *Hint:* Consider the canonical homomorphism $\pi : \mathcal{P}(\omega) \to \mathcal{P}(\omega)/Fin$ and use Theorem 20.2.

Let T be a first-order theory, $\mathfrak{A}$ a model of T. We say that $\mathfrak{A}$ is κ-*universal* for T if every model $\mathfrak{B}$ of T with universe B of cardinality $\leq \kappa$ can be embedded in $\mathfrak{A}$. For example, a slight modification of the proof of Theorem 4.16 shows that the natural order of the rationals $\langle \mathbb{Q}, \leq \rangle$ is $\aleph_0$-universal for the theory of linear orders, and Corollary 25.16(d) implies that $Intalg(\mathbb{Q}^+)$ is $\aleph_0$-universal for the theory of Boolean algebras.

Let κ be an infinite cardinal, let T be a first-order theory, and let $\mathfrak{A}$ be a model of T. If X is a subset of the universe of $\mathfrak{A}$, then let $(X)_\mathfrak{A}$ denote the submodel of $\mathfrak{A}$ generated by X, i.e., the smallest submodel of $\mathfrak{A}$ whose universe contains X. We say that $\mathfrak{A}$ has the κ-*extension property* if the following holds: Whenever $\mathfrak{B}$ is a model of T, X is a subset of the universe of $\mathfrak{B}$ with $|X| < \kappa$, h is an embedding of

$(X)_\mathfrak{B}$ into $\mathfrak{A}$, and b is an element of the universe of $\mathfrak{B}$, then h can be extended to an embedding of the structure $(X \cup \{b\})_\mathfrak{B}$ into $\mathfrak{A}$.

EXERCISE 25.34 (R). Suppose $\mathfrak{A} = \langle A, \ldots \rangle$ and $\mathfrak{B} = \langle B, \ldots \rangle$ are models of a first-order theory T, $|A| = |B| = \kappa$, and both $\mathfrak{A}$ and $\mathfrak{B}$ have the κ-extension property. Show that $\mathfrak{A} \cong \mathfrak{B}$.

Now we want to show that $\mathcal{P}(\omega)/Fin$ is $\aleph_1$-universal for the theory of Boolean algebras. We need some preliminaries.

Let B be a Boolean algebra. We say that B has the *strong countable separation property* if B is infinite and for all $X, Y \in [B]^{\leq \aleph_0}$ such that $\forall F \in [X]^{<\aleph_0} \forall G \in [Y]^{<\aleph_0} (\sum_{a \in F} a < \prod_{b \in G} b)$ there exists $c \in B$ such that $\forall a \in X \forall b \in Y (a < c < b)$.

EXAMPLE 25.9. The strong countable separation property implies the countable separation property, but not vice versa. To see the latter, assume that a is an atom in B. Let $X = \{0\}$, $Y = \{a\}$. Since there is no $c \in B$ with $0 < c < a$, we see that every Boolean algebra with the strong countable separation property must be atomless. So, in particular, the complete Boolean algebra $\mathcal{P}(\omega)$ has the countable separation property but does not have the strong countable separation property.

An element $a \in B$ is called a *countable limit* if there exists $X \in [B]^{\aleph_0}$ such that $\sum_{b \in X} b = a$ and $\forall F \in [X]^{<\omega} (\sum_{b \in F} b < a)$.

EXERCISE 25.35 (PG). Prove that if B is a Boolean algebra with the strong countable separation property, then

(i) B has no countable limits; and
(ii) Every infinite maximal disjoint family in B is uncountable.

LEMMA 25.27. *The Boolean algebra $\mathcal{P}(\omega)/Fin$ has the strong countable separation property.*

PROOF. Let $X, Y \in [\mathcal{P}(\omega)]^{\leq \aleph_0}$ be such that $\forall F \in [X]^{<\aleph_0} \forall G \in [Y]^{<\aleph_0} (\bigcup F \subset^* \bigcap G)$. We show that there exists $c \subseteq \omega$ such that $\forall a \in X \forall b \in Y (a \subset^* c \subset^* b)$.

We only consider the case where $|X| = |Y| = \omega$; the proof in the remaining cases is analogous. Let $X = \{a_i : i < \omega\}$, $Y = \{b_i : i < \omega\}$. We construct recursively a sequence $(n_i)_{i<\omega}$ of pairwise distinct natural numbers such that $n_i \in \bigcap_{k<i} b_k \setminus \bigcup_{k<i} a_k$. Let $d = \{n_i : i < \omega\}$ and $e = \{n_{2j} : j < \omega\}$. Then $|d \cap a_i| < \omega$ and $e \subset^* b_i$ for each $i < \omega$. We define:

$$c = e \cup \bigcup_{n<\omega} (\bigcup_{i<n} a_i \cap \bigcap_{i<n} b_i).$$

This c is as required. □

THEOREM 25.28. *Every Boolean algebra with the strong countable separation property is $\aleph_1$-universal.*

PROOF. In order to prove Theorem 25.28, we need the following:

LEMMA 25.29. *Let A, B be Boolean algebras such that A is countable and B has the strong countable separation property. If $\pi : A \to B$ is a homomorphic embedding and $A(a)$ is a simple extension of A, then π can be extended to a homomorphic embedding $\pi^+ : A(a) \to B$.*

PROOF. Let $A(a)$ be a simple extension of A. If $a \in A$ there is nothing to prove, so assume that $a \notin A$. Consider the cut $\langle S_a, T_a \rangle$. Since $a \notin A$, $\sum_{b \in F} b < \prod_{b \in G} b$ for all finite $F \subseteq S_a$ and $G \subseteq T_a$. Since π is a homomorphic embedding, it follows that $\langle \pi[S_a], \pi[T_a] \rangle$ is a cut in B, and $\sum_{b \in F} b < \prod_{b \in G} b$ for all finite $F \subseteq \pi[S_a]$ and $G \subseteq \pi[T_a]$.

Let $J = \{s + (-t) : s \in S_a, t \in T_a\}$. Then J is an ideal in A. For $b \in A \setminus J$ let $Y'(b) = \{b \cdot -j : j \in J\}$. Then $Y'(b)$ is closed under $\cdot$ and $0_A \notin Y'(b)$. Let $X = \{0_B\}$ and $Y(b) = \pi[Y'(b)]$. Then $X < Y(b)$ and there exists $c_b \in B$ such that $0 < c_b < Y(b)$. Let $C = \{c_b : b \in A \setminus J\}$. By the Balcar-Vojtáš Theorem and Exercise 25.35(ii), we may assume without loss of generality that the c_b's are pairwise disjoint. For each $b \in A \setminus J$ we choose disjoint elements d_b, e_b such that $0 < d_b, e_b < c_b$. The sets

$$U = \pi[S_a] \cup \{d_b : b \in A \setminus J\},$$
$$V = \pi[\{-t : t \in T_\alpha\}] \cup \{e_b : b \in A \setminus J\}$$

are countable and $U \perp V$. Hence there exists $x \in B$ that separates U and V. Let $\pi^+ : A \to B$ be the extension of π given by the formula $\pi^+(a) = x$. By Exercise 25.18, this π^+ is a homomorphic embedding. □

EXERCISE 25.36 (PG). Derive Theorem 25.28 from Lemma 25.29. □

COROLLARY 25.30 (CH). *Every Boolean algebra B of cardinality $\aleph_1$ that has the strong countable separation property is isomorphic to $\mathcal{P}(\omega)/Fin$.*

PROOF. Since the Boolean algebra $\mathcal{P}(\omega)/Fin$ has the strong countable separation property, Lemma 25.29 applies. In other words, $\mathcal{P}(\omega)/Fin$ is a model of the theory of Boolean algebras that has the $\aleph_1$-extension property. By CH, Exercise 25.34 applies, and Corollary 25.30 follows. □

Let B be a Boolean algebra, $X \subseteq B$. The set X is called *irredundant* if $a \notin \langle X \setminus \{a\} \rangle$ for all $a \in X$.

THEOREM 25.31 (McKenzie). *Every maximal irredundant subset of a Boolean algebra generates a dense subalgebra.*

PROOF. Let B be a Boolean algebra, $X \subseteq B$ an irredundant subset. Let $a \in B^+$ be such that $B|a \cap \langle X \rangle = \{0\}$. We show that $X \cup \{a\}$ is irredundant: Suppose $X \cup \{a\}$ is not irredundant. Then there is $b \in X$ with $b \in \langle (X \cup \{a\}) \setminus \{b\} \rangle$. Let $D = \langle X \rangle$, $Y = X \setminus \{b\}$. Since $b \in \langle Y \rangle(a)$, there are $y, z \in \langle Y \rangle$ such that $b = y \cdot a + z \cdot -a$. Then $b \Delta z \leq a$ (since $b \cdot -a = z \cdot -a$), and since $b, z \in D$, we also have $b \Delta z \in D$. Moreover, $z \in \langle Y \rangle$, but $b \notin \langle Y \rangle$, since X was assumed irredundant. Thus $b \neq z$; hence $b \Delta z \neq 0$ and $B|a \cap D \neq \{0\}$, which contradicts the choice of a. □

Let B be a Boolean algebra. The π-*weight* of B, denoted by πB, is the smallest cardinal κ such that there exists a dense subset $A \subseteq B^+$ of cardinality κ. A subset $X \subseteq B$ is called *incomparable* if its elements are pairwise incomparable. We define $Inc\, B = \sup \{|X| : X$ is an incomparable subset of $B\}$.

EXERCISE 25.37 (G). Show that the π-weight of a Boolean algebra is equal to the π-weight of its Stone space.

We conclude this chapter with a theorem that establishes a connection between πB and $Inc\, B$. First we need a lemma.

LEMMA 25.32. *Let B be a Boolean algebra with $\pi B = \omega$. Then B can be embedded in $\mathcal{P}(\omega)$.*

PROOF. It is not hard to see[2] that $Intalg\,(\mathbb{Q}^+)$ can be embedded in $\mathcal{P}(\omega)$. It follows from Corollary 25.16(d) that every countable Boolean algebra can be embedded in $\mathcal{P}(\omega)$.

Now let A be a countable dense subalgebra of B, and let $f : A \to \mathcal{P}(\omega)$ be an embedding. By Sikorski's Theorem there exists an extension to a homomorphism $f^+ : B \to \mathcal{P}(\omega)$. Since f is injective and A is dense in B, f^+ is also injective, and thus f^+ is a homomorphic embedding of B into $\mathcal{P}(\omega)$. □

THEOREM 25.33 (Baumgartner und Komjáth). *Assume B is a Boolean algebra with $Inc\,B \leq \aleph_0$. Then $\pi B \leq \aleph_0$, and hence B can be embedded in $\mathcal{P}(\omega)$.*

PROOF. Suppose towards a contradiction that B is a Boolean algebra with $Inc\,B \leq \aleph_0$ and $\pi B \geq \aleph_1$. Then there exists a sequence $(b_\alpha)_{\alpha < \omega_1}$ of elements of B^+ such that for each $\alpha < \omega_1$ there is no $d \in B_\alpha$ with $0 < d < b_\alpha$, where $B_\alpha = \langle \{b_\beta : \beta < \alpha\} \rangle$. Let

$$Z = \{\alpha < \omega_1 : \exists b \in B_\alpha \, (b \cdot b_\alpha \neq 0 \wedge -b \cdot b_\alpha \neq 0)\}.$$

Case 1: Z is stationary in ω_1.

Let $(c_\alpha)_{\alpha < \omega_1}$ be an enumeration of B_{ω_1}. For each $\alpha \in Z$ let $f(\alpha)$ be the smallest ordinal β such that $b_\alpha \cdot c_\beta \neq 0 \neq b_\alpha \cdot -c_\beta$. Let

$$C = \{\alpha < \omega_1 : B_\alpha = \{c_\beta : \beta < \alpha\}\}.$$

Then C is *club*, and thus $Z \cap C$ is stationary. By the Pressing Down Lemma (Theorem 21.12), f is constant on some stationary $S \subseteq Z \cap C$.

Let $c = c_{f(\alpha)}$ for $\alpha \in S$. For $\alpha \in S$ let $d_\alpha = c \cdot b_\alpha + -c \cdot -b_\alpha$. We show that $\{d_\alpha : \alpha \in S\}$ is an incomparable set, which contradicts our assumption about B. Let $\alpha, \beta \in S$ be such that $\alpha < \beta$. If $d_\alpha \leq d_\beta$, then $c \cdot b_\alpha \leq b_\beta$, which contradicts the choice of b_β. If $d_\beta \leq d_\alpha$, then $-c \cdot -b_\beta \leq -c \cdot -b_\alpha$ and thus $-c \cdot b_\alpha \leq -c \cdot b_\beta \leq b_\beta$. Again we get a contradiction with the choice of b_β.

Case 2: Z is not stationary.

Let $Y = \omega_1 \setminus Z$. Then $|Y| = \omega_1$, and for $\alpha, \beta \in Y$ with $\alpha < \beta$ either $b_\alpha \cdot b_\beta = 0$ or $b_\beta \leq b_\alpha$. Let

$$W = \{\alpha \in Y : |\{\beta \in Y : b_\beta \leq b_\alpha\}| = \aleph_1\}.$$

Suppose $|W| \leq \aleph_0$. Then we can recursively construct a sequence $(b_{\alpha_i})_{i < \omega_1}$ such that $\alpha_i \in Y \setminus W$ and $b_{\alpha_i} \not\leq b_{\alpha_k}$ for $i < k < \omega_1$. But then $\{b_{\alpha_i} : i < \omega_1\}$ is an uncountable incomparable set, contradicting the choice of B. Thus $|W| = \aleph_1$.

Suppose that $b_\beta \leq b_\alpha$ for all $\alpha, \beta \in W$ with $\alpha < \beta$. Then $\{b_\alpha : \alpha \in W\}$ is a chain. Let $(\alpha_i)_{i < \omega_1}$ be an enumeration of W in strictly increasing order. For $i < \omega_1$ let $d_i = b_{\alpha_i} \cdot -b_{\alpha_{i+1}}$. Then $\{d_i : i < \omega_1\}$ is an uncountable incomparable set, contradicting the choice of B.

Thus there exist $\alpha, \beta \in W$ with $\alpha \neq \beta$ such that $b_\alpha \cdot b_\beta = 0$. We pick functions

$$f : \omega_1 \to \{\gamma \in Y : b_\gamma \leq b_\alpha\},$$
$$g : \omega_1 \to \{\gamma \in Y : b_\gamma \leq b_\beta\}$$

such that $f(i), g(i) < f(k), g(k)$ for $i < k < \omega_1$. For $i < \omega_1$ let

$$d_i = b_{f(i)} + b_\beta \cdot -b_{g(i)}.$$

---

[2] Do we still have to remind you that here is an EXERCISE?

If $i < k$, then d_i and d_k are incomparable; hence $\{d_i : i < \omega_1\}$ is an uncountable incomparable set. This contradiction shows that B must contain a countable dense subalgebra. □

Mathographical Remark

The three-volume *Handbook of Boolean Algebras*, edited by J. D. Monk and R. Bonnet, Elsevier Science Publishers, 1989, gives a comprehensive account of the current state of knowledge about Boolean algebras.

CHAPTER 26

Appendix: Some General Topology

The main purpose of this appendix is to provide the reader with a convenient source for looking up the concepts and theorems from general topology that are used elsewhere in this book. We also include some proofs that we consider particularly good illustrations of certain set-theoretic techniques. For a serious study of general topology we recommend the encyclopaedic treatise *General Topology*, by R. Engelking, Heldermann Verlag, Berlin, 1989. In particular, all the proofs that are skipped in this appendix can be found in Engelking's book.

Let X be a set, $\tau \subseteq \mathcal{P}(X)$. We say that τ is a *topology on X*, if $\emptyset, X \in \tau$ and τ is closed under finite intersections and arbitrary unions (i.e., if $F \in [\tau]^{<\aleph_0}$ and $G \subseteq \tau$, then $\bigcap F \in \tau$ and $\bigcup G \in \tau$). A *topological space* is a pair $\langle X, \tau \rangle$, where τ is a topology on X. The elements of X are called the *points* and the elements of τ are called the *open sets* of this space. If τ is implied by the context, we simply speak of the topological space X.

A subset A of a topological space X is *closed* if its complement $X \backslash A$ is open. Let $B \subseteq X$. The largest open subset of X that is included in B is called the *interior of B* and is denoted by $int_\tau(B)$. The smallest closed superset of B is called the *closure of B* and is denoted by $cl_\tau(B)$. We simply write $int\, B$ and $cl\, B$ if τ is implied by the context. A point x is called an *accumulation point of B* if $x \in cl(B \backslash \{x\})$. The set of all accumulation points of B is denoted by B' and called the *derived set of B*. One can show that $cl\, B = B \cup B'$. If x is not an accumulation point of X, then x is said to be *isolated*. A subset of X is called *clopen* if it is simultaneously closed and open. Unions of countably many closed sets are called F_σ-sets; intersections of countably many open sets are called G_δ-sets.

The notion of a topological space provides an abstract framework for the study of continuity. Suppose $\langle X, \tau \rangle, \langle Y, \rho \rangle$ are two topological spaces and $f : X \to Y$. Then f is *continuous* if $f^{-1}V \in \tau$ for every $V \in \rho$. The function f is a *homeomorphism* if it is a bijection and both f and its inverse f^{-1} are continuous. The spaces $\langle X, \tau \rangle$ and $\langle Y, \rho \rangle$ are *homeomorphic* (i.e., topologically the same) if there exists a homeomorphism $f : X \to Y$.

Let $\langle X, \tau \rangle$ be a topological space, $x \in X$. We say that $N \subseteq X$ is a *neighborhood of x* if there exists $U \in \tau$ such that $x \in U \subseteq N$. As we mentioned in Example 13.3, the family $\mathcal{N}_x$ of all neighborhoods of x is a proper filter of subsets of X, called the *neighborhood filter of x* or the *neighborhood system at x*. If $\mathcal{B}_x$ is a filter base of $\mathcal{N}_x$, i.e., if $\mathcal{B}_x \subseteq \mathcal{N}_x$ and for each $N \in \mathcal{N}_x$ there exists $B \in \mathcal{B}_x$ with $B \subseteq N$, then $\mathcal{B}_x$ is called a *neighborhood base at x*. A neighborhood base at x that consists of open sets is also called a *local base at x*. A *base* for τ is a family $\mathcal{B} \subseteq \tau$ such that

$$\forall x \in X \, \forall N \in \mathcal{N}_x \, \exists U \in \mathcal{B} \, (x \in U \subseteq N).$$

Equivalently, $\mathcal{B}$ is a base for τ iff $\tau = \{\bigcup \mathcal{A} : \mathcal{A} \subseteq \mathcal{B}\}$. Similarly, if we are given a family of neighborhood bases $\{\mathcal{B}_x : x \in X\}$, then $U \in \tau$ iff $\forall x \in U \, \exists B \in \mathcal{B}_x \, (B \subseteq U)$. Thus topologies can be defined by specifying a base or a family of neighborhood bases.

Now let us look at some important examples.

EXAMPLE 26.1. Let X be an arbitrary nonempty set, $\tau = \{\emptyset, X\}$. This topology is called the *antidiscrete topology* on X.

EXAMPLE 26.2. Let X be an arbitrary nonempty set, $\tau = \mathcal{P}(X)$. This topology is called the *discrete topology* on X. In a discrete space, all points are isolated. Clearly, two discrete spaces X and Y are homeomorphic if and only if $|X| = |Y|$. If κ is a cardinal, then $D(\kappa)$ denotes the discrete space with κ points.

EXAMPLE 26.3. Let $\langle L, < \rangle$ be a strict l.o., and let $a, b \in L$. We define:

$$(a, b) = \{c \in L : a < c < b\};$$
$$(-\infty, a) = \{c \in L : c < a\};$$
$$(b, \infty) = \{c \in L : b < c\}.$$

The sets defined above are called *open intervals in L*. The family of open intervals forms a base of a topology $\tau(<)$ on L, called the *order topology on L*. A topological space X is called a *linearly ordered space* if it is homeomorphic to $\langle L, \tau(<) \rangle$ for some strict l.o. $\langle L, < \rangle$.

Here are some special cases of the above construction:

(a) If $<$ is the usual inequality of reals, then $\tau(<)$ is the standard topology on the real line.

(b) If α is an ordinal and no other topology on α is specified, then we consider the order topology $\tau(\in)$ as the default option. Note that in this topology, the family $\{(\gamma, \beta] : \gamma < \beta\}$ forms a local base for each $\beta < \alpha$. Thus, if $\alpha \leq \omega$, then $\langle \alpha, \tau(\in) \rangle$ is homeomorphic to the discrete space $D(\alpha)$, but if $\omega < \alpha$, then the space $\langle \alpha, \tau(\in) \rangle$ is not discrete.

EXAMPLE 26.4. Let κ be an infinite cardinal, and let $\mathcal{F}$ be a nonprincipal filter on κ. We define a topology $\tau^{\mathcal{F}}$ on $\kappa + 1$ by specifying neighborhood bases for all points $\alpha \in \kappa + 1$. If $\alpha < \kappa$, then we let $\mathcal{B}_x = \{\{\alpha\}\}$ be a neighborhood base for α. In this way each $\alpha < \kappa$ is declared an isolated point of the space $\langle \kappa + 1, \tau^{\mathcal{F}} \rangle$. A neighborhood base for κ is defined as $\mathcal{B}_x = \{\{\kappa\} \cup X : X \in \mathcal{F}\}$. Of course, the properties of the space $\langle \kappa + 1, \tau^{\mathcal{F}} \rangle$ depend on the choice of κ and $\mathcal{F}$. For example, if $\mathcal{F}$ is the filter of all cofinite subsets of ω, then $\langle \omega + 1, \tau^{\mathcal{F}} \rangle = \langle \omega + 1, \tau(\in) \rangle$.

EXAMPLE 26.5. Let $\beta\omega$ denote the set of all ultrafilters on ω. For each $a \subseteq \omega$, let $U_a = \{\mathcal{F} \in \beta\omega : a \in \mathcal{F}\}$. The family $\{U_a : a \subseteq \omega\}$ is a base for a topology τ on $\beta\omega$. The space $\langle \beta\omega, \tau \rangle$ is called the *Stone-Čech compactification of ω*. Note that for each $n \in \omega$ we have $U_{\{n\}} = \{\mathcal{F}_n\}$, where $\mathcal{F}_n = \{a \subseteq \omega : n \in a\}$. It follows that each principal ultrafilter is an isolated point of the space $\beta\omega$.

Let X be a nonempty set. A *metric on X* is a function $d : X \times X \to [0, \infty)$ such that for all $x, y, z \in X$:

(i) $d(x, y) = 0$ iff $x = y$;
(ii) $d(x, y) = d(y, x)$;
(iii) $d(x, z) \leq d(x, y) + d(y, z)$.

Given a metric d on X, a point $x \in X$, and a positive real ε, one defines the *open ball with center x and radius ε* by:

$$B(x, \varepsilon) = \{y \in X : d(x, y) < \varepsilon\}.$$

The *diameter* of a set $A \subseteq X$ is defined as $diam(A) = \sup\{d(x, y) : \{x, y\} \in [A]^2\}$. Note that (iii) implies that the diameter of an open ball $B(x, \varepsilon)$ is at most 2ε. For $x \in X$, let $\mathcal{B}_x = \{B(x, \varepsilon) : \varepsilon > 0\}$. The family $\{\mathcal{B}_x : x \in X\}$ is a family of neighborhood bases for a topology τ_d on X, called the *topology induced by the metric d*. A topological space $\langle X, \tau \rangle$ is *metrizable* if there exists a metric d on X such that $\tau = \tau_d$.

Let us look at some specific instances. The function $M : \mathbb{R}^2 \to [0, \infty)$ defined by $M(x, y) = |x - y|$ is a metric that induces the usual topology on $\mathbb{R}$. This shows that the space $\langle \mathbb{R}, \tau(<) \rangle$ of Example 26.3(a) is metrizable. Similarly, if κ is a cardinal and the function $dd : \kappa \times \kappa \to [0, \infty)$ is defined by:

$$dd(\alpha, \beta) = \begin{cases} 0, & \text{if } \alpha = \beta; \\ 1, & \text{if } \alpha \neq \beta, \end{cases}$$

then dd is a metric such that τ_{dd} is the discrete topology on κ. It follows that all spaces $D(\kappa)$ are metrizable.

Given a metric d on a set X, we say that a sequence $(x_n)_{n \in \omega}$ of points in X *converges* (in the sense of d) to $x \in X$ if $\lim_{n \to \infty} d(x_n, x) = 0$. We say that $(x_n)_{n \in \omega}$ is a *(d-)Cauchy-sequence* if $\lim_{n \to \infty} \sup_{m > n} d(x_n, x_m) = 0$. Every convergent sequence is a Cauchy sequence, but not necessarily vice versa. We say that a metric d on a set X is *complete* if every d-Cauchy sequence converges with respect to d to some $x \in X$. A topological space $\langle X, \tau \rangle$ is *completely metrizable* if τ is induced by a complete metric on X. For example, the metric dd is complete by default since there are no nontrivial dd-Cauchy sequences. The metric M is complete on $\mathbb{R}$, but neither the restriction of M to $(0, 1)$ nor its restriction to $\mathbb{Q}$ is complete. However, the space $\langle (0,1), \tau_{M \restriction (0,1)^2} \rangle$ is homeomorphic to $\langle \mathbb{R}, \tau_M \rangle$ and thus is completely metrizable. In contrast, one can show that the space $\langle \mathbb{Q}, \tau_{M \restriction \mathbb{Q}^2} \rangle$ is not completely metrizable.

There are several standard constructions of new spaces from old ones. In this text we need only two of them, namely subspaces and product spaces.

Let $\langle X, \tau \rangle$ be a topological space, $Y \subseteq X$. Define $\tau \restriction Y = \{U \cap Y : U \in \tau\}$. The topology $\tau \restriction Y$ is called the *subspace topology on Y induced by τ*, and $\langle Y, \tau \restriction Y \rangle$ is called a *subspace of X*.

EXAMPLE 26.6. (a) $\langle \mathbb{Q}, \tau_{M \restriction \mathbb{Q}^2} \rangle$ is a subspace of $\langle \mathbb{R}, \tau_M \rangle$.
(b) If $\alpha \in \beta \in \mathbf{ON}$, then $\langle \alpha, \tau(\in) \rangle$ is a subspace of $\langle \beta, \tau(\in) \rangle$.
(c) If $\kappa < \lambda$, $\mathcal{F}$ is a nonprincipal filter on κ, and $\mathcal{G}$ is a nonprincipal filter on λ, then $\langle \kappa + 1, \tau^{\mathcal{F}} \rangle$ is *not* a subspace of $\langle \lambda + 1, \tau^{\mathcal{G}} \rangle$.
(d) Let ω^* be the set of all nonprincipal ultrafilters on ω. Then $\langle \omega^*, \tau \restriction \omega^* \rangle$ is a subspace without isolated points of the space $\langle \beta\omega, \tau \rangle$ defined in Example 26.5.

Let $\{\langle X_i, \tau_i \rangle : i \in I\}$ be a family of topological spaces. The *product* of these spaces is the space $\langle \prod_{i \in I} X_i, \tau \rangle$, where the *product topology* τ is generated by the base of all sets of the form $\prod_{i \in I} U_i$ such that $U_i \in \tau_i$ for all $i \in I$ and $U_i \neq X_i$ for only finitely many $i \in I$.

EXAMPLE 26.7. We occasionally refer to the "standard" or "product" topologies on $^\omega 2$ and $^\omega\omega$. These topologies are defined as the product topologies on the spaces $\prod_{i\in\omega} D(2)$ and $\prod_{i\in\omega} D(\aleph_0)$ respectively. For $s \in {}^{<\omega}\omega$, we let $[s] = \{f \in {}^\omega\omega : s \subset f\}$. Note that the family $\{[s] : s \in {}^{<\omega}\omega\}$ forms a base for the product topology on $^\omega\omega$. It can be shown that $^\omega\omega$ is homeomorphic to the set of irrationals with the subspace topology inherited from $\langle \mathbb{R}, \tau_M \rangle$.

Clearly, $^\omega 2$ can be considered a subspace of $^\omega\omega$. We often identify $^\omega 2$ with $\mathcal{P}(\omega)$ as follows: For $a \subseteq \omega$, let $\chi_a : \omega \to 2$ be defined by:

$$\chi_a(n) = \begin{cases} 1, & \text{if } n \in a; \\ 0, & \text{if } n \notin a. \end{cases}$$

We call χ_a the *characteristic function of a*. The function $F : \mathcal{P}(\omega) \to {}^\omega 2$ that assigns χ_a to $a \subseteq \omega$ is a bijection. Thus, if τ denotes the product topology on $^\omega 2$, then the family $\rho = \{F^{-1}U : U \in \tau\}$ is a topology on $\mathcal{P}(\omega)$ such that the spaces $\langle \mathcal{P}(\omega), \rho \rangle$ and $\langle {}^\omega 2, \tau \rangle$ are homeomorphic. We refer to the topology ρ as the *standard topology on $\mathcal{P}(\omega)$*.

General topology is the study of properties of topological spaces. Among the most important of these properties are *separation properties*, such as Hausdorffness and regularity. A topological space $\langle X, \tau \rangle$ is a *Hausdorff space* or a T_2-*space* if

$$\forall \{x, y\} \in [X]^2 \exists U, V \in \tau \, (x \in U \wedge y \in V \wedge U \cap V = \emptyset).$$

We say that $\langle X, \tau \rangle$ is *regular* if it is Hausdorff[1] and

(26.1) $\quad \forall x \in X \, \forall A \subseteq X \, (x \notin cl\, A \to \exists U, V \in \tau \, (x \in U \wedge cl\, A \subseteq V \wedge U \cap V = \emptyset)).$

All metrizable spaces are regular. Another important class of regular spaces are the zero-dimensional Hausdorff spaces, where a space is called *zero-dimensional* if it has a base consisting of clopen sets. For example, if α is an ordinal, then intervals of the form $(\gamma, \beta]$, where $\gamma < \beta < \alpha$, are clopen in the space $\langle \alpha, \tau(\in) \rangle$, and it follows that this space is regular. Similarly, the spaces $\beta\omega$ and ω^* are zero-dimensional and hence regular.

Many properties of metrizable spaces generalize to Hausdorff spaces (or regular spaces), but not to topological spaces in general. For example, if $(x_n)_{n \in \omega}$ is a sequence of points in a topological space $\langle X, \tau \rangle$, then $x \in X$ is considered a limit of this sequence if each neighborhood of x contains all but finitely many elements of the sequence. Note that if τ is the antidiscrete topology of Example 26.1, then every point of X is a limit of every sequence in X. Such an anomaly cannot happen if τ is a T_2-topology: In a Hausdorff space, each sequence has at most one limit.

Here is another important property of metrizable spaces that generalizes to Hausdorff spaces:

THEOREM 26.1. *Let X be a Hausdorff space, and let $f : X \to X$ be a continuous function. Then the set $\{x \in X : f(x) = x\}$ of fixed points of f is closed.*

Other important topological properties are connected with *cardinal functions*. Let us give a few important examples of cardinal functions. Let $\langle X, \tau \rangle$ be a topological space. The *character of a point x in X* is defined as:

$$\chi(x, X) = \min\{|\mathcal{B}_x| : \mathcal{B}_x \text{ is a local base at } x\}.$$

---

[1]Most topologists require only condition (26.1) for regularity, and use the name T_3-*spaces* for what we call regular spaces.

The *character of the space* X is defined as:
$$\chi(X) = \sup\{\chi(x,X) : x \in X\} + \aleph_0.$$
The *weight* of X is defined as:
$$w(X) = \min\{|\mathcal{B}| : \mathcal{B} \text{ is a base for } \tau\} + \aleph_0.$$
The π-*weight* of X is defined as:
$$\pi w(X) = \min\{|\mathcal{B}| : \mathcal{B} \text{ is a } \pi\text{-base for } \tau\} + \aleph_0,$$
where $\mathcal{B}$ is a π-*base for* τ if $\mathcal{B} \subseteq \tau\setminus\{\emptyset\}$ and $\forall U \in \tau\setminus\{\emptyset\} \exists V \in \mathcal{B}\, (V \subseteq U)$.

If $D \subseteq X$ is such that $\operatorname{cl} D = X$, then we say that D is *dense* in X. Equivalently, D is dense in X iff $D \cap U \neq \emptyset$ for every nonempty open subset U of X. The *density* of X is defined as:
$$d(X) = \min\{|D| : D \text{ is a dense subset of } X\} + \aleph_0.$$
A family $\mathcal{U} \subseteq \tau$ is an *open cover* of X if $\bigcup\mathcal{U} = X$. A *subcover of* $\mathcal{U}$ is a family $\mathcal{V} \subseteq \mathcal{U}$ that is itself an open cover of X. The *Lindelöf degree* of X is defined as:
$$L(X) = \min\{\kappa : \text{every open cover of } X \text{ has a subcover of cardinality } \leq \kappa\} + \aleph_0.$$
The *Suslin number* of X is defined as:
$$S(X) = \min\{\kappa : \forall \mathcal{U} \in [\tau\setminus\{\emptyset\}]^\kappa \exists\{U,V\} \in [\mathcal{U}]^2\, (U \cap V \neq \emptyset)\}.$$

Each cardinal function has its corresponding *countability property*. Spaces with countable character are called *first countable*, spaces with countable weight are called *second countable*, spaces with countable density are called *separable*, and spaces with countable Lindelöf degree are called *Lindelöf spaces*. If $S(X) = \kappa$, then X satisfies the κ-*chain condition* or κ-*c.c.*, and if $S(X) \leq \aleph_1$, then X satisfies the *countable chain condition* or *c.c.c.* A space is said to have a property *hereditarily* if every subspace of it has this property. First and second countability are always inherited by subspaces, but countable π-weight, separability, Lindelöfness, and the c.c.c. are not hereditary.

EXAMPLE 26.8. (a) Every metrizable space is first countable. If κ is infinite, then $w(D(\kappa)) = \kappa$. Thus not every metrizable space is second countable.

(b) If $\mathcal{F}$ is a nonprincipal ultrafilter on ω, then $\langle \omega + 1, \tau^\mathcal{F}\rangle$ is a countable, not first countable space.

(c) Let α be an ordinal and consider the space $\langle \alpha, \tau(\in)\rangle$. If $\beta \in \alpha$, then $\chi(\beta, \alpha) = cf(\beta)$. Thus α is a first countable space iff $\alpha \leq \omega_1$. Moreover, α is second countable iff $\alpha < \omega_1$. Every dense subset of α must contain all successor ordinals $< \alpha$. Thus α is separable iff $\alpha < \omega_1$. It is not hard to show that α is Lindelöf iff $cf(\alpha) \leq \omega$. Thus $\langle \omega_1 \cdot \omega, \tau(\in)\rangle$ is an example of a regular Lindelöf space that is not separable. However, its subspace ω_1 is not Lindelöf, and thus $\omega_1 \cdot \omega$ is not hereditarily Lindelöf.

(d) Every open cover of $\beta\omega$ contains a finite, hence countable, subcover; the same is true for ω^*. Thus both of these spaces are Lindelöf. The set of isolated points is a dense subset of $\beta\omega$, and the family $\{U_{\{n\}} : n \in \omega\}$ is a π-base in this space. It follows that $\beta\omega$ is separable, satisfies the c.c.c., and has a countable π-base. On the other hand, this space is not first countable and hence not second countable. Its subspace ω^* is not separable, does not have a countable π-base, and does not satisfy the c.c.c.

For metrizable spaces, most of the countability properties discussed above coincide. In particular, we have the following.

THEOREM 26.2. *Let X be a metrizable space. Then the following are equivalent:*

(i) X *is second countable;*
(ii) X *is Lindelöf;*
(iii) X *is hereditarily Lindelöf;*
(iv) X *is separable;*
(v) X *is hereditarily separable.*

As Examples 26.8(c) and (d) show, the equivalence between (i), (ii), and (iv) of Theorem 26.2 does not generalize to the class of all topological spaces, not even to the class of all regular spaces. The question whether the equivalence between (iii) and (v) generalizes is more complex. A space X is called an *S-space* if it is regular, hereditarily separable, but not Lindelöf, and an *L-space* if it is regular, hereditarily Lindelöf, but not separable. There exist various consistent constructions of S-spaces and L-spaces; for example, the Kunen line constructed in Chapter 17 with the help of CH is an S-space. However, it is relatively consistent with ZFC that there are no S-spaces. At the time of this writing it is still an open problem whether ZFC proves the existence of an L-space.

Let us return to Example 26.8(d). The spaces $\beta\omega$ and ω^* satisfy a very strong version of Lindelöfness, namely:

(26.2) Every open cover of X has a finite subcover.

Topological spaces that satisfy 26.2 are called *compact spaces*. Compact Hausdorff topologies are minimal topologies in the following sense:

THEOREM 26.3. *Suppose $\langle X, \tau_1 \rangle$ is a compact Hausdorff space, and $\tau_0 \subseteq \tau_1$ is a topology on X such that the space $\langle X, \tau_0 \rangle$ is Hausdorff. Then $\tau_0 = \tau_1$.*

For a proof of Theorem 26.3, see Corollary 3.1.14 of Engelking's book.

If τ_0, τ_1 are two topologies on the same set X such that $\tau_0 \subseteq \tau_1$, then we say that τ_1 is a *refinement* of τ_0.

EXAMPLE 26.9. (a) An ordinal $\alpha > 0$ is compact iff it is a successor ordinal.

(b) The real line $\mathbb{R}$ is not compact, but its subspace $[0,1]$ is. More generally, every closed interval $[a,b]$ of reals is compact.

A Hausdorff space X is called *locally compact* if every point x of X has a neighborhood that is a compact subspace of X. Example 26.9(b) shows that the real line is a locally compact, noncompact space. Similarly, every ordinal is a locally compact space.

The most conspicuous difference between intervals in the real line and ordinals is that the former contain no isolated points, whereas in the latter spaces nonisolated points are few and far between. Let us capture this intuition in a formal definition. A subset A of a topological space X is *perfect* if it is closed and every point of A is an accumulation point of A; i.e., A is perfect iff $A = A'$. For example, each closed nontrivial interval is a perfect subset of the real line.

The set X' of all nonisolated points of a space X is called the *Cantor-Bendixson derivative of X*. For example, if α is an ordinal with the order topology, then α'

is the set of all nonzero limit ordinals below α. The Cantor-Bendixson derivative of $\beta\omega$ is ω^*. As in calculus, the operation of taking derivatives can be iterated. By recursion over $\xi \in \mathbf{ON}$ we define:

$$X^{(0)} = X;$$
$$X^{(\xi+1)} = (X^{(\xi)})';$$
$$X^{(\delta)} = \bigcap_{\xi<\delta} X^{(\xi)} \text{ if } \delta \text{ is a limit ordinal } > 0.$$

If X is Hausdorff, then the set $\bigcap_{\xi \in \mathbf{ON}} X^{(\xi)}$ is either perfect or empty. In the latter case the space X is said to be *scattered*. All ordinals are scattered spaces. The next theorem gives a sufficient condition for being scattered. We say that a space X is *locally countable* if every point $x \in X$ has a countable neighborhood.

THEOREM 26.4. *Every locally countable, locally compact Hausdorff space is scattered.*

A space is *connected* if it is not possible to decompose X into two disjoint nonempty open subsets. A *continuum* is a connected compact Hausdorff space. The unit interval $[0,1]$ is an example of a continuum. If a Hausdorff space with more than one point contains an isolated point, then X is not connected. In particular, no scattered Hausdorff space with more than one point is connected. The space $^\omega 2$ is an example of a disconnected compact Hausdorff space without isolated points.

Compact metrizable spaces are *Polish spaces*, i.e., they are completely metrizable and separable. The spaces $D(\aleph_0)$, $\mathcal{P}(\omega)$, $^\omega\omega$, and $\mathbb{R}$ are examples of noncompact Polish spaces.

Recall the definition of meager sets from Chapter 13. Let us call a subset C of a topological space X *comeager* if $X\setminus C$ is meager.

THEOREM 26.5 (Baire Category Theorem). *In a Polish space, every comeager set is dense.*

If Y is a dense subspace of a compact Hausdorff space X, then X is called a *compactification of Y* and the subspace $X\setminus Y$ is called the *remainder* of this compactification. Each noncompact, locally compact Hausdorff space Y has a (unique up to homeomorphism) *one-point compactification,* i.e., a compactification X such that the remainder $X\setminus Y$ is a singleton. For example, if κ is an infinite cardinal and $\mathcal{F}$ is the filter of cofinite subsets of κ, then $\langle \kappa+1, \tau^{\mathcal{F}}\rangle$ is the one-point compactification of $D(\kappa)$. In particular, $\langle \omega+1, \tau(\in)\rangle$ can be regarded as the one-point compactification of $D(\aleph_0)$.

One-point compactifications are the smallest possible ones. On the other end of the spectrum are Stone-Čech compactifications. The *Stone-Čech compactification βY of a space Y* is a compactification of Y such that for every compactification X of Y there exists a continuous surjection $f : \beta Y \to X$ such that $f\lceil Y$ is the identity. The Stone-Čech compactification of Y, if it exists[2], is unique up to homeomorphism. If we identify each natural number n with the principal ultrafilter $\mathcal{F}_n$ on ω, then the space $\beta\omega$ of Example 26.5 is the Stone-Čech compactification of the discrete space $D(\aleph_0)$, and ω^* is the remainder of this compactification.

---

[2]One can show that Y has a Stone-Čech compactification if and only if Y is a Tychonoff space, where the Tychonoff property is a slightly stronger separation property than regularity.

Let us conclude this chapter with some proofs that illustrate how set-theoretic arguments can be used as tools in general topology.

THEOREM 26.6 (Hewitt-Marczewski-Pondiczery). *Let κ be an infinite cardinal. If $\{X_i : i \in I\}$ is a family of topological spaces such that $d(X_i) \leq \kappa$ for all $i \in I$, and if $|I| \leq 2^\kappa$, then $d(\prod_{i \in I} X_i) \leq \kappa$. In particular, a product of at most $2^{\aleph_0}$ separable spaces is separable.*

PROOF. Let κ, I be as in the assumptions. Since taking the product of a space X with any number of copies of the one-point space results in a homeomorphic space, we may without loss of generality assume that $|I| = 2^\kappa$, or even more specifically, that $I = {}^\kappa 2$. So, let $\{X_f : f \in {}^\kappa 2\}$ be a family of nonempty topological spaces of density $\leq \kappa$ each. For each $f \in {}^\kappa 2$, let D^f be a dense subset, and enumerate D^f, possibly with repetitions, as $\{d^f_\alpha : \alpha < \kappa\}$. For $H \in [\kappa]^{<\aleph_0}$, let $\mathcal{G}_H$ be the set of all functions from ${}^H 2$ into κ. Let $\mathcal{G} = \bigcup_{H \in [\kappa]^{<\aleph_0}} \mathcal{G}_H$. Clearly, $|\mathcal{G}| = \kappa$. For every $H \in [\kappa]^{<\aleph_0}$ and $G \in \mathcal{G}_H$, we define a function $e_G \in \prod_{f \in {}^\kappa 2} X_f$ by letting $e_G(f) = d^f_{G(f \restriction H)}$ for all $f \in {}^\kappa 2$. Since $\mathcal{G}_{H_0} \cap \mathcal{G}_{H_1} = \emptyset$ for $H_0 \neq H_1$, this definition is unambiguous.

We show that the set $\{e_G : G \in \mathcal{G}\}$ is dense in $\prod_{f \in {}^\kappa 2} X_f$. For this, it suffices to show that for every nonempty basic open set U of the product space there exists $G \in \mathcal{G}$ such that $e_G \in U$. So let $U = \prod_{f \in {}^\kappa 2} U_f$ be nonempty basic open, and let $F \in [{}^\kappa 2]^{<\aleph_0}$ be such that $U_f = X_f$ for all $f \in {}^\kappa 2 \backslash F$. Since any two different functions in F differ on some coordinate $\alpha \in \kappa$, there exists $H \in [\kappa]^{<\aleph_0}$ such that $f \restriction H \neq g \restriction H$ whenever $\{f, g\} \in [F]^2$. Fix such H and define $G : {}^H 2 \to \kappa$ as follows: If $\varphi = f \restriction H$ for some (unique by the choice of H) $f \in F$, then let $G(\varphi)$ be the smallest α such that $d^f_\alpha \in U_f$. If $\varphi \in {}^H 2$ is not of this form, let $G(\varphi) = 0$. This definition of G implies that $e_G(f) \in U_f$ for all $f \in {}^\kappa 2$, and hence $e_G \in U$. □

In a certain sense, the result of Theorem 26.5 is the best possible one.

THEOREM 26.7. *Suppose κ is an infinite cardinal, $|I| > 2^\kappa$, and $\{X_i : i \in I\}$ is a family of Hausdorff spaces such that $|X_i| > 1$ for all $i \in I$. Then $d(\prod_{i \in I} X_i) > \kappa$.*

PROOF. Let $\{X_i : i \in I\}$ be as in the assumptions, and let D be a subset of $\prod_{i \in I} X_i$ of cardinality $\leq \kappa$. We show that D is not dense in $\prod_{i \in I} X_i$. For $i \in I$, pick $\{x_i, y_i\} \in [X]^2$ and two disjoint nonempty open sets $U_i, V_i \subset X_i$ such that $x_i \in U_i$ and $y_i \in V_i$. Let $D = \{d_\alpha : \alpha < \lambda\}$, where $\lambda \leq \kappa$. For each $i \in I$, define a function $f_i : \lambda \to 2$ as follows:

$$f_i(\alpha) = \begin{cases} 1, & \text{if } d_\alpha(i) \in U_i; \\ 0, & \text{if } d_\alpha(i) \notin U_i. \end{cases}$$

Since $|I| > 2^\lambda$, the Pigeonhole Principle implies that there exist $i_0, i_1 \in I$ such that $i_0 \neq i_1$ and $f_{i_0} = f_{i_1}$. Consider the set $W = \prod_{i \in I} W_i$, where $W_{i_0} = U_{i_0}$, $W_{i_1} = V_{i_1}$, and $W_i = X_i$ for $i \in I \backslash \{i_0, i_1\}$. Then W is a nonempty open subset of $\prod_{i \in I} X_i$ with $W \cap D = \emptyset$. □

The last two theorems of this appendix illustrate the use of cardinal invariants of the continuum in topology. More material on this topic can be found in the articles *The integers and topology*, by E. K. van Douwen, in *Handbook of Set-Theoretical Topology*, K. Kunen and J. E. Vaughan, eds., North-Holland, 1984, 111–167; and

Small Uncountable Cardinals and Topology, by J. E. Vaughan, in *Open Problems in Topology*, J. van Mill and G. M. Reed, eds., North-Holland, 1990, 195–216.

A topological space X is called *sequentially compact* if every sequence of points in X has a convergent subsequence. Every compact metrizable space is sequentially compact. In particular, the space $D(2)$ is sequentially compact.

THEOREM 26.8. *Suppose $\kappa < \mathfrak{t}$ and $\{\langle X_\alpha, \tau_\alpha \rangle : \alpha < \kappa\}$ is a family of sequentially compact topological spaces. Then the space $\prod_{\alpha<\kappa} X_\alpha$ is also sequentially compact.*

PROOF. Let κ and $\{\langle X_\alpha, \tau_\alpha \rangle : \alpha < \kappa\}$ be as in the assumptions, and let $(f_n)_{n\in\omega}$ be a sequence of elements of $\prod_{\alpha<\kappa} X_\alpha$. By recursion over α, we construct $x_\alpha \in X_\alpha$ and $b_\alpha \in [\omega]^{\aleph_0}$ such that for all $\alpha < \beta < \kappa$:

(i) $\forall U \in \tau_\alpha\, (x_\alpha \in U \to |\{n \in b_\alpha : f_n(\alpha) \notin U\}| < \aleph_0)$;
(ii) $b_\beta \subseteq^* b_\alpha$.

Sequential compactness of X_α can be used to assure (i), and at limit stages β one uses the assumption that $\kappa < \mathfrak{t}$ to assure (ii). The latter assumption also implies that there exists $b \in [\omega]^{\aleph_0}$ such that $b \subseteq^* b_\alpha$ for all $\alpha < \kappa$. Fix such a b, and let $f \in \prod_{\alpha<\kappa} X_\alpha$ be such that $f(\alpha) = x_\alpha$ for all $\alpha < \kappa$. A straightforward verification shows that the sequence $(f_n)_{n\in b}$ converges to f. □

THEOREM 26.9. *The topological space $(D(2))^{\mathfrak{s}}$ is not sequentially compact.*

PROOF. Let $S = \{b_\alpha : \alpha < \mathfrak{s}\} \subseteq \mathcal{P}(\omega)$ be a splitting family. Define a sequence $(f_n)_{n\in\omega}$ of elements of ${}^{\mathfrak{s}}2$ as follows:

$$f_n(\alpha) = \begin{cases} 1, & \text{if } n \in b_\alpha; \\ 0, & \text{if } n \notin b_\alpha. \end{cases}$$

We show that $(f_n)_{n\in\omega}$ has no convergent subsequences. Let $b \in [\omega]^{\aleph_0}$. Since S is a splitting family, there exists $\alpha < \mathfrak{s}$ such that $|b \cap b_\alpha| = \aleph_0 = |b \setminus b_\alpha|$. Fix such α, and let $U = \{f \in {}^{\mathfrak{s}}2 : f(\alpha) = 1\}$ and $V = \{f \in {}^{\mathfrak{s}}2 : f(\alpha) = 0\}$. Then U and V are disjoint open sets in the product space ${}^{\mathfrak{s}}2$, and the sets $\{n \in b : f_n \in U\}$ and $\{n \in b : f_n \in V\}$ are both infinite. Thus the sequence $(f_n)_{n\in b}$ cannot be convergent. □

Index

F_σ-set, 207
G_δ-set, 207
$\mathbb{Q}$-embedding of a tree, 41
T_2-space, 210
T_3-space, 210
♣-principle, 144
♣-sequence, 144
♢-principle, 141
♢-sequence, 142
↾-principle, 145
$\aleph_1$-separation property, 201
Δ-system, 67
κ-Aronszajn tree, 36
κ-Kurepa tree, 30
κ-chain condition
  in a Boolean algebra, 19
  in a l.o., 44
  in a p.o., 7
  in a topological space, 211
κ-closed p.o., 6
κ-directed closed p.o., 6
κ-extension property, 201
κ-scale, 76
κ-tree, 36
κ-universal model, 201
$\langle \kappa, \lambda^* \rangle$-gap, 117
$\langle \kappa, \lambda^* \rangle$-pregap, 117
λ-independent family, 84
π-base, 211
π-weight
  of a Boolean algebra, 203
  of a topological space, 211
σ-centered p.o., 91
σ-field, 13
σ-linked p.o., 91
Łoś' Theorem, 9

absolute formula, 168
accumulation point, 207
additivity
  of a measure, 147
  of an ideal, 74
almost contained in, 102
almost disjoint families, 119
almost disjoint family, 82, 127
always first category set, 76
amoeba forcing, 92
antichain
  in a p.o., 6
  in a tree, 40
antidiscrete topology, 208
Aronszajn line, 43
Aronszajn tree, 36
atom
  of a Boolean algebra, 188
  of a measure, 153
atomic formula, 159
atomic measure, 153
atomless measure, 153

Baire Category Theorem, 213
Baire property, 14
Balcar-Vojtáš Theorem, 200
base
  of a topology, 207
  of an ideal, 72
Baumgartner-Komjáth Theorem, 204
Boolean algebra, 14
  atomic, 188
  atomless, 188
  canonical homomorphism, 188
  complement, 16
  complete, 19

disjoint elements, 199
disjoint subset, 199
dual, 16
incomparable subset, 203
κ-complete, 19
separated subsets, 201
superatomic, 197
trivial, 14
weakly homogeneous, 190
weakly κ-homogeneous, 190
Boolean combination, 33
Boolean order, 15
Boolean ring, 187
Boolean space, 18
Borel Conjecture, 74
Borel set, 13
bounded subset
of an ordinal, 123
of $^{\omega}\omega$, 76
bounding number, 76
branch, 29

canonical coloring, 49
Cantor discontinuum, 18
Cantor set, 18
Cantor space, 18
Cantor-Bendixson characteristics, 198
Cantor-Bendixson derivative, 197, 212
Cantor-Bendixson height, 198
cardinal function, 210
cardinal invariant, 74
Cauchy sequence, 209
cellularity, 199
centered set, 5, 91
character
of a point, 210
of a space, 211
characteristic function, 210
clopen set, 14, 207
closed
in **ON**, 131
under a function, 124
closed subset
of a topological space, 207
of an ordinal, 123
of $[X]^{<\kappa}$, 133
closure, 207
closure under functions, 132
$CLUB$, 123
CLUB, 133
$club$, 123
club, 133
cocountable set, 1
cofinal branch, 29
comeager set, 213
compact space, 212
compactification, 213
compactness argument, 31
compatible elements, 3

complete metric, 209
completely metrizable space, 209
completion
of a Boolean algebra, 22
of a l.o., 19
cone, 134
connected space, 213
continuous function, 207
continuum, 213
convergent sequence, 209
countability property, 211
countable chain condition
in a p.o., 7
in a topological space, 211
countable limit, 202
countable separation property, 201
cover, 211
cut, 21, 192

de Morgan's Laws, 16
Dedekind cut, 19
definability, 172
dense set
in a Boolaen algebra, 19
in a l.o., 43
in a p.o., 7
in a topological space, 211
density, 211
derived set, 123, 207
diagonal intersection, 125
diagonal union, 127
diameter, 209
discrete space, 18
discrete topology, 208
disjoint refinement, 200
dominating number, 76
dual ideal, 1

elementary embedding, 24, 159
elementary equivalence, 160
elementary submodel, 159
embedding, 159
elementary, 159
Erdős cardinal, 60
Erdős-Dushnik-Miller Theorem, 60
Erdős-Rado Theorem, 55
Erdős-Sierpiński Theorem, 72
Erdős-Tarski Theorem, 199
eventual dominance, 75
extension
finite, 191
simple, 191
Extension Property, 178

factor algebra, 190
Fichtenholz-Kantorovitch-Hausdorff
Theorem, 83
field of sets, 12
σ-complete, 13

filter
  countably complete, 3
  generated by a family, 2
  in a Boolean algebra, 17
  in a p.o., 3
    κ-closed, 6
    generated by a subset, 5
  κ-closed, 3
  κ-complete, 3
  of closed sets, 3
  on a set, 1
  proper, 1
  uniform, 1
filter base, 2
  in a p.o., 4
finite intersection property, 2
fip, 2
first countable space, 211
Fodor's Lemma, 127
forest, 28
full binary tree, 28
Fundamental Theorem on Ultraproducts, 9

Galvin-Mycielski-Solovay Theorem, 78
gap, 19, 117, 192
generic blueprint, 140
greatest lower bound, 5

Hausdorff gap, 118
Hausdorff space, 210
height
  of a node, 28
  of a tree, 28
hereditary property, 211
Hewitt-Marczewski-Pondiczery
  Theorem, 214
homeomorphic spaces, 207
homeomorphism, 207
homogeneous set, 49
homomorphism, 159

ideal
  in a p.o., 4
  of null sets, 147
  on a set, 2
  proper, 2
idempotent, 187
incompatible elements, 3
independent family, 83
indiscernible elements, 160
induced coloring, 50
initial part of a tree, 29
interior, 207
interval
  half-open, 194
interval algebra, 194
irredundant set, 203
isolated point, 207

kernel
  of a Δ-system, 67
  of a homomorphism, 188
Knaster Property, 93
König's Lemma, 31
Kunen line, 81
Kurepa Hypothesis, 31, 143
Kurepa tree, 31, 143

L-space, 212
lattice, 14
level of a tree, 29
Lévy collapse, 199
Lévy's partial order, 199
lexicographical order, 54
Lindelöf degree, 211
Lindelöf space, 211
linearly ordered continuum, 47
linearly ordered space, 208
linked set, 91
local base, 207
locally compact space, 212
locally countable space, 213
lower bound, 3
lower semilattice, 5
Luzin gap, 121
Luzin set, 71
  generalized, 74

mad family, 88
maximal almost disjoint family, 88
maximal antichain, 7
  below an element, 189
McKenzie's Theorem, 203
meager set, 14, 72
measurable cardinal, 3, 12, 150
measure
  atomic, 153
  atomless, 153
  κ-additive, 150
  two-valued, 147
  vanishing on the singletons, 147
meet, 5
metric, 208
metrizable space, 209
Miller-Rado Theorem, 62

Negative Stepping-Up Lemma, 57
neighborhood, 207
neighborhood base, 207
neighborhood filter, 1, 207
neighborhood system, 1, 207
network, 176
network weight, 176
node, 27
  high, 46
  immediate successor, 44
  splitting, 28
nonstandard natural number, 12

normal function
  on an ordinal, 124
  on $[X]^{<\kappa}$, 136
normal functional class, 131
normal ultrafilter, 151, 184
nowhere dense set, 13
null set, 72

OCA, 64
one-point compactification, 213
open ball, 209
Open Coloring Axiom, 64
open cover, 211
open interval, 208
open partition, 64
open set, 207
open subset of a p.o., 4
order topology, 208
Ostaszewski principle, 146

P-point, 3, 110
Paris-Harrington Theorem, 51
partition, 49
partition calculus, 50
partition of unity, 196
path, 29
perfect set, 212
perfectly meager set, 76
Polish space, 213
power set algebra, 12
predense set, 7
pregap, 117
Pressing Down Lemma, 127
Prime Ideal Theorem, 6
probability measure, 147
product of Boolean algebras, 190
product space, 209
product topology, 209
Property K, 93
pseudo-intersection, 102

Q-point, 110
quasi-normal ultrafilter, 115
quotient Boolean algebra, 188

ramification system, 194
  generating, 194
  linear, 195
  linear generating, 195
Ramsey number, 53
Ramsey Theory, 50
Ramsey ultrafilter, 114
Ramsey's Theorem, 50
Rasiowa-Sikorski Lemma, 87
real-valued measurable cardinal, 150
reduced product, 8
reduct, 164
refinement, 212
Reflection Property, 178

regressive function
  on an ordinal, 127
  on $[X]^{<\kappa}$, 135
regular cut, 21
regular open set, 22
regular space, 210
regularization, 22
relativization, 168
remainder, 213
root
  of a Δ-system, 67
  of a tree, 27

S-space, 212
saturated ideal, 150
scattered space, 213
second countable space, 211
selective ultrafilter, 110
semi-Q-point, 112
separable space, 211
separated families, 119
separation property, 210
separative p.o., 20
sequentially compact space, 215
set algebra, 12
  κ-complete, 13
sfip, 102
side constraints, 105
Sierpiński set, 72
  generalized, 74
Sikorski's Extension Theorem, 192
Silver's Theorem, 129
simple product, 42
Skolem functions, 164
Skolemized model, 164
Solovay's Lemma, 95
special Aronszajn tree, 41
specializing function, 41
Specker's Theorem, 63
splitting family, 80
splitting node, 28
splitting number, 80
splitting tree, 28
standard topology on $\mathcal{P}(\omega)$, 210
stationary in **ON**, 131
stationary set
  in $[X]^{<\kappa}$, 135
  of ordinals, 126
  with respect to a filter, 1
stick-principle, 145
Stone Duality Theorem, 18
Stone space, 18
Stone-Čech compactification, 213
  of ω, 208
strong countable separation property, 202
strong finite intersection property, 102
strong limit cardinal, 30
strong measure zero, 74
strongly independent family, 83

strongly λ-independent family, 84
strongly meager set, 78
subcover, 211
subfield, 13
submodel
  elementary, 159
  generated by a set, 132
subspace topology, 209
substructure generated by a set, 132
subtree, 27
Suslin Hypothesis, 40
Suslin line, 45
Suslin number, 211
Suslin Problem, 47
Suslin tree, 40
symmetric difference, 187

Tarski-Vaught criterion, 161
tidy formula, 33
topological space, 207
topology, 207
topology induced by a metric, 209
totally disconnected space, 18
transversal, 127
tree, 27
  bushy, 44
  full binary, 28
  lexicographical ordering, 42
  splitting, 28
  tall, 46
tree property, 36
Tychonoff's Theorem, 6

Ulam matrix, 148
Ulam measurable, 153
ultrafilter
  in a p.o., 4
    fixed, 4
    free, 4
    principal, 4
  normal, 151, 184
  on a set, 1
  principal, 1
  quasi-normal, 115
  selective, 110
  uniform, 150
ultrapower, 11
ultraproduct, 8
unbounded in **ON**, 131
unbounded subset
  of an ordinal, 123
  of $^\omega\omega$, 76
  of $[X]^{<\kappa}$, 133
uniform ultrafilter, 150
universal measure zero set, 78

Vaught relation, 196
Vaught's Theorem, 196

weakly compact cardinal, 39, 58
weight, 211
well-met p.o., 5

zero-dimensional space, 210

Index of Symbols

$\nleq$, 3
$\perp$, 3, 119, 201
$\prod_{\xi \in X} \mathfrak{A}_\xi / \mathcal{U}$, 8
$\prod_{\xi \in X} A_\xi / \mathcal{U}$, 8
$\bar{\in}$, 11
$-X$, 13
$\mathbf{\Sigma}^0_\alpha$, 13
$\mathbf{\Pi}^0_\alpha$, 13
$\leq_B$, 15
$\prod X$, 19
$\sum X$, 19
$\subseteq^*$, 37, 102
$=^*$, 37, 102
$\leq_e$, 40
$\hat{t}$, 20, 27
$\hat{t}(\beta)$, 42
$\otimes^s$, 42
$\kappa \to (\lambda)^\rho_\sigma$, 49
$\kappa \not\to (\lambda)^\rho_\sigma$, 50
$n \xrightarrow{*} (m)^k_\ell$, 51
$\kappa \to (\lambda)^{<\omega}_\sigma$, 60
$\kappa \to (\lambda_i)^n_{i<\sigma}$, 60
$\lambda \to (\beta_i)^n_{i<\sigma}$, 62
$\lambda \to (\beta)^n_\sigma$, 62
$\lambda \to (\beta)^{<\omega}_\sigma$, 62
$\omega \to (\mathcal{F})^n_m$, 114
$<^*$, 75
$y + X$, 78
$\subset^*$, 102
A', 123, 207
$\Delta_{\alpha<\gamma} C_\alpha$, 125
$\Delta_{x \in X} \mathcal{C}_x$, 135
$\nabla_{\alpha<\kappa} A_\alpha$, 127
$\check{A}$, 134
$\diamondsuit^\circ(C)$, 140
$\diamondsuit$, 141
$\diamondsuit^-$, 142
$\diamondsuit^*$, 143
$\diamondsuit^+$, 143
$\clubsuit$, 144
$\diamondsuit(E)$, 146
$\mathfrak{A} \prec \mathfrak{B}$, 159
$\mathfrak{A} \equiv \mathfrak{B}$, 160
$\varphi/\mathbf{X}$, 168
$a\Delta b$, 187

$\mathfrak{B}|a$, 190
$\mathfrak{A} \times \mathfrak{B}$, 190
$\langle X \rangle$, 191
$B(a)$, 191
$\sim_I$, 188
$[a, b)$, 194
$(X)_\mathfrak{A}$, 201
$\prod_{i \in I} X_i$, 209
$[s]$, 210

$\mathfrak{a}$, 90
$add(\mathcal{I})$, 74
$\mathbb{A}(\varepsilon)$, 92
$At(B)$, 188

$\mathfrak{b}$, 76
B^+, 16
$\mathfrak{B}^*$, 16
$\beta\omega$, 208
βY, 213
$b(\mathfrak{R})$, 188
$\mathcal{B}_x$, 207
$B(x, \varepsilon)$, 209

$c(a)$, 199
$c(B)$, 199
cB, 22
c.c.c., 7, 211
$cFr(\kappa)$, 189
χ_a, 210
$\chi(x, X)$, 210
$\chi(X)$, 211
$cl\, B$, $cl_\tau(B)$ 207
$Clop(X)$, 14
$club$, 123
$CLUB(\gamma)$, 123
club, 133
$CLUB([X]^{<\kappa})$, 133
$cof(\mathcal{I})$, 79
$cov(\mathcal{I})$, 79

D, 18
$\mathfrak{d}$, 75
dd, 209
$\Delta(s, t)$, 42
$diam(A)$, 209

26. INDEX OF SYMBOLS

$D(\kappa)$, 18
$d(X)$, 211

$\exp_n(\kappa)$, 55

$\mathcal{F}^*$, 1
$\mathcal{F}_a$, 1
$\mathbb{F}_a$, 4
$\mathbb{F}(B)$, 4
$Finco(A)$, 13
fip, 2
$Fix(f)$, 124
$Fix(\mathbf{F})$, 131
$F_{<\kappa}(A)$, 13
$flt(\mathcal{A})$, 2
$Fn(F,J)$, 4, 91
$Form_L^n$, 164
$f_\mathcal{P}$, 49

glb, 13.3
$g_{/\mathcal{U}}$, 8
g^{id}, 183
g^z, 183
gR, 194

$h(A)$, 198
$ht(t)$, 28
$ht(T)$, 28

$\mathcal{I}^*$, 2
$I(A)$, 197
$I_\alpha(A)$, 197
$\mathcal{I}_\mu$, 147
$Inc B$, 203
$Intalg(L)$, 194
$int B$, 207
$int_\tau(B)$, 207

$j_\mathcal{U}$, 24

$\kappa(\alpha)$, 62
κ-c.c., 7, 19, 211
KH, 31
$\kappa(\lambda)$, 60
$ker(\pi)$, 188
$\mathcal{K}(X)$, 3

lgR, 195
lub, 15
$L(X)$, 211

MA, 90
MA_{Ba}, 97
MA_{cBa}, 97
$MA(\kappa)$, 90
$MA_{Property\ K}$, 94
$MA_{\sigma-linked}$, 94
$MA_{\sigma-centered}$, 94
MA_{top}, 97
$MA^-(\kappa)$, 98, 132
$\mathcal{M}$, 72

$\mathcal{M}_g$, 124
$\mathcal{M}_G$, 125
$\mu_\mathcal{U}$, 147
$M(x,y)$, 209

$\mathcal{N}$, 72
$n(A)$, 198
$neg\ U$, 23
$\mathcal{N}_x$, 1, 207
$nw(X)$, 176

OCA, 64
ω^*, 209

$\mathfrak{p}$, 102
PA, 54
πB, 203
π_I, 188
$\pi w(X)$, 211

rA, 22
$rank(x)$, 179
$r(\mathfrak{B})$, 187
$R(m,k,\ell)$, 52
$R^*(m,k,\ell)$, 53
$RO(X)$, 22

$\mathfrak{s}$, 80
sfip, 102
SH, 40
SK, 163
$sk(L_m)$, 165
$sk(\mathfrak{B})$, 165
SQ, 47
$st B$, 18
$S(X)$, 211

$\mathfrak{t}$, 103
$T(\alpha)$, 29
$T_{(\alpha)}$, 29
τ, 207
τ_d, 209
$\tau^\mathcal{F}$, 208
τ^M, 176
$\tau\lceil Y$, 209
$\tau(<)$, 208
$\tau(\in)$, 208

U_a, 17
$Ult B$, 17
$\mathcal{U}_\mu$, 147

vWCT, 32
vWTY, 32

WCT, 32
WTY, 32
$w(X)$, 211

$^XV_\kappa/\mathcal{U}$, 11
$^X\mathbf{V}/\mathcal{U}$, 24